DIGITAL SENSORY S(

DIGITAL SENSORY SCIENCE
Applications in New Product Development

Edited by

J. BEN. LAWLOR
Nestlé NPTC, Singen, Germany

JEAN A. MCEWAN
Jean A McEwan Consulting Ltd., Bishop's Stortford, United Kingdom

DAVID LABBE
Société des Produits Nestlé S.A. Route du Jorat, Lausanne, Switzerland

ELSEVIER

WP
WOODHEAD
PUBLISHING
An imprint of Elsevier

Woodhead Publishing is an imprint of Elsevier
50 Hampshire Street, 5th Floor, Cambridge, MA 02139, United States
The Boulevard, Langford Lane, Kidlington, OX5 1GB, United Kingdom

ISBN: 978-0-323-95225-5 (print)

ISBN: 978-0-323-95226-2 (online)

For information on all Woodhead Publishing publications
visit our website at https://www.elsevier.com/books-and-journals

Publisher: Nikki P. Levy
Acquisitions Editor: Megan R. Ball
Editorial Project Manager: Lena Sparks
Production Project Manager: Prem Kumar Kaliamoorthi
Cover Designer: Christian J. Bilbow

Typeset by MPS Limited, Chennai, India

Working together
to grow libraries in
developing countries

www.elsevier.com • www.bookaid.org

Contents

2

How digital is creating new opportunities for sensory science

8. Sensory self-service — digitalisation of sensory central location testing   83

Sven Henneberg, Annika Ipsen, Manuel Rost, Femke W.M. Damen and Torsten Koch

3

Digitalization in instrumental, neurological, psychological and behavioural methods: Current applications and opportunities

9. Predicting sensory properties from chemical profiles, the ultimate flavour puzzle: a tale of interactions, receptors, mathematics and artificial intelligence   95

Andrew J. Taylor

10. Electronic noses and tongues: current trends and future needs   117

Gianmarco Gabrieli, Michal Muszynski and Patrick Ruch

11. Leveraging neuro-behavioural tools to enhance sensory research   135

Kathryn Ambroze and Michelle M. Niedziela

4

Immersion technologies, context and sensory perception

16. Multisensory immersive rooms: a mixed reality solution to overcome the limits of contexts studies 249

Adriana Galiñanes-Plaza, Agnès Giboreau and Jacques-Henry Pinhas

17. Voice-activated technology in sensory and consumer research: a new frontier 259

Tian Yu, Janavi Kumar, Natalie Stoer, Hamza Diaz and John Ennis

5

How to tell powerful digital sensory stories

18. Shining through the smog: how to tell powerful, purposeful stories about sensory science in a world of digital overload 271

Sam Knowles

Index 285

List of contributors

Kathryn Ambroze HCD Research, Flemington, NJ, USA

J. Ben. Lawlor Nestlé NPTC, Singen, Germany

Nathalie Boireau Danone Nutricia Research, Utrecht, The Netherlands

Cristina Botinestean Food Industry Development Department, Teagasc Food Research Centre, Ashtown, Dublin, Ireland

Derek Victor Byrne Food Quality Perception and Society Science Team, iSENSE Lab, Department of Food Science, Faculty of Technical Sciences, Aarhus University, Aarhus, Denmark; Sino-Danish College (SDC), University of Chinese Academy of Sciences, Beijing, P.R. China

Emily Crofton Food Quality and Sensory Science Department, Teagasc Food Research Centre, Ashtown, Dublin, Ireland

Victoire Dairou Danone Global Research & Innovation Center, Paris, France

Femke W.M. Damen isi GmbH, Goettingen, Germany

René A. de Wijk Wageningen Food & Biobased Research, WUR, Wageningen, The Netherlands

Thibault Delafontaine MMR Sensory Science Centre, Wokingham, Berkshire, United Kingdom

Hamza Diaz Aigora, Richmond, VA, United States

John Ennis Aigora, Richmond, VA, United States

Christopher J. Findlay C-K Findlay Consultants Inc., Guelph, ON, Canada

Rebecca Ford Sensory Science Centre, Division of Food, Nutrition and Dietetics, School of Biosciences, University of Nottingham, Sutton Bonington Campus, Loughborough, Leicestershire, United Kingdom

Gianmarco Gabrieli IBM Research Europe, Säumerstrasse 4, Rüschlikon, Switzerland

Adriana Galiñanes-Plaza Reperes Insights, Paris, France; The Lab in the Bag, Paris, France

Agnès Giboreau Research & Innovation Centre, Institut Lyfe (ex Institut Paul Bocuse), Lyon, France

Chantalle Groeneschild Danone Nutricia Research, Utrecht, The Netherlands

Sven Henneberg isi GmbH, Goettingen, Germany

Annika Ipsen isi GmbH, Goettingen, Germany

Frances Jack The Scotch Whisky Research Institute, Research Avenue North, Edinburgh, United Kingdom

Martin J. Kern SAM Sensory and Consumer Research, Munich, Germany

Sam Knowles Chief Data Storyteller, The Insight Agents, Lewes, East Sussex, United Kingdom

Torsten Koch isi GmbH, Goettingen, Germany

Janavi Kumar General Mills, Minneapolis, MN, United States

David Labbe Société des Produits Nestlé S.A. Route du Jorat, Lausanne, Switzerland

Anne-Sophie Marcelino Danone Global Research & Innovation Center, Paris, France

Jean A. McEwan Jean A McEwan Consulting Ltd., Bishop's Stortford, United Kingdom

Phiala Mehring MMR Sensory Science Centre, Wokingham, Berkshire, United Kingdom

Michal Muszynski IBM Research Europe, Säumerstrasse 4, Rüschlikon, Switzerland

Michelle M. Niedziela HCD Research, Flemington, NJ, USA

Lucas P.J.J. Noldus Section Neurophysics, Donders Centre for Neuroscience, Radboud University, Nijmegen, The Netherlands

Danni Peng-Li Food Quality Perception and Society Science Team, iSENSE Lab, Department of Food Science, Faculty of Technical Sciences, Aarhus University, Aarhus, Denmark; Sino-Danish College (SDC), University of Chinese Academy of Sciences, Beijing, P.R. China; Neuropsychology and Applied Cognitive Neuroscience Laboratory, CAS Key Laboratory of Mental Health, Institute of Psychology, Chinese Academy of Sciences, Beijing, P.R. China

Jacques-Henry Pinhas The Lab in the Bag, Paris, France

Imogen Ramsey Sensory Science Centre, Division of Food, Nutrition and Dietetics, School of Biosciences, University of Nottingham, Sutton Bonington Campus, Loughborough, Leicestershire, United Kingdom

Manuel Rost isi GmbH, Goettingen, Germany

Patrick Ruch IBM Research Europe, Säumerstrasse 4, Rüschlikon, Switzerland

Natalie Stoer General Mills, Minneapolis, MN, United States

Andrew J. Taylor Flavometrix Limited, Loughborough, United Kingdom

Qian Janice Wang Food Quality Perception and Society Science Team, iSENSE Lab, Department of Food Science, Faculty of Technical Sciences, Aarhus University, Aarhus, Denmark; Sino-Danish College (SDC), University of Chinese Academy of Sciences, Beijing, P.R. China; Department of Food Science, University of Copenhagen, Frederiksberg, Denmark

Qian Yang Sensory Science Centre, Division of Food, Nutrition and Dietetics, School of Biosciences, University of Nottingham, Sutton Bonington Campus, Loughborough, Leicestershire, United Kingdom

Huizi Yu MMR Sensory Science Centre, Pleasantville, NY, United States

Tian Yu Aigora, Richmond, VA, United States

Foreword

It is hard not to notice. Digital technologies are ubiquitous in our everyday lives, from lightbulbs to watches and cars, from restaurant menus to train tickets, and from medical records to the route directions to our next appointment. Scientific research and business life of course are no exception. Today's most advanced phones are more powerful than 5-year old laptop computers, and they allow us to generate and to treat ever-growing amounts of data. With this evolution, sensory and consumer science have deeply transformed and will continue to do so. It has never been so easy to capture sensory information. Likewise, collected sensory and consumer data have never been so rich and complex. This book is being published in the right time and gives multiple examples of the opportunities brought by digitalization. New methods have emerged, and conversely, practices and methods being used 20 years ago are no longer thinkable. Who would use a joystick for time-intensity measurements? Who would mail a paper form for a consumer survey? Who would use a ruler to measure a reported intensity on a visual scale? Who would look at statistical tables to determine if the result of a test is significant?

Technological evolution also comes with an evolution of consumption and usage patterns. We have seen an extreme diversification of consumer typologies to which the consumer goods industry is increasingly responsive, thanks to the digital tools that allow us to closely monitor trends and analyse consumer experience. As a result, new and customized products are launched at an ever-faster pace, which, in turn, contributes to more diverse consumption habits. Phones and apps have also transformed the way people shop and consume goods. One can select a makeup shade using a selfie, buy shoes using the 3D scan of their feet and adjust their shopping list thanks to open-food databases and customized dietary guidelines. All these functionalities allow consumers to make more informed decisions although they are unlikely to be more rational than they used to be (which perhaps questions the abundance of information available to consumers).

From an educator's perspective, the question then arises on how to accompany this evolution and what are the skills of today's and tomorrow's sensory scientist? One would immediately think about computer-related skills and mastery of advanced technologies. Programming and data science skills are certainly an advantage. However, digital objects, apps, and even statistical software are immensely more intuitive than they used to be. Speaking from experience, any fifth-grade child can now easily launch an online survey and analyse their data using simple but beautiful and telling charts. College students who pursue a degree in sensory and consumer science (or in food science with a significant sensory science component) have access to about the same tools. Now those college students are expected to have more extensive knowledge of what makes a good questionnaire, how to limit biases and how to recruit relevant participants. The same goes with sensory tests. Anyone could pick a method and run it reasonably well. Dedicated software and web-based applications have made the job easy. They allow us to set up proper test designs and, thereby, certainly prevent many mistakes. Yet, sensory scientists are expected to wisely choose which method to apply, to understand the principles of sensory perception and of human–product interactions and to make the results of the test meaningful to others. From this perspective, the situation today is not so different from 1989, when David Thomson wrote in a paper entitled *Structuring the education of sensory analysts in the age of computers*: 'A sensory analyst who is capable of working towards realistic project objectives, designing and implementing a study in practice, analyzing and interpreting data and finally presenting the results concisely and accurately in a form that is readily understood by non-experts, is a rare and much sought after professional'. I strongly believe that such qualities are even more important today, precisely because collecting sensory (and nonsensory) data and analysing them with advanced techniques are now very easy. As a result, it is also very easy to make mistakes and to collect nonsensical data. This echoes Thomson's claim, 'familiarity and a "feel" for raw data is essential when interpreting and drawing inferences from statistical analyses'. I would double down and add that a 'feel' for the sensory measure itself is nowadays even more essential. Among the basics of any sensory evaluation comes the fact that human beings are different that they perceive things differently and like different things. This is inseparable from the fact that sensory measures result from an interaction between objects (the products that are being tested) and human beings (who use

or consume the products). By definition, sensory measures are therefore subjective. Besides, this product–subject interaction takes place in a given context — experimental or natural — that could affect the sensory and behavioural response. Digital tools offer exciting perspectives to better approach this reality thanks to richer measures, videos, sounds, images, biosensors, etc. Likewise, data science and artificial intelligence should help extracting more insightful and more actionable information from that complexity.

By contrast, there would be a risk to multiply the type of measures with richer content and more appealing media without considering interindividual differences and the subjectivity of sensory measures. This would inevitably lead to oversimplification, possibly to misinterpretation and eventually wrong decisions. Sensory scientists should always keep in mind the specificities of working with human subjects that are not mere measuring tools but have an opinion and a personal history with the products that they evaluate. Last, and certainly not the least, they think and have their own views about the study they participate to. Thus experimenters should always consider participants' perspective when implementing new methods or new technologies, and they should weigh the opportunities offered by those methods or technologies with the risk of introducing biases to their study design.

As for context, sensory data acquisition apps that can be remotely accessed from tablets and phones (and maybe soon from connected watches or glasses) have clearly helped taking consumer tests out of the sensory booth (boosted with a 2-year pandemic). Conversely, virtual reality headsets and immersive spaces are efficient ways to bring elements of context to the lab, thereby making product evaluation more meaningful. Thanks to digital technologies, the wealth of information that is potentially accessible to sensory and consumer scientists leads them to reach aspects of human interactions with their food (or with any other goods) that were previously confined to very different fields such as anthropology, sociology, or design. Therefore sensory science will certainly continue to cross-fertilize with human sciences. Likewise, sensory and consumer scientists in the age of digitalization must have knowledge of these fields and receive some training in psychology and cognitive sciences. In the same line of thinking, the democratization of sensory science and sensory evaluation methods enabled by digital technologies opens the way to broader studies and to citizen science. This opens further perspectives not just to sensory and consumer scientists but to sensory and consumer science as a field and to the populations that may benefit from this evolution.

This book brings different and complementary perspectives to digital sensory science as it gathers contributions from a wide range of authors, all from very different horizons. I am grateful to J. Ben. Lawlor, Jean A. McEwan, and David Labbe for having put it together, and I am certain it will strongly benefit the readers, should they be students, or experienced sensory scientists and practitioners, in the academia and in the industry.

Julien Delarue
University of California Davis, Davis, CA,
United States

CHAPTER 1

Introduction

J. Ben. Lawlor[1], Jean A. McEwan[2] and David Labbe[3]

[1]Nestlé NPTC, Singen, Germany [2]Jean A McEwan Consulting Ltd., Bishop's Stortford, United Kingdom [3]Société des Produits Nestlé S.A. Route du Jorat, Lausanne, Switzerland

Any sufficiently advanced technology is indistinguishable from magic. **Arthur C. Clarke**

1.1 Introduction

1.1.1 Why this book now?

It is well-recognised that effective and efficient leveraging of digital technologies is critical for consumer *value capture, creation and delivery*. The arrival of digital sensory science, an area that has broadly been defined here as *the use of technology to measure, explain and/or predict sensory perception*, reflects our sensory mindset of constant reinvention. The rapid advances in technologies driving digitalisation in sensory and consumer science have meant that more than ever we can now truly put consumer/user experience at the heart of the equation to *create value for*, and *be valued by*, our consumers. However, despite the buzz and excitement that digitalisation creates, there remains a lot of debate and confusion around exactly what digitalisation means for sensory science, how it can be implemented and leveraged and how it will affect our ways of working, our perspectives, etc. Where do we start? Our goal with this book was to demonstrate how sensory science is, in fact, already fully engaged and on its digital transformational journey, to illustrate the opportunities and new options it has led to in our field and, of course, to point out some of the challenges that are still ahead of us. In this way, we hope to bring this much sought-after clarity and how digital will evolve the global sensory science community's worldview. To achieve this, we contacted several key opinion leaders across the sensory community and asked them to contribute a chapter of their current work/perspective on digital sensory. We consciously chose a wide spectrum of contributors from research, agency and industry backgrounds as we hoped to capture a more wide-lens view, which would be reflective of the dual research and practice nature and focus of sensory. To add relevance to the reader, each chapter comes with at least one case study in which authors clearly illustrate their topic in a research or practical application.

1.2 Organisation and macro summary of content

We have organised the chapters into five themed sections. All chapters come with different topics, ideas and examples that complement and build on each other by looking at digital through different aspects. Digitalisation is about connecting the dots — to do so we must first collect them. What we want to do in this short introduction here is to summarise, very briefly and at a macro level, some (as there are many more) of the main ideas that each author brings as we know for sure that the authors can express their focus areas far better than we could ever hope to.

1.2.1 Setting the scene

Following a forward from Julien, Section 1.2.1 begins with Chris (Chapter 2, The emergence of digital technologies and their impact on sensory science) and Martin (Chapter 3, The arrival of digitisation in sensory research and its further development in the use of artificial intelligence) setting the foundation for us in terms of detailing how we got to where we are today (a journey from pencil and paper to wearable smart devices). Chris points out the opportunities that digitalisation has brought to *data entry, statistics, method application and user experience* and notes that without the benefits of digital transformation, sensory science would be seriously limited in both its scope and effectiveness. Martin (Chapter 3) then very neatly demonstrates how digitalisation can provide both *agility* and *efficiency*. For instance, Martin details possibilities for the digitalisation of the sensory lab furnishings resulting in considerable work facilitation and quality improvements, and discusses how digitalisation is enabling a 'flexibilization' of research. An example is given through *Target Oriented Sensory Analysis* — an innovative approach that can tailor the sensory panel set-up to meet the objective and thus save on time, cost and general resources.

1.2.2 How digital is creating new opportunities for sensory science

Section 1.2.2 is about how digitalisation is changing how we think about, plan and perform sensory science with panels. Annika, Sven, Torsten and Manuel (Chapter 4, Digital toolbox - ways of increasing efficiency in a descriptive panel) and Sven, Annika, Manuel, Femke and Torsten (Chapter 8, Sensory self-service - digitalisation of sensory central location testing) demonstrate how digitalization can be leveraged in both descriptive analysis and consumer testing via the creation of an *automation process* driven by the creation of *digital platforms*, and how this can result in increased *efficiency, lower turnaround times and cost* while maintaining a no compromise mindset on data quality. In Chapter 5 ('Hybridisation' of physical and virtual environments in sensory design), Anne-Sophie, Chantalle, Nathalie and Victoire discuss new ways of working, enabled via application of digital technologies, including the enablement of *remote sensory panels* (remote meaning performing tasks out of the sensory lab but with participants connected centrally in real-time), *remote sensory trainings* and *remote team tastings*. All are facilitated by a 'phigital' (a hybridisation of both Physical and Digital worlds, seen as a continuum) way of thinking and working. For instance, by asking 'what panel tasks do we need to do in person and what panel tasks can be done remotely?' a hybridised solution with increased efficiency and equal effectiveness can be readily arrived at. The concept of using *phigital environments* is compelling and opens a myriad of new agile hybridised sensory solutions for research and practice. Navigating sensory into new territory, Phiala, Thibault and Huizi

(Chapter 6, Sensory evolution: deconstructing the user experience of products) discuss their concept 'Digital Trained User Panels' (*think like a consumer — report as a panellist*), the novel thinking being that in addition to the use of the trained panel in the sensory lab, we move to *the use of product users as panellists*. The shift here is about *moving from a pure product focus to an entire product 'deconstruction' process to understand multiple consumer touch points*, an approach that *brings us closer to understanding sensory from a more valid consumer perspective*. Phiala and colleagues also point out that success in the application of new technologies in sensory science will also demand striking the *right balance between academic standards for methods on the one hand with a strong sense of commercial viability on the other* (a similar point is made by Kathryn and Michelle in Chapter 11, Leveraging neuro-behavioural tools to enhance sensory research). In Chapter 7, Using digital tools to understand individual differences in flavour perception which impact on food preferences, Frances discusses the benefits of digitalisation to sensory panel data by facilitating a deeper understanding of *individual differences between panellists* in terms of *sensory sensitivity to different aroma compounds*. Consumers *(including panellists) are not homogenous in their perception* (also see Chapter 9, Predicting sensory properties from chemical profiles, the ultimate flavour puzzle: a tale of interactions, receptors, mathematics and artificial intelligence, by Andy) and this has important implications in how we should look at and interpret data in areas including sensory quality control and profiling studies for product development. There are also implications for the panel make-up where Frances recommends a panel with 'enough members with a range of sensitivities to cover all off-notes of interest'. Frances provides the reader with clear case study examples of how leveraging digital to understand individual panellist sensitivity contributes to more actionable and powerful business insights. For anyone working in taint/off-note detection, this chapter should prove exceedingly thought-provoking.

Throughout this book, the reader will observe that several authors have commented on how the increased digitalisation of sensory will affect our sensory *skill-set* requirements (see Chapters 2, 6, 7 and 9 of Chris, Phiala, Thibault and Huizi, Frances and Andy, for instance). One universal conclusion is that it will be *impossible for every individual practitioner to have all the requested skills in all the disciplines around digitalisation*. Success in digitalisation will only be achieved by leveraging the skills of scientists working in other disciplines — or as Danni, Qian and Derek (Chapter 12, Emerging biometric methodologies for human behaviour measurement in applied sensory and consumer science) put it, will just entail more 'teaming-up' of different disciplines. Such an approach to new ways of working will need to be more integrative in nature (e.g. an *interdisciplinary* or a *transdisciplinary approach*) to synthesise elements of different disciplines, including their perspectives, knowledge, concepts, theories, methods, etc. Diversity, meaning targeted diversity, of team skills will be essential. However, despite this, skills of sensory scientists (and panellists) will still require some degree of 'digitally focused levelling up'. The question then becomes whether the need is for sensory to integrate with different disciplines either on a permanent team basis or on a part-time, individual project basis. There is no one-size-fits-all solution to this and team skills and technologies should be rooted in business and research objectives. It is also worth mentioning that the reader should consider *both hard and soft skill needs*. The digitalisation challenge is not just about evolving / learning and practicing more advanced technical knowledge; it is also about how we will teach our teams to be more creative, agile, better communicators and influencers and ultimately, be purpose-ready sensory leaders in a digital age.

1.2.3 Digitalisation in instrumental, neurological, psychological and behavioural methods: current applications and opportunities

This section begins with Andy (Chapter 9) and Gianmarco, Michal and Patrick (Chapter 10, Electronic noses and tongues: current trends and future needs) detailing the current usage of instrumental measures (gas chromatography mass spectrometry (GC-Ms), electronic nose (e-nose) and tongue (e-tongue)) to explain and/or predict sensory flavour perception. Definitions of Artificial Intelligence (AI) and Machine Learning (ML), their goals and current uses in instrumental analysis are provided. Gianmarco, Michal and Patrick note that instrumental technologies such as e-nose and e-tongue are designed to act as *'digital analogues'* of human biological sensing functions to *simulate human behaviours* and *solve complex problems* using trained models. Output yields an *estimate of human perception*. Andy (Chapter 9, Predicting sensory properties from chemical profiles, the ultimate flavour puzzle: a tale of interactions, receptors, mathematics and artificial intelligence) discusses the feasibility and readiness of instrumentation by detailing three main areas that are hindering the development of predictive flavour models, namely, the *complexity of the flavour perception mechanisms* and the *need for high-quality data in both chemical and sensory analyses*. The challenges of moving from an understanding of flavour perception in aqueous systems containing single or mixtures of tastants and odourants to an understanding in real foods with eating and drinking behaviours with consumers that vary (genetically, culturally, physiologically) in their perceptions are surely complex. Andy emphasises that the *success of AI and ML will depend on the quality of the chemical and sensory data used to establish the model*. For instance, for sensory data, the choice is either to use expert panel data, a dataset that represents a very small proportion of the population but with high-quality data, or to use a wide sampling of the consumer population but in the knowledge that variance will be higher and the model may not be very accurate. In Chapter 10, Electronic noses and tongues: current trends and future needs, Gianmarco and colleagues discuss the use of e-nose and e-tongue that leverage ML and highlight current applications in case studies of sensory science in food safety and innovation. *Technological advances in miniaturisation of sensors, greater portability of ML models and software and affordability* have meant that these tech tools are starting to facilitate *decentralised analysis* in more *consumer-relevant environments* (also see Chapter 13, Added value of implicit measures in sensory and consumer science, of René and Lucas for similar comments on tools in a different field). Nevertheless, as consumer acceptance/liking is the main objective in industry, *e-noses and e-tongues are never expected to completely displace sensory panels but could be effectively used to support and complement sensory panels* in their mission by offering an objective platform that can be trained with sensory knowledge. On this point, the reader is referred to Chapter 7 of Frances, who gives an example of musty taint in wine (2,4,6-trichloroanisole) that requires sensory evaluation for its detection as current instrumental limits of detection around 10 ppt and assessors detecting at the nose at much lower levels (with different sensitivities reported). For digitalisation to be successful, Gianmarco, Michal and Patrick point to the importance and need for *'a common framework'* of instrument best practice in terms of *interoperability, training, validation and benchmarking* to promote their wider use. Authors cite two important *paradigm shifts* that digitalisation will bring to us: (1) facilitate the outsourcing of some sensory tasks to automated, high-throughput systems (therefore saving resource, etc.); and (2) cause a shift from chemical analysis targeting recognition of specific molecules to a *data-driven analysis* of complex matrixes.

The next chapters (Chapters 11−13) discuss the application of methods from neuroscience, psychology and behavioural sciences in sensory science with case studies focused on digitised methods. Traditionally, sensory science leverages self-report methods to capture explicit/conscious consumer feedback on product performance. However, two points are important here: (1) the use of self-report methods, with metrics such as liking, has mostly failed to predict long-term product success in the market. (2) Such self-report methods are generally product-centric and, therefore, unable to capture other key consumer touch points that contribute to holistic user experience and determine long-term product market value (see Chapter 11, Leveraging neuro-behavioural tools to enhance sensory research, of Michelle and Kathryn). As René and Lucas (Chapter 13) put it, 'if explicit tools were successful in the development of food products there would be no need to add more tests to this repertoire. Unfortunately, this is not the case'. Authors in all three chapters give excellent overviews of methods from neuroscience, psychology and behavioural sciences, detail what each method measures, their opportunities and challenges and how digitalisation is affecting their use in sensory via important case studies. For instance, in Chapter 12, Danni, Qian and Derek show the application of, and technology behind, measures including eye-tracking to capture visual attention, electrodermal activity for implicit emotions/arousal and electroencephalography to measure cognitive processes. Authors write how technological advancements have clearly facilitated some of these *technologies becoming more price competitive, democratised their use into other disciplines and have mobilised methods out of the lab to measure in naturalistic settings and in-the-moment user/consumer experiences*. It is important to point out to the reader that authors emphasise that the methods they describe in these chapters are not superior to traditional (explicit self-report) tools but rather that it depends on what needs to be measured and, possibly, where. As Michelle and Kathryn (Chapter 11) write 'choose the right (combination of), tools for the right research questions'. In the final chapter of this section (Chapter 13), René and Lucas argue that implicit measures are not just a more expensive and a more complex equivalent of traditional explicit tools but that both types of measures provide *complementary information*. Explicit measures are more introspective, reflective and *focused on the product*. Implicit measures go beyond the product, particularly when used outside of the lab, and capture the *whole product experience* − so a broader food experience in real-life conditions. This has an important implication on how to best capture holistic user experience: *Location matters*. Ideally, as René and Lucas write, implicit measures would be *plug-and-play* regarding data recordings, processing, analysis and interpretation with the capacity to operate in real-life environments. Currently, this is not fully the case but encouraging steps in sensor miniaturisation and software development are being made. Another point is the development in immersion technologies, which when combined with implicit measures in a lab can offer a solution to some current shortcomings of implicit measures.

1.2.4 Immersion technologies, context and sensory perception

Section 1.2.4 discusses how digital immersion technologies are facilitating more context-relevant studies to be performed (for definitions of context and immersion, see Chapters 15 and 16, respectively). Immersive technologies can be nondigital or digital. Nondigital can include imagined scenarios or physical rooms created to mimic particular environments such as a coffee bar (for an overview, see Chapter 14, Using virtual reality as a context-enhancing technology in sensory science, of Emily and Cristina). Digital can include Virtual Reality (VR) which immerses participants within a computer-generated virtual environment, Augmented Reality (AR) that enhances views of the real world by overlaying what participants see with computer-generated information and Mixed Reality (MR) which sits between AR and VR (see Chapter 15, Next-generation sensory and consumer science: data collection tools using digital technologies, of Rebecca, Imogen and Qian). In Chapter 14, Using virtual reality as a context-enhancing technology in sensory science, Emily and Cristina discuss types of VR systems and how immersive VR technologies are being used to simulate diverse and different consumer environments (restaurant, busy street or countryside) for the sensory assessment of food, thus adding agility to challenges around study location (e.g. study cost, time as well as sample shipping and preparation, etc.) and the validity/relevance of data collected in a lab compared to a 'real-life' environment. Like René and Lucas, Emily and Cristina illustrate in their case studies that *environment matters (to attributes and liking), adds meaning, validity to real-life* and increases *consumer engagement*. However, authors point out that the effect of immersion and context may vary by factors including *context-sensitivity of individual foods and consumers, situational appropriateness and congruency*, and of course, how hedonic responses are captured using an immersive VR device. Using a novel case study, Rebecca, Imogen and Qian (Chapter 15) demonstrate the added value of using MR technologies to *merge both real and digital worlds together*, an approach that facilitates the embedding of questionnaires and the consumption experience into the immersion experience itself, thus overcoming some of the shortfalls of using VR technology. In Chapter 16, Multisensory immersive rooms: a mixed reality solution to overcome the limits of contexts studies, Adriana, Agnès and Jacques-Henry list the detailed requirements of a *Mixed Reality Immersive Room* using the one at Institut Lyfe (formerly Institut Paul Bocuse) as a reference — key information for any researcher hoping to recreate such an immersive space. As noted earlier, immersive technologies also offer a solution of how to use implicit methods that require control in uncontrolled 'real-life' environments. Using 'immersion in the lab', we can effectively achieve this, thus opening up the possibility of measuring user experience in a lab (see Chapter 13 of René and Lucas). In the final chapter (Chapter 17, Voice-activated technology in sensory and consumer research: a new frontier) of this section, Tian, Janavi, Natalie, Hamza and John detail the development of voice-activated technologies and how technology running on the *Internet of Things* (also see Chapter 10) will pave the way for smart speaker research. Advantages include a better fit for products (e.g. shampoos) and experiences (e.g. cooking) and capture of in-the-moment feedback. Challenges/future opportunities are also listed, and in their novel case study, Tian and colleagues illustrate that smart speakers can capture closed-ended quantitative data similar to an online/traditional survey but may have limitations for open-ended questions. Given time, and perhaps as consumers get more familiar with these methods, some of these challenges can be overcome.

1.2.5 How to tell powerful digital sensory stories

What better way to end the book than to talk about storytelling. Till now authors shared their original thoughts and real examples on how exactly digitalisation is transforming sensory science. But how do we now bring the new digital sensory stories to our stake-holders? With the emergence of new technologies fuelling new opportunities and options, an increasingly important part of the sensory scientist's job will be communicating our new stories in engaging and influencing ways. In Chapter 18, Shining through the smog: how to tell powerful, purposeful stories about sensory science in a world of digital overload, Sam details how we can achieve this task — of bringing together the data and the narrative to tell compelling stories. Sam neatly outlines the six golden rules of storytelling and parcels up the book rather neatly by noting that Analytics + Storytelling = INFLUENCE.

1.3 Final words

As existing digital technologies evolve and new ones emerge, sensory science will continue to advance in terms of both research capabilities and everyday practical applications. In this book, the authors have detailed some of the main areas in research and practice where digital is currently enabling positive change and of course, listed the current challenges. It is clear that digitalisation is enabling new metrics to be brought to the sensory testing table, metrics that have the power to complement, or fully displace in some instances, more traditional metrics such as liking for potentially more successful product development, but this will also require more interdisciplinary ways of working. Digitalisation is an enabler to a deeper understanding of the complexity and intertangled nature of global consumer/user experience and can lead to greater value creation. However, the true genius of digitalisation lies not in the technologies but in the curiosity, razor-sharp focus and consistent effort of the researchers and practitioners driving it forward. In wrapping up here, the final word must go to acknowledge and thank the authors. It was their contributions that made this book grow from an idea to a reality, and for this, we are grateful beyond words.

S E C T I O N 1

Setting the scene

2

The emergence of digital technologies and their impact on sensory science

Christopher J. Findlay

C-K Findlay Consultants Inc., Guelph, ON, Canada

2.1 The beginning

Before the introduction of digital technologies, sensory testing methods were chosen for their simplicity. By keeping methods simple, they provided answers that were useful and there were no great barriers to their adoption. The triangle test (Standard Test Method for Sensory Analysis—Triangle Test, 2018) could be conducted using sample sets randomised by using a deck of playing cards, modified into groups of 6 that represented the 6 possible presentation orders. At the end of the test, the results would be analysed by using a simple table that could be found at the back of most sensory publications. The ASTM E-18 Standard Method provides a clear description of the test and the tables to interpret the outcome. Books such as *'Laboratory Methods for Sensory Evaluation of Food'* by Elizabeth Larmond (Larmond, 1977), and her colleagues at Agriculture Canada, gave the beginner a framework of a sensory programme that covered difference testing, consumer preference and basic descriptive analysis. This relatively short publication found its way into hundreds of new sensory labs globally. No complicated calculations were required, and no challenging statistics were needed to interpret the results.

Then, the electronic calculator hit the market. Complex calculations were now within the reach of any scientist or technician who wanted a precise result. Early models were limited to basic arithmetic, but in no time at all, models that managed complex calculations and multiple step equations became common and affordable. It was not long before anyone with a bit of mathematical knowledge could programme their own equations. There was no easy user interface to shepherd neophytes through statistical analysis, but for anyone prepared to learn to programme the machine, previously inaccessible methods were available. A major impact on sensory science was felt in the blossoming field of sensory statistics.

In sensory evaluation courses, a typical lab exercise was to calculate the analysis of variance, manually, to help understand the steps involved in the analysis of variance (ANOVA) and then to do the same again using a calculator. For a modest-sized experiment with four products, evaluating five attributes on a ballot, with ten assessors, with three replicates there are 150 data points for each attribute. To do the ANOVA calculation by hand seems to take forever, with lots of opportunity for error. The calculator makes the process much faster, but there's still plenty of room for error. The use of spreadsheet programs automated the calculations by using a template or a wizard to do the heavy lifting. Of course, by the time fully computerised sensory analysis systems came along, the calculations all took place in the background, instantly. That instant process is easy to take for granted and serves as a reminder of the power of the digital transformation. Not only was a lot of time saved, but statistical analysis that was once too much effort became routine.

2.2 Photocopiers

Another digital electronic innovation that had a profound effect was the photocopier. The copier and the computer printer made the production of ballots and the dissemination of information quick and simple. Nowadays, we take the use of these tools for granted. The earlier technology for duplicating and printing paper forms was slow and difficult. Customising ballots became trivial, allowing tests to be tailored to actual needs. Although, the reliance on paper ballots would continue to restrict the scope and flexibility of sensory evaluation.

2.2.1 Digital consumers

A major digital innovation that exploded globally, starting in the late 1960s was the Automated Teller Machine (ATM or ABM). This introduced consumers to routinely interacting with computerised devices. The interface needed to be simple enough for anyone to use and secure enough to deter robbery. The success and wide adoption of the ATM has paved the way for a broad range of automated point-of-sale devices.

Video games also introduced people to interacting with digital devices. Children of the 1980s all grew up with a comfortable familiarity with computer-based machines for entertainment, education and communication. By the 1990s, mobile handheld phones became common and widely accepted. In addition to being a phone, these machines were quickly expanded in function to be music players, cameras and walking encyclopaedia. All of these innovations made the adoption of computerised sensory analysis viable.

2.2.2 Scanning and digitising

Paper is a very reassuring medium for business purposes. Anyone can pick up a sheet of information and interact with it spontaneously. A pencil allows the user to erase and change what has been recorded. A pen requires greater commitment to permanence and is less forgiving if an error is made. Transferring information from a ballot into a data table required a lot of work. This task was made more difficult when the data came in the form of scores from a line scale. The method applied to making this task manageable was to use a standard line size and a ruler to measure the value. Of course, if your line length was a convenient metric value, like 10 or 15 cm, it was easier to record the score. If the data could be recorded as an integer, less work for input was needed.

A great benefit of human judgement is the ability to look at a ballot and decide which marks are valid. A digitizer tablet allowed the user to place the ballot on the digitizer surface and mark the location with an electronic stylus. In this way, the user could compensate for damaged or distorted ballots quite easily. In food and beverage testing, it was not unusual for the ballot to be stained with the product being tested. A practiced analyst could process several hundred ballots in an hour, such as scanner speeds, but still a boring and tedious job at best.

2.2.3 Digitising

Digitising from paper forms emerged in several formats. The first was the electronic card reader (ECR). The ECR allows a fill-in-the-box data entry on a standard printed card that is machine readable. ECR works very well, with dedicated equipment and structured tests. Similarly, the digital scanner uses photocopying technology to take an image of the paper form and extract data from programmed locations on the sheet. Most people think of paper as a relatively stable medium. That's almost true, but when you need to handle and process hundreds of forms, wrinkles and misalignment create a real problem. Paper also expands and contracts with changes in humidity. Scanning large numbers of sheets requires automatic document feeders that can jam and mangle any sheet of paper that's not perfect. To ensure smooth handling, a paper press would be used to smooth and align the sheets, requiring yet another step in the process of scanning. Hypothetically, a scanner could process several thousand pages in an hour. In practice it was often only several hundred.

2.2.4 Paper ballots

Several sensory software products were developed to create paper ballots and to scan the results for analysis. The products were limited in their scope and required skill and patience to use, but they provided a steppingstone to fully automated systems. In addition, they were considerably cheaper than the early personal computer-based direct data collection systems. The relative simplicity, flexibility and low price led to a wide adoption of these systems, some of which are still in operation 30 years after their introduction. However, the continued requirement for paper ballots limited the benefits of digital technology. It was difficult, if not impossible, to manage complex incomplete block designs and temporal measures.

2.2.5 Sample identity

Another byproduct of scanning technology was experienced in sample labelling. Getting a blinding code on to a test sample seems to be a minor task, however, the test is only as good as the certainty of product identity. Typically, three-digit or character identifiers are used. When it is easy to print labels and prepare sample cups or plates in advance, that task becomes less onerous. In fact, if you can print bar codes or QR codes to identify samples, assessors can't even guess the code. Imagine having sample glasses etched with unique bar codes and being able to dispense samples by scanning the code just before serving. This technique has been used in practice for decades at major breweries. That way, appropriate glasses can be used without having to apply and remove labels. Glasses can be easily cleaned and sanitised and sample identity remains secure. In the same manner, it is possible to inkjet print codes directly on products, such as individual cigarettes, to ensure their identity. The same technology that allows self-checkout in retail stores can be used to have assessors record their own samples with certainty.

2.2.6 Direct data entry

Before wireless and the internet, the personal computer was described as the equivalent of a car that sat in your driveway with nowhere to go. There were no data highways to travel on. PCs had a lot of power, but limited connectivity. Local area networks that were being marketed in the 1980s were not built to a single standard. The hardware and software to create the LAN were both expensive and complicated. For early adopters of direct data entry, the cost of hardware was 80% of the cost of implementation. This also required the active assistance of the organisation's computing services department, at a time when many business-critical activities were demanding the same resources. Consequently, it was quite common for researchers from both industry and academia to build their own networks. This photo shows an early sensory system of networked PCs for trained assessors in an open area (Fig. 2.1).

Many researchers built their own software to do exactly what they wanted it to do (Brady et al., 1985). Most programs were written in Basic and ran in the Microsoft DOS

FIGURE 2.1 Networked PCs in an open sensory data collection room. Descriptive sensory panel leader prepares study for trained assessors. *Source: C.J. Findlay.*

(disk operating system) environment. Because they were built for a specific purpose, they lacked flexibility and the software tended to be largely undocumented, both for the end user and by the programmer. This led to a serious problem when the creator of the system changed jobs and was no longer available to maintain the system. The challenge was also exacerbated by the frequent upgrades in operating systems. In the period between 1985 and 1995 there was a new DOS operating system launched almost every year and twice in some years. These releases operating system upgrades fixed bugs and introduced some new features, but at the same time required rewriting existing sensory software code. Many updates of early sensory software were made for the sole purpose of ensuring DOS compatibility, not to enhance software functionality. Sensory software makers felt that they were constantly developing their programs just to keep up with the technology. There was also concern expressed about whether or not the computer collected the same data as on paper ballots. Was the process itself a source of bias and influenced the quality of results. Several studies concluded that there was no significant difference between paper ballots and direct entry computerised ballots (Swaney-Stueve & Heymann, 2002).

The keyboard became the default system for direct data entry since all personal computers required a keyboard. However, for people who had no experience with keyboards or computers, it was difficult to enter their responses. Friendly and intuitive user interfaces were not part of the early systems used to operate computers. To make it easier, data entry was restricted to as few keys as possible. In some case, paper templates were used to cover the unnecessary keys. Although trained panellists could be taught how to work with the systems, no matter how counterintuitive, a lot of effort was required to make it work. Matters were made worse when researchers tried to get naive consumers to enter their own responses.

2.2.7 The light pen

Long before Luke Skywalker waved his light sabre in Star Wars, the light pen (Fig. 2.2) provided a digital input device for video screens. The operation was simple, the pen tip was a light detector with a switch that was triggered by touching the CRT screen. The

FIGURE 2.2 **Light pen for personal computer.** A simple direct data entry device used to respond directly on screen.

location on the screen was detected by the position of the raster. Although resolution was limited to the pixel level, fortunately, the data collected did not require great accuracy. For example, to mark a score on a line scale anchored at zero and one hundred which was 60 pixels in length, the smallest unit that could be recorded would about 1.3. The biggest advantage to the use of the light pen was its intuitive similarity to a pencil on paper. The biggest drawback was that the light pen had to be installed and wired to each computer. Although this tethering stopped consumers from walking off with the device. The computer mouse was in the process of being introduced, it was still a great novelty and anecdotes at the time described numerous challenges due to eye-to-hand coordination difficulty on the part of assessors (Kuesten et al., 1994).

2.2.8 Personal digital assistants

Personal digital assistants (PDAs) such as the GriDpad and the PalmPilot were introduced around 1990 as electronic notepads. They used a stylus to input directly on their screens and became a simple and direct way of getting data entry. The devices were not cheap and came with several drawbacks, including the ease with which any panellist could walk off with a stylus. Tests were created and uploaded to each device individually. Data were collected locally on each device and then downloaded and synchronised at the end of the test. Many sensory labs found these devices effective, regardless of their limitations. They provided the benefit of portability and remote data collection. It became possible to transport a sensory test to a field location. In this photo a technician is using a tablet to input sensory quality data (Fig. 2.3).

Battery life and fragility were part of the cost of working with these devices, but they provided a tantalising glimpse of the future of data collection.

In the 1990s, there was an explosion of graphic capabilities in personal computing. The mouse soon became the ubiquitous input device and screen resolutions increased to photographic quality. This progress provided sensory software with greater flexibility and

FIGURE 2.3 **Technician evaluating sensory quality using a digital tablet.** Portable pen-based tablets make data collection very flexible. *Source: C.J. Findlay.*

more functionality. Microsoft rapidly developed its Windows operating systems, and the graphic user interface became much more user friendly and design standards helped reduce learning time for new applications. Now, most households have a home computer, which may be a laptop or desktop, and internet connection. This provides researchers with potential access to a large population of appropriate consumers. Once a consumer database is established, recruitment and screening that took days of telephone calls happens in minutes, thanks to the internet.

2.2.9 Touchscreens

Apple touchscreen devices were introduced in 2007, with the first iPhone, and in 2010 with the first iPad. Smartphones and smart tablets could be networked easily, and software became cloud-based, liberating data collection from the original locally installed software. This fundamental change has revolutionised data collection. Literally any internet connected device can be used for data entry. There are hundreds of devices that operate using Android or iOS systems. A question that has been raised about the comparison of data collected on vastly different sized screens (Fig. 2.4).

It was found that for unstructured line scales, although there is a small difference, that the effect of device size is minimal (Fisher et al., 2016). There is no appreciable difference with structured scales that use boxes to capture discrete responses.

FIGURE 2.4 **Flexible data input devices of various sizes.** Every internet device that has a browser can be used to collect data. *Source: C.J. Findlay.*

2.3 Statistics unleashed

Statistical packages have had a major impact on sensory science. The ability to handle large datasets with sophisticated mathematics has opened the doors to deeper understanding of innumerable connections that assist in the interpretation of results and provide startling insights. Most sensory software automatically produces extensive statistical reports including impressive graphic output. The presentation of results has transitioned from numerical summary tables to dynamic graphical representations. Multivariate statistics (Dijksterhuis, 2008; Meullenet et al., 2008; Multivariate analysis of data in sensory science, 1996) are typically displayed in the form of a biplot. The first two dimensions or principal components provide a sensory map of the products tested. These plots are great conceptual tools and can lead to interesting insights into the success or failure of products. The caveat has always been to account for enough variance that the picture created resembles the truth. An illustration of a conventional principal component analysis (PCA) of 15 products can be found in Fig. 2.5.

A more accessible PCA of whole grain breads uses photos to identify the products and their arrangement on the biplot (Fig. 2.6).

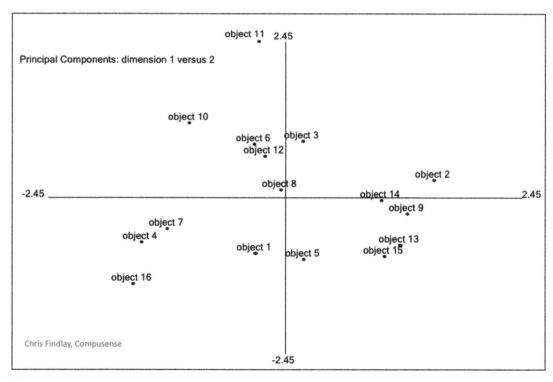

FIGURE 2.5   Conventional principal component analysis biplot of 15 samples identified by number. Principal component analyses have become a standard tool for displaying products in a sensory space. *Source: C.J. Findlay.*

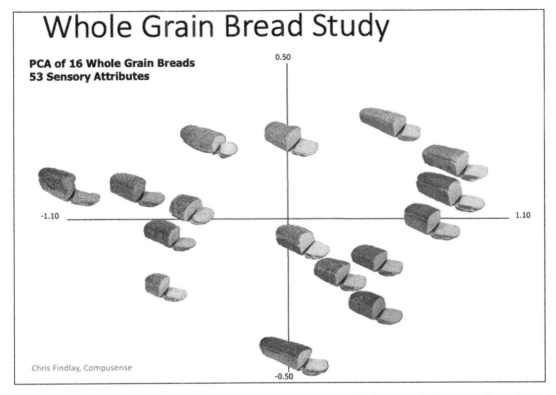

FIGURE 2.6 **Enhanced principal component analysis using graphic images of 16 test products.** Images improve the ease of interpretation of results for clients. *Source: C.J. Findlay.*

It is easier to identify the visual attributes that contribute to the organisation of the samples. If they were only points and numbers on the plot, the similarities and differences between products would not be as apparent.

Often, final reports take the form of computerised files that are presented as slide shows, commonly known as decks. Where it was once a major task to prepare a slide show presentation, so much of the process is automated. It has become easy to create a branded and polished format suitable for presentation to your top management. What once required a week of painstaking graphic manipulations is now done automatically in minutes, providing more time for the sensory scientist to interpret the results. Of course, that has also raised the expectations of clients for fast and professional results.

2.4 Temporal methods

The original collection of time—intensity data was achieved through recording a line on a moving strip chart with a pen (Guinard et al., 1985). The curves were tedious to record

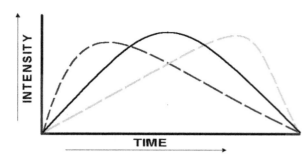

FIGURE 2.7   **Time−intensity curves modelling early, intermediate and late temporal intensity peaks.** Continuous data collection is easy in a digital system. *Source: C.J. Findlay.*

and challenging to extract data and analyse. Computerised data collection of temporal data has become an essential tool in the sensory scientist's kit (Dijksterhuis & Piggott, 2000). Many products are characterised by their temporal properties. You can consider when beers of the same bitterness deliver that sensory experience. Temporal effects may also be observed with the heat of different peppers. Single attribute changes may be tracked using continuous time intensity (Fig. 2.7).

Time−intensity curves with similar total area under the curve (AUC) demonstrate early, intermediate and late temporal peak intensity that differentiate the products. Additional attributes are more challenging and have led to methods such as temporal dominance of sensations (TDS) or temporal check-all-that-apply (TCATA) (Ares et al., 2015). The collection and analysis of these types of data would be virtually impossible were it not for digital technology.

2.5 Summary

Digital technology has had a major impact in three important areas of sensory science: data entry, statistics and the application of computer-enabled methods.

Data entry has been transformed from paper-based ballots that required extensive time to create and much more time to score and tabulate, to a process that's both quick and easy. According to Findlay et al. (1986), typical conventional paper sensory descriptive analysis could take a week to turn around a set of results. The integrated computerised system could accomplish the same in less than an hour. The time savings had the impact of permitting more sensory studies to be conducted with the same resources. A major paper products company estimated (anecdotally) that when they implemented a computerised system, their volume of work increased from 100 to 600 studies a year. In addition to the rapid work cycle, issues such as last-minute changes to a ballot became trivial. Because results were available rapidly, sensory data became a key factor in business-critical decisions. This ended the dependence on subjective decisions being made pragmatically by management.

The statistical treatment of sensory data benefitted greatly from direct data entry as there was less opportunity for transcription errors and guaranteed consistency in data format. By standardising data formats, analysis was much easier since preparation of data files was automatic. The increased numerical processing capacity of personal computers also made the calculations themselves much faster. This increase in computing power also opened the door to more rigorous regression analysis and modelling. Routine application of multivariate analysis, such as Generalised Procrustes Analysis (GPA) or PCA (Dijksterhuis, 2008; Meullenet et al., 2008; Multivariate analysis of data in sensory science, 1996) provided value-addition in the interpretation of complexity in product sensory properties and consumer diversity.

The computer-enabled methodological opportunities that were not available to a paper-based study. Interactive and adaptive techniques utilising conditional ballot branches and entry criteria became well accepted. Complex experimental designs, particularly incomplete blocks, could be generated reliably with balance and proper order rotation as they were required. Temporal measures became much more accessible, and methodology expanded from single-attribute time-intensity to temporal dominance of sensations and TCATA and their many permutations.

Without the benefits of digital transformation, Sensory Science would be seriously limited in both its scope and effectiveness. Sensory methods have become an integral part of User Experience Design (UXD). New product development in almost every commercial field is being shaped and led by the application of UXD. Better design has reduced the costly failure rate of new product introductions. After all, if a developer can anticipate consumer's needs and desires, it is much easier to deliver a product that meets consumer expectation. Digital technology has driven the growth of sensory science by making sensory tools accessible, rapid and affordable.

2.6 Quo Vadis

The future is always an interesting place. Our literature is full of utopian and dystopian narratives that project our possibilities. There are many pundits who paint dramatic visions of digital worlds dominated by machines and managed by artificial intelligence. Since I am not a pundit, I will reserve my predictions to the logical extension of where we really are. Wisdom tells me that humans are actually 50,000-year-old hardware trying to run 21st century software. So although we have great aspirations, we must still nourish and develop our physical bodies as well as our minds. There will continue to be demand for unique and individual product experience. We all enjoy our differences, even if we try to predict what "everyone" will like. The Metaverse will provide an environment for the mind that will evolve in diversity and complexity. It will not replace the "real" world. To understand how we can spend our planets finite resources in the most beneficial way for the greatest number of people in a diverse world, product creators will come to rely on the insights of sensory and consumer scientists. To generate those insights, sensory practitioners must become adept in using the digital tools that are available.

The word of caution that I must add is that it is impossible for every practitioner to have state-of-the-art skills in all of the disciplines that combine to deliver outstanding sensory and consumer outcomes. There is no doubt that early adopters of the automobile, the computer or the internet had to understand the inner workings of their novel technologies. The same may be said for the management of big data and the application of artificial intelligence. Consequently, effective future sensory scientists must know enough of each of the disciplines that they have to manage to execute any project using a combination of technologies. You do not have to know everything yourself. You must know who you can call on and trust when you need their assistance. In addition to the core skills that sensory and consumer scientists already require, I would add the following. Advanced project management, behavioural economics, data visualisation, virtual and augmented reality, cultural anthropology, mathematical modelling, data mining.

Be prepared for a lifetime of learning. The future is very exciting and fortune favours those who are prepared.

References

Ares, G., Jaeger, S. R., Antúnez, L., Vidal, L., Giménez, A., Coste, B., Picallo, A., & Castura, J. C. (2015). Comparison of TCATA and TDS for dynamic sensory characterization of food products. *Food Research International*, 78, 148–158. Available from https://doi.org/10.1016/j.foodres.2015.10.023; http://www.elsevier.com/inca/publications/store/4/2/2/9/7/0.

Brady, P. L., Ketelsen, S. M., & Ketelsen, L. J. (1985). Computerized system for collection and analysis of sensory data. *Food technology*.

Dijksterhuis, G. B. (2008). *Multivariate data analysis in sensory and consumer science*. John Wiley & Sons.

Dijksterhuis, G. B., & Piggott, J. R. (2000). Dynamic methods of sensory analysis. *Trends in Food Science and Technology*, 11(8), 284–290. Available from https://doi.org/10.1016/S0924-2244(01)00020-6.

Findlay, C. J., Gullett, E. A., & Genner, D. (1986). Integrated computerized sensory analysis. *Journal of Sensory Studies*, 1(3–4), 307–314. Available from https://doi.org/10.1111/j.1745-459X.1986.tb00180.x.

Fisher, C. M., King, S. K., Castura, J. C., & Findlay, C. J. (2016). Does data collection device affect sensory descriptive analysis results? *Journal of Sensory Studies*, 31(4), 275–282. Available from https://doi.org/10.1111/joss.12210; http://www.blackwellpublishing.com/journal.asp?ref = 0887–8250&site = 1.

Guinard, J., Pangborn, R. M., & Shoemaker, C. F. (1985). Computerized procedure for time-intensity sensory measurements. *Journal of Food Science*, 50(2), 543–544. Available from https://doi.org/10.1111/j.1365-2621.1985.tb13450.x.

Kuesten, C. L., Mclellan, M. R., & Altman, N. (1994). Computerized panel training: Effects of using graphic feedback on scale usage. *Journal of Sensory Studies*, 9(4), 413–444. Available from https://doi.org/10.1111/j.1745-459X.1994.tb00257.x.

Larmond. (1977). *Laboratory methods for sensory evaluation of food*. Research Branch, Canada Department.

Meullenet, J. F., Xiong, R., & Findlay, C. J. (2008). *Multivariate and probabilistic analyses of sensory science problems*. John Wiley & Sons.

Multivariate analysis of data in sensory science. Elsevier, 1996.

Standard Test Method for Sensory Analysis—Triangle Test. 100 (2018).

Swaney-Stueve, M., & Heymann, H. (2002). A comparison between paper and computerized ballots and a study of simulated substitution between the two ballots used in descriptive analysis. *Journal of Sensory Studies*, 17(6), 527–537. Available from https://doi.org/10.1111/j.1745-459X.2002.tb00363.x; http://www.blackwellpublishing.com/journal.asp?ref = 0887-8250&site = 1.

3

The arrival of digitisation in sensory research and its further development in the use of artificial intelligence

Martin J. Kern

SAM Sensory and Consumer Research, Munich, Germany

3.1 Digitisation is already old

When we started collecting data with a computer in the late 1970s, it was the beginning of digitisation, which has been spreading ever faster and ever more comprehensively ever since and never stops. Year after year, digitisation made new steps and penetrated into more dimensions. In this very dynamic process of digitisation, we can identify several leaps that have been significantly and paradigmatically changing the way we work:

- Computer technology was the first and earliest step in digitisation: it made it possible to measure intensity using digital techniques, replacing paper and pencil and the manual measurement of intensity.
- Internet availability − the launch of the internet ushered a second wave of digital penetration, allowing data collection programs to be available online and measurements to be made regardless of location. Software development has become agile and traditional licensed deployment to the end user is now almost completely replaced by the use of SAAS (Software as a Service), which is updated daily − and whenever we use it, we have the latest version.
- Digitised communication in combination with audio visualisation/videoconferencing was another big accelerator to take digitisation to a further and deeper level − and digitised communication/video conferencing got a huge boost with the 2020/2021 pandemic situation (Fig. 3.1).

FIGURE 3.1   Development of digitisation in sensory and consumer research. *Source: SAM Sensory and Consumer Research.*

In parallel, and in the background, there have been many other developments that support digitisation and increase the speed of invasion of digitised techniques and tools, such as:

1. Increased capabilities of computer performance.
2. Worldwide availability of servers and the ability to temporarily set up virtual servers in cloud solutions such as Azure (Ms), Amazon Web Services (AWS), Google Cloud and others for data collection — and to close them down again after conclusion of the work. It was this step that made centralised project management and global outbound from it possible, extremely fast and very highly efficient.
3. The digitisation of speech, fully digitised speech recognition with everything that goes with it complete the possibilities of digitisation. This includes smart speakers such as Siri (Apple), Alexa (Amazon), Google Assistant (Google), Bixby (Samsung), Cortana, etc.
4. The enormous development of smartphones and tablets as personal devices, which is now a standard for everyone worldwide, ensures that digitisation is available everywhere and in every area.

As we write this article in early 2022, we can see that the level of digitisation in all areas is simply overwhelming. On the contrary, users of sensory and consumer research are finding it difficult to recognise and take advantage of all the opportunities available to them. Very often, the way companies do their work falls far short of the digital possibilities — and the speed of development makes it hard to keep up.

One important aspect of digitisation is that it is not so important to get the best software or the best technical equipment to exploit all the possibilities on offer today. Much more decisive is the combination of the individual components with each other in order to have the most powerful solution for the particular company-specific situation.

This article therefore attempts to give a picture of the solutions that are possible today — without claiming to be complete.

3.2 State-of-the-art sensory analysis

The way we can conduct Sensory Analysis today has gained an enormous lift in flexibility, ushered by COVID-19 and the requirement to adapt many workflows to work from home (WFH), including panel training and product preparation.

Best practice is to conduct sensory analysis in designated rooms with professional sensory equipment. The premises include a training room that allows for moderated creation of the attribute list and experience of the sensory universe of the test set, a measurement room with professional sensory booths and a separate preparation room. It should go without saying that all rooms are air-conditioned and have uniformly defined lighting conditions so that work can be done regardless of weather conditions. Ideally, there is a slight negative pressure on the preparation side and a slight positive pressure on the test subject side, so that odour from the preparation side can be safely prevented from reaching the test subject side.

Depending on the particular needs of a specific product category, the equipment can also be considerably more expensive, for example, controlled humidity may be required, spittoons or even washbasins with running hot and cold water with a very precisely defined temperature range or even, in the case of fragrance measurement, completely enclosed cabins with laminar overflow.

3.3 Digitised furniture for sensory evaluation

Even if the basic requirements for the sensory equipment have not changed, there are already possibilities for digitisation of the furnishing, which results in considerable work facilitation and quality improvements, as will be shown below:

- Connecting workstations to a LAN network and/or W-LAN network is today's standard. It is the basic prerequisite for exploiting all the possibilities of working with a tablet or laptop. The associated advantages are a considerable simplification and increase in efficiency in all individual steps of the sensory analysis: training, moderation, coordination amongst the test persons, visualisation and data acquisition itself in connection with a powerful sensory software.
- The aforementioned requires individual workplaces to be equipped with power sockets, LAN connection (K-Stone and/or USB modules), USB charger, HDMI connection and more to benefit optimally from the use of the workplaces.
- The room lighting as well as the lighting of the sensory workplace in the booth is LED. The light colours are daylight white, for normal sensory analysis; LED strings in RGB (Red, Green, Blue) allow any other colour, thus standardising the colour of product samples and making them less detectable to the subject during assessment. This means that the assessment can be carried out independently of the product colour. Today's possibilities of digitalisation should also be taken into account for the lighting. Lighting is controlled without switches via light-management systems such as Casambi, Gira, Gigaset or similar and can be set more precisely via those than via switches.

3.4 Room management — smart networks via KNX-control

The control of the many components that determine the working conditions for sensory analysis can be done individually (conventional switching), or via a so-called smart network, which is used by many companies, and especially hotels. A smart network brings enormous advantages because it allows central control of all relevant issues that are responsible for space management, such as light, air conditioning, humidity control, availability of water, ventilation, signalisation, just to name a few. The standard for such a network is KNX (KNX is an abbreviation for the word Konnex). KNX is based on three technologies: EIB (European Installation Bus), EHS (European Home Systems Protocol) and BatiBUS.

Fig. 3.2 shows the digital components of a room with sensory cabins for performing sensory analysis.

3.5 High-performance sensory tables: intelligent, digitised and modular for sensory evaluation

High-Performance Sensory Table systems are developed for classical conference and executive usage, training purposes and of course, its core function as a Sensory Booth for Descriptive Analysis. The elevating partitioning walls transforming the table within a few seconds from a conference table with training purposes to a sensory booth with evaluation purposes for descriptive analysis.

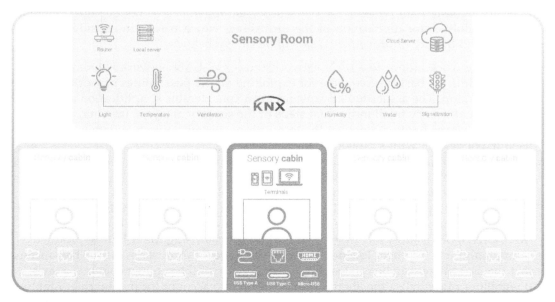

FIGURE 3.2 Digital components of rooms with for performing sensory analysis. *Source: SAM Sensory and Consumer Research.*

In the original state (partitions retracted), the table functions as a normal conference table to be used independently for sensory tests/descriptive analysis as well as for the pre- and post-meetings and result-discussions of sensory sessions (or whatsoever). At the press of a button, the conventional conference room table transforms into a valuable piece of professional sensory equipment with its smoothly and quickly elevating partitioning walls. The three use cases can be described as follows:

- Sensory Booths

  Partition walls form a complete Sensory Booth with full separation and protection of each sitting respondent. Each booth is equipped with power socket, one LAN connection, USB connections, etc. A high-power LED lighting is integrated, which homogeneously illuminates each working place allowing daylight as well as the entire colour spectrum RGB.
- Training Tool

  The perfect combination of the Sensory Table used for panel training optimises the use of rooms increasing the overall working efficiency of each descriptive panel.
- Conference Table

  The sensory table is clean in design and highly resistant in material, with intelligent media technology integrated. Thus it is ideally used also as a conference table (Fig. 3.3).

FIGURE 3.3 SAM Sensign Sensory Table. *Source: SAM Sensory and Consumer Research; Sensign® Sensory Furniture.*

1. Setting the scene

3.6 The software

The use of powerful software is of elementary importance for the degree of digitisation that can be applied. Good software for sensory analysis makes it significantly easier to conduct training and data collection, analyses and report. And it provides all of this in one: In addition to pure data entry, it also includes tools for visualisation and panel performance. When data collection is completed, raw data set, graphics and panel performance are directly available. During a single-panel training session, it allows for instance to collect data, analyse them and provide a thorough feedback to each individual panellist. This has a powerful and positive impact on panel performance.

Most packages today are offered as SaaS (Software as a Service). This is a great advantage as the user always has access to the latest version of the application. Table 3.1 shows the different providers.

Using a sensory software also increases the flexibility of the sensory analysis.

3.7 Performance characteristics of sensory analysis software

What should a modern sensory software offer and do today? The below list gives some ideas, although it does not claim to be complete:

- Intuitive programming.
- The availability of different scales.
- Offers an option to create an even more complex presentation design.
- Can provide result-outputs in different formats to allow further analysis. Raw data also should be available in different formats.
- Can integrate time management and time sensitive approaches, such as progressive profiling, TDS, TCATA, break for palate cleansing, etc.
- Can allow for the implementation of pictures and videos (e.g., to explain protocol of testing).

TABLE 3.1 Overview of the different software providers for data collection.

Name	Origin	Website
Compusense	Canada	https://compusense.com/
Eyequestion	The Netherlands	https://eyequestion.nl/
Fizz/Biosystem	France	https://www.biosystemes.com/en/fizz-software.php
isiSensorySuite	Germany	https://www.isi-insights.com/en/sensoryservices
Redjade	USA	https://redjade.net/
SenseCheck/Cara	Great Britain	https://senscheck.com/
SensoTASTE	Switzerland	https://sensoplus.ch/sensotaste/

- Provides a comprehensive and meaningful panel performance after completion of data collection (either in a pilot measurement or normal measurement), consisting of the following:
  - Feedback at global panel level as well as feedback at individual level.
  - Comprehensive analysis on discrimination, repeatability, consensus (crossover), scale usage, contribution of each panellist to the panel.
  - Provides access to different type of analysis (significance levels, ANOVA models, post hoc tests, etc.).

3.8 Target-oriented sensory analysis

Conducting sensory analysis as best practice undoubtedly delivers very robust and valuable results on the products under investigation. However, sensory analysis conducted as best-practice is time-consuming and expensive and some objectives do not necessarily demand a descriptive analysis that fully meets the requirements of science. In many cases, the objective of the work allows for an adjustment of the conditions under which the descriptive work is carried out so that the results are less accurate, but much easier, cheaper and faster to obtain. This makes descriptive analysis a powerful partner for agile product-development and -management. Fig. 3.4 gives an overview of the most frequently used approaches in the descriptive analysis.

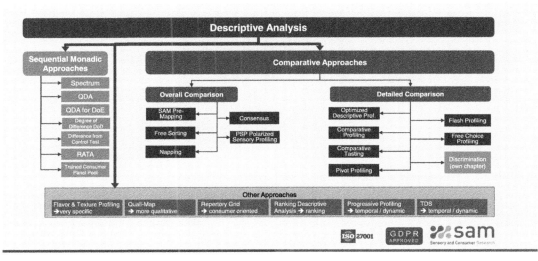

FIGURE 3.4 Overview of the most frequently used approaches in descriptive analysis. *Source: SAM Sensory and Consumer Research.*

3.9 Agile elements in sensory analysis

Digitisation allows for the integration of agile elements in the procedure of product characterisation by trained consumers. There are several basic ideas, how to make the approach adaptive to different aims of the research:

1. Vary the number of panellists — Fig. 3.5: → reduce to 5—6 and stay narrative → increase to 30 and shorten the procedure.
2. Reduce number of measurements from 3 measurements (best practice), down to 2 measurements (saving time and cost), or even 1 measurement (in case of high number of products or obvious differences between the products or lower risk to take a wrong decision), (Fig. 3.6).
3. Use RATA (rate all that apply) to focus on main differences and exclude attributes which do not apply (Fig. 3.7).
4. Modify the overall setting @ facility, @ home, online training (Fig. 3.8).
5. Use references for faster alignment of the panel (Fig. 3.9).
6. Adjust the training and attribute list to the requirements (Fig. 3.10).

3.10 Flexibilisation of research

Digitisation, and even more the digitisation of the communication via videoconferencing, allows the flexibilisation of sensory analysis. The basic requirement is that all participants have all products and references (if applicable), available. If this is the case, then

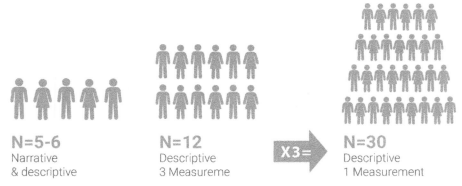

N=5-6
Narrative
& descriptive

N=12
Descriptive
3 Measureme

X3=

N=30
Descriptive
1 Measurement

FIGURE 3.5   Vary number of panellists. *Source: SAM Sensory and Consumer Research.*

FIGURE 3.6   Vary number of measurements. *Source: SAM Sensory and Consumer Research.*

Best practice:
3 Measurements

Adaptation:
2 Measurements
Saving time & cost

Specific case:
1 Measurement
High frequency
of samples

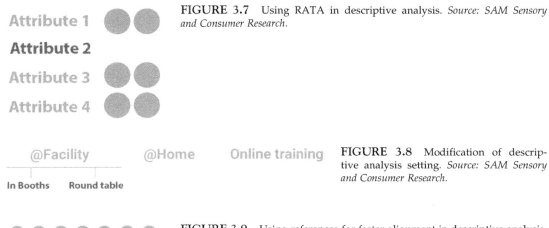

FIGURE 3.7 Using RATA in descriptive analysis. *Source: SAM Sensory and Consumer Research.*

FIGURE 3.8 Modification of descriptive analysis setting. *Source: SAM Sensory and Consumer Research.*

FIGURE 3.9 Using references for faster alignment in descriptive analysis. *Source: SAM Sensory and Consumer Research.*

Adjust the training to requirements

Comprehensive list · Focus on specific sensory dimensions

FIGURE 3.10 Adjusting training requirements in descriptive analysis. *Source: SAM Sensory and Consumer Research.*

there are almost no limits to the implementation of sensory analysis: The training and creation of the attribute list can be done at home. The moderator can centrally moderate the different participants and conduct the voting via internet. The data collection is done via the internet and the collected data is evaluated via the used software (see Fig. 3.11).

3.11 Artificial intelligence: how does it work?

Although this article is not about defining artificial intelligence (AI), we should start by clarifying what AI is all about. Mainly there are three topics or targets:

1. **Higher level of explanation**: AI claims to better, and more accurately, explain human behaviour. In the case of sensory analysis this should mean a better and deeper understanding of the preference and rejection of products from the data collected on them.
2. **Replacing work**: AI makes it possible to replace some of the work done in sensory analysis today, that is, to make it obsolete, as the computational tools do the same or even more with less data.

		Digitized & Offline	Digitized & Online	Digitized, Online, high Datavolume
		Digital Data Entry	Internet Availability	Audio-Visualization & Videoconferencing
Training	@ facility	X	X	X
	@ home		X	X
	@ different locations			X
Measurement	Paper & Pencil	X	X	X
	Digital entry @ facility	X	X	X
	Digital entry @ home		X	X
	Digital entry @ different locations			X
Panel Performance	Manually	X	X	X
	Automatically		X	X

FIGURE 3.11 Flexibilisation of descriptive analysis/sensory research. *Source: SAM Sensory and Consumer Research.*

3. Self-learning ability: AI is capable of learning. A finding from a data set can be expanded or supplemented at a later point in time and this supplementation is an automatic process of AI described by 'machine learning'.

The above sounds promising, but like everything else, it does not happen all by itself. It needs preparation and it requires a guided process. AI can only be powerful if three conditions apply: (1) new computational tools are used that (2) examine a larger amount of data than before with (3) a far greater computer power compared to our previous (or current) approach.

Keep in mind that with AI, more is true than with classical research approaches: if you put rubbish in, rubbish will come out (Fig. 3.12).

3.12 Data warehouse — a fundamental prerequisite

The elementary prerequisite for a meaningful application of AI is the availability of sufficient data material. If this is missing, even the most modern and powerful AI tools are of no help. The workload reduction and additional insights provided by AI must therefore first be 'deserved' by means of targeted additional work: Data on products from different sources must be collected in order to have a larger data set, which is AI conform. In the case of sensory analysis this means:

- All sensory profiles of all products of the same product category from different time periods and in different test sets.
- All chemical—physical parameters available on the products.

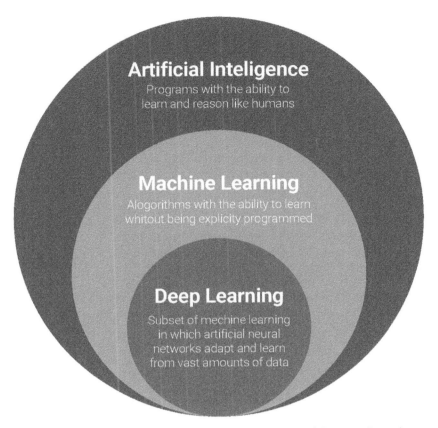

FIGURE 3.12   The model of artificial intelligence. *Source: SAM Sensory and Consumer Research.*

- Hedonic data on product evaluation by consumers.
- All information available on the respective data set and on the corresponding product and product category.

With all data, it is important that they are entered into the data warehouse in a way that can be digitally analysed. Without this, the information is not compatible with the computational tools of AI.

3.13 Artificial intelligence: what can be expected from artificial intelligence in sensory analysis?

One thing is certainly true: AI is not a no-brainer. But the more we engage with AI, the more will come back — and what we get back is unpredictable. Nevertheless, we can make some basic assumptions:

- *Behavioural patterns*: AI will help us to identify reasons for consumer behaviour when choosing products. Understanding what is beyond liking and rejection. For this, we will have to rethink and expand our questionnaires — more hedonic questions and answers are important to identify what is really relevant in product choice and what is not.
- *Elimination of attributes*: It is quite plausible that the measurement of some attributes may be rendered redundant by the measurement of others and therefore their measurement can be dispensed with.
- *Replacement of attributes by chemical—physical parameters*: There is certainly potential here, but this will not be possible through a simple correlation of individual parameters. Instead, it requires a thorough treatment of the data. This data treatment alone involves a great deal of experimental procedure, itself costing time and effort and requiring a willingness to do so. Two directions are possible here:
- *Indexing*: The development of indexes from different chemical—physical parameters can then better explain an issue and result in a correlation than a simple parameter alone.
- *Data Aggregation*: Aggregated data, in the same way as indexes, can have a higher explanatory potential than individual data. An example from a product optimisation is the physical parameter 'conductivity' of sparkling wine. Conductivity is a parameter that results from the concentration of anions and cations in the liquid. The analysis of individual parameters (potassium, sodium, phosphate, sulphate, etc.), showed no impact on product acceptance in any of the cases. Conductivity, on the other hand, was a strong negative driver: the higher the conductivity, the lower the acceptance.

3.14 Conclusion

Digitisation and the path to artificial intelligence are the working methods of our time, as are sensory and consumer research, and are a fundamentally important pillar of efficient and agile product development. There should be no doubt at all about that. The approach is difficult and requires an engagement with several disciplines at the same time. The reward, however, are very important insights that contribute significantly to the development of products with strong competitive advantages in today's increasingly complex world. Embark on the journey, it is worth it!

How digital is creating new opportunities for sensory science

CHAPTER
4

Digital toolbox — ways of increasing efficiency in a descriptive panel

Annika Ipsen, Sven Henneberg, Torsten Koch and Manuel Rost

isi GmbH, Goettingen, Germany

4.1 Introduction

Descriptive analysis is a valuable tool for R&D, quality assurance and for technical consumer insights teams as well as marketing insights. Carried out by a specially trained panel under standardised conditions (ideally in a sensory laboratory), objective sensory profiles can (1) provide insights into sensory changes after a recipe adjustment, (2) screen competitor products and explore the market/category from a sensorial point of view, (3) compare production facilities and assess the influence of different batches, packaging materials or storage conditions or (4) support marketing in validating sensory claims. For the collection of data, well-selected and well-trained panellists and standardised tasting conditions are of key importance.

During a laboratory profiling session, and especially when running in-home projects, digital tools can make the training sessions more efficient, facilitate the measurements, minimise sources of error. Overall, they can significantly simplify the job of the panel leader and the panellists, thus saving costs and time and significantly increase data quality.

isi has developed a constantly-growing platform, the *isiSensorySuite* (isi Sensory Suite, 2022), specifically tailored to meet increasing demands on efficiency during descriptive analysis. It combines a series of different tools such as the *isiTrainingTool*, *isiSensoryProfiling* and *isiPanelManagement* and is constantly being extended and individually customised to fit clients' needs.

This chapter will introduce the several tools that can be used by panel leaders, panellists and project leaders to increase efficiency and eliminate sources of error in the panel training, data collection, project management, panellist management and analysis.

4.2 **Excursus: descriptive analysis under Covid-19 restrictions**

During the Covid-19 pandemic, when the world suddenly came to a standstill with contact restrictions, curfews and lockdowns, many projects still had to go on and creative solutions had to be found to continue the panel work using in-home profiling sessions. In addition to the difficulties of getting the products to the panellists' homes and creating an environment that is as standardised as possible, the importance of digital implementation of panel sessions once again came to the fore. Panel training and data collection had to take place online and was implemented with the help of digital tools to ensure high data quality despite less-than-ideal circumstances.

When the pandemic started many sensory departments were closed for external traffic or completely shut down for several months. However, ongoing or already scheduled projects had to continue, as important business decisions depended on the results, and, therefore, creative solutions had to be found quickly. isi is a sensory market research company in Germany which had about 20 sensory panels running at that time. Creative solutions had to be found quickly to ensure that most of these projects could continue. Many panels were transferred into in-home profiling within 1–2 weeks. As most panellists were on reduced working hours in their regular jobs and all leisure activities were suspended, the panellists showed high motivation to continue panel activities from home, even if this meant some additional effort for them. The samples were prepared, labelled, and safely packed in the isi labs and, together with all needed equipment and palate cleaners, sent to their homes via post or courier or picked up daily by the panellists. Contact-free pick-up stations were set up outside the building, with specific timeslots for the panellists to guarantee minimal contact between panellists and staff (Fig. 4.1).

Beforehand, all panellists submitted a list of their available devices at home, such as laptops, cameras and microphones. Panellists who had no such devices available were equipped with tablets or other devices. Additional requirements were a stable home internet connection and the ability to run communications software such as *Microsoft Teams* or *Zoom*.

During an initial session, a technical check was performed, and all panellists were introduced to the programs. Etiquette guidelines were established, such as only activating the microphone when speaking, raising one's "hand" when wanting to speak, avoiding background noises, limiting distractions and preparing all materials needed for the session in advance. From our experience these are very important steps when implementing in-home profiling. We recommend running a quick technical check when transferring projects into in-home profiling, especially when there are panellists in the group who are not experienced with computers and web calls.

Web links for training tools or programs such as *GoogleSheets* (Google Sheets, 2022) or *MiroBoard* (Miro, 2022) were shared during the sessions when needed via the chat function.

For profiling sessions, the link for the questionnaire was shared via e-mail with the panellist.

For certain product categories, where in-home profiling was not possible because samples needed specific preparation or immediate tasting (such as coffee, whipped cream, Bolognese), at isi we had the fortunate situation that we have multiple mobile cabin booths at our facility that can be used for individual profiling sessions. Panellists were invited separately for individual time slots and samples were prepared by the lab staff and served in the cabin booths. The booths were in separate entrance areas with access to sinks for

FIGURE 4.1 Profiling during Covid-19 Samples were prepared in the lab and packed for panellists including all required material, then picked-up at a contact-free station for panellists to set up a profiling testing area at home.

hand washing and disinfection stations. When the situation became more relaxed after a few months, isi was able to open the labs again for lab profiling. In addition to the regular cabin booth, the group discussion rooms were equipped with mobile cabin booths to split panels into different rooms to create maximum social distancing. Each booth is equipped with standard technical equipment. Even, after 2 years of living with the pandemic, all lab training sessions are still performed online.

Digitalisation would have progressed even without Covid-19, but the need became especially clear in the pandemic and has significantly accelerated the development of appropriate tools.

4.3 isiSensorySuite

4.3.1 Panel training

Before a descriptive panel can deliver high-quality data, qualified panellists must pass an extensive training to learn the methodologies, create an attribute list of objective terms, define a common understanding of these terms and tasting standards, be familiarised with references to train to a common scale usage and, overall, to be familiarised with the test products. Typically, the training sessions are held in a group discussion room where all panellists are present in person and the panel leader leads the discussions.

In general, it is possible for the panellists to complete all tasks using paper-and-pencil questionnaires. However, during training, the data collection of profiling scores is a time-consuming and error-prone task. After the panellist scores each product individually on a paper questionnaire, each scale needs to be measured with a ruler to determine the intensity score in preparation for the discussion. During the discussion, each score needs to be collected orally and a mean score is roughly estimated by the panel leader. Errors often happen, as attributes are missed during profiling and scales are measured or transmitted incorrectly. Furthermore, there may be a risk of overlooking quieter participants in the discussions and losing their opinions and objections. On top of this, after training a lot of time is required by the panel leader to summarise all notes.

It is much more efficient to supply the panellists and the panel leader with PCs or tablets to collect, measure and share the results directly digitally via a beamer or monitor in the panel room.

There are a variety of simple tools available on the market that are suitable for a wide range of tasks to be used during training sessions. Methods and tasks can be explained via PowerPoint presentations visually by the panel leader. Simple exercises for attribute generation can be easily done with individual or shared tables (e.g., *GoogleSheets, MiroBoard*), and then discussed in the entire group. Simultaneous transmission via beamer with subsequent discussion saves time and avoids duplication of comments during the discussion. Corrections are made quickly without the often-confusing flipchart outputs.

In addition, many digital solutions specifically developed for descriptive analysis are available, including the *isiTrainingTool, EyeQuestion* (EyeQuestion, 2022) or *Compusense* (Compusense, 2022). These cover all kinds of training exercises such as basic taste identification, threshold tasks, discrimination tests (triangle, ranking pairwise, R-index) and profiling exercises (Spectrum, QDA). Panellists can individually rate their products and results are shown immediately on the screen and can also be discussed as a group when shared by the panel leader via monitor. Furthermore, the panel leader has the option to download and process the results after the training for record keeping without transcription errors and time-consuming counting of individual worksheets.

The use of digital tools is also a must-have for in-home profiling sessions. Trainings that take place digitally need a platform, such as *Microsoft Teams* or *Zoom* for group discussions. Cameras should be used by all participants to strengthen the group feeling, to develop and monitor tasting standards and to be able to show product characteristics visually. All of the digital software solutions already mentioned are internet-based and can also be used during in-home sessions.

Based on our experience, training conducted with digital tools is up to 50%—80% more time-efficient, especially when conducting and discussing test profiling. Long paper—and-pencil questionnaires are no longer necessary; digital solutions save time and after initial investment they can save costs and manpower on the long run.

4.3.2 Business case study: a training tool for sensory profiling

In sensory profiling, methods such as Spectrum or Qualitative Descriptive Analysis (QDA) are used in which products are scored individually by each panellist on an unstructured line scale (translated into 10 points), going from left to right with increasing intensity (e.g., low to high, weak to strong). Anchor points on the scale for orientation during the profiling are commonly used. Depending on the study background, it is sometimes necessary to include intensity scores of a reference product on the scale for orientation during profiling. References can be products, a food reference or watery solution for basic taste attributes.

The *isiTrainingTool* is an online programme specifically designed to use for profiling practices during a descriptive panel training.

In preparation of the sessions the panel leader can easily upload an Excel datafile to the programme that includes the information, such as the project name, number of samples, attribute names, description of the anchor intensities (low to high, weak to strong), and optionally the reference scores.

The number of uploaded data files is unlimited, which allows profiling several sample sets during one training session or working on several studies at the same time with multiple panels.

During the training sessions, the panellist can open the programme via weblink, select the project and after entering the panellist-ID and the sample code start entering their individual scores. Attributes belonging to one sensory dimension are shown on one page, and optionally reference scores are indicated on the scale. Scores are entered by clicking on the slider and can be additionally fine-tuned by clicking on the arrow buttons. The reference scale with the anchors on intensity 1 and 9 are displayed on the very bottom of the page (Fig. 4.2).

For a later discussion, the panel leader opens a new interface, selects the project and can share the results via screen with the entire group.

The individual scores of each panellist are displayed in different colours on the scale of each product. This allows each panellist to monitor their own performance in comparison to the group, as well as agreement among the panel and the discrimination between the profiled products. In addition, the mean score is automatically determined and shown on the scale. A small table indicates how many panellists rated which sample with a significantly higher intensity. For a deep dive into the scores, the panel leader can also open a separate overview with the numerical scores of each panellist for the particular attribute. A drop-down selection allows jumping between attributes during the discussion if attributes are to be discussed by category (e.g., sour taste together with sour aftertaste) instead of the order of the questionnaire (Fig. 4.3).

After the training session, the panel leader can download the captured data in Excel format for further analysis and record keeping.

4.3.3 Data collection

For data collection paper-and-pencil questionnaires are occasionally still used however, most companies have already implemented computer-based collection software. Online-based

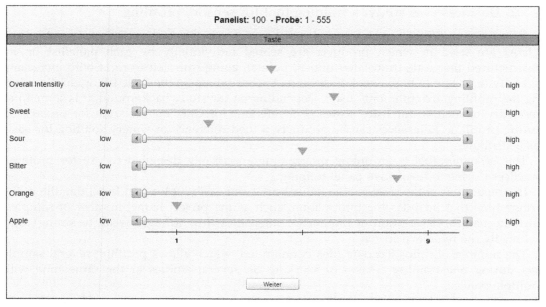

FIGURE 4.2   isi TrainingTool The Tool allows panellists to score products individually on an unstructured scale with increasing intensity. Reference scores can be displayed optionally.

programs allow a flexible, virtually hardware-independent, comfortable and accurate collection of sensory data.

There are numerous different programs on the market (see Chapter 3, The arrival of digitisation in sensory research and its further development in the use of artificial intelligence by Martin J. Kern), for creating quick, standard tests in a user-friendly way; for example, discrimination tests, CATA questions, ranking tests, or profiling scaling questionnaires, but also complex methods such as time intensity, Temporal Dominance of Sensation (TDS), flash profiling, napping or implicit-association tests can be used. Design codes and rotations plans are automatically generated by most programs.

With some programs such as *isiSensoryProfiling*, the panel leader can monitor the progress of data collection via dashboards and view results in real time. For maximum efficiency once data collection is completed analysis is automatically executed. Upon specification of statistical analysis parameters and level of significance, automated reports can be generated including standard evaluations, such as significance testing, mean score tables, line or spider plots, PCAs or MFAs. Results are automatically charted in PowerPoint presentation templates and even an automated output of panel performances parameters is available.

This saves an enormous amount of time compared to paper-and-pencil questionnaires or manual analysis and data handling since missing data due to unticked scales is eliminated and the manual transfer of data is no longer necessary. The collected data is available immediately after the panellists have finished profiling and can be further processed by the panel leader.

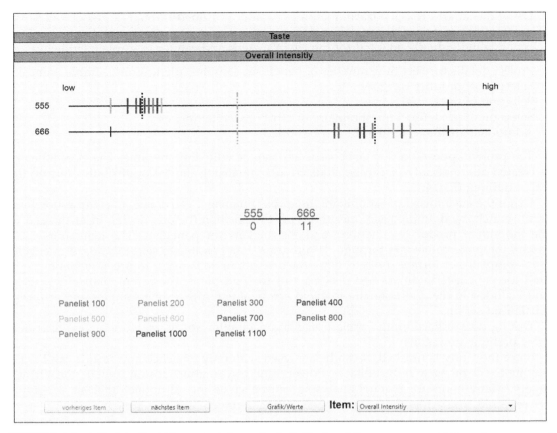

FIGURE 4.3 isiTrainingTool The output shows the individual and the mean scores of the entire group and allows a fast and meaningful discussion.

4.3.4 Business case study: sensory profiling programme

isiSensoryProfiling is a valuable programme combining a project management tool, a questionnaire generator, automated analysis and panel performance checks. The programme is specifically tailored by isi for individual needs, that is unlike classical questionnaire software's it is very individual and optimised into users' workflows to achieve maximum efficiency rather than a jack-of-all-trades solution that is more flexible but also less adaptable to existing workflows and needs.

A dashboard displays an overview of all ongoing projects. The project leader can create, manage and archive projects easily.

When creating a new project, the project leader has two options for generating the questionnaire. Either by a smart Project Wizard or by specifically customising the questionnaire to specific needs.

2. How digital is creating new opportunities for sensory science

Using the smart Project Wizard, the project leader is guided through a predefined series of customised selections to quickly set up a project and generate a questionnaire. This selection is individually adapted for the client's needs and can for example include:

- Project details (project title, number of panellists, samples, replications, sample names)
- Selection of product category
- Selection of method(s)
- Selection of predefined questionnaire or attribute list templates
- Additional attributes of interest can be added

The project appears in the dashboards and can be edited at any time. The questionnaire is created automatically or can be adjusted after finalising the list of attributes during the panel training (Fig. 4.4).

Using the customised questionnaire, the project manager can create a new questionnaire from scratch, by defining each question type, the position (before, during or at the end of the individual product evaluation), and instructions for panellists. The generator easily guides the project manager through the progress. It is possible to select standard questions from templates or former projects. Once the questionnaire is ready to use, a weblink is created for panellists to enter. Test designs and rotation plans can be automatically created and included in the questionnaire to avoid errors of entering wrong samples and ensuring a balanced set up.

During data collection, the project leader can monitor study progress in real time and view preliminary results.

Once the study is finished, it can be analysed with only one click and results can be put out in Excel or as a fully editable PowerPoint presentation customised in the client's layout. Many hours for manual analysing can be saved, no additional analysis programs are required, and errors that may happen during a manual analysis are eliminated.

FIGURE 4.4 isiSensoryProfiling Questionnaires are easily generated by using a Smart Wizard from templates or by customising each question according to the desired needs.

4.3.5 Panellist management

Not only can training and data collection be implemented digitally, but using a panellist management programme can save precious time. Without such a programme, panellists need to be contacted for project invitations, panellist information needs to be tracked and updated regularly in different files and attendance during each session needs to be captured (including holidays and other absences). Also, panellist incentives need to be managed and panel performance development must be monitored. To avoid multiple Excel and data files, software to monitor and track all information in panel management, specifically designed for use in descriptive analysis, is available (such as *isiPanelManagement* or *EyeQuestion*). These programs are created to relieve the panel leader of organisational tasks that often cost a lot of time.

4.3.6 Business case study: panel management tool

isiPanelManagement is a programme that is specifically tailored to individual user needs (i.e., departments, teams, business units, etc.). The layout can be adapted to the organisation's layout and desires. It contains elements for panellist and project management, a link to accounting to facilitate the invoicing process, as well as a link to a booking planner for facilities and can support the recruitment process of new panellists. All features are optional and can be adapted individually.

4.3.6.1 *Panellist profile*

In one database all contact details, account information, attendance checks and holiday dates of panellists can be recorded. Panellists have individual accounts where they can log in from home to view upcoming projects, change contact information and account data, accept or cancel project participations, enter vacations or contact the panel leader. Furthermore, they can monitor their incentives and access questionnaire links during in-home profiling.

Static panels or dynamic panel pools can be created for ensuring that the appropriate panellists are invited to projects. Especially helpful for long-term participation of panellists, a panellist history can be viewed to check on participation in general training sessions and in former panel activities. Further, data privacy protection sheets, consent forms and declarations of confidentially can be captured in the panellist profile, and additional printouts or e-mails can be collected.

Panellist attendance can be monitored automatically, for example, by linking the tool to the data collection tool or by manually input during or after the session by the panel leader or lab staff.

Invoices for incentives, as well as annual tax certificates can automatically be created based on the attendance lists and shared with panellists and the accounting department. A dashboard also informs the panellist of previous incentive amounts.

A minor but highly appreciated feature is an e-mail reminder for panel leaders when panellists' birthdays are coming up. After all, panellists' motivation is one of the key drivers of the success.

4.3.6.2 Project management

The panel leader can schedule and plan projects in the management tool. Invitations are automatically shared with panellists and by logging in to their profile, the panellists can accept or decline project attendance. If multiple slots are offered, panellists can self-schedule their attendance. Quotas on the timeslots avoid overrecruiting and back-up lists can be created. In case of cancellations, panellists from the back-up list can be invited automatically. Reminder e-mails prior to the project start are automatically sent out.

By linking the panel management tool to the data collection programme, panel performance information can be automatically stored and monitored and all results from previous studies are saved in one database.

4.3.6.3 Recruitment

Recruiting new panellists is always a time-consuming process. Using the panel management tool can make panel leaders' lives a bit easier, saving time and costs.

Potential candidates can register via a shared weblink or a permanent link on the client's website. The *isiPanelManagement* can help setting up a screener questionnaire for interested candidates. The questionnaires are easy to create and fully customised to the companies' layout and logo. For each new candidate, a new profile is automatically generated, and the panel leader can easily select candidates for an upcoming screening.

Questionnaires can be saved for future needs and are easily adapted. For upcoming screenings, the panel leader can easily send out invitations for a screening test via the programme and the candidates can self-register for screening dates. At all times, the panel leader has an overview of registrations and booked screening slots.

4.3.6.4 Booking planner

The booking planner is an additional add-on where a sensory lab or discussion room booking can managed. A synchronisation with Outlook calendars or other programs is possible and the Planner provides an overview of all booked facilities. This is especially helpful when managing multiple panels or if facilities are used for multiple purposes. This avoids double bookings, provides a quick overview of capacities and upcoming projects and allows efficient and effortless capacity management with as little unnecessary downtimes.

4.3.7 Business case study: isiSensorySuite

With the *isiSensorySuite*, isi developed a multifunctional sensory programme to combine all necessary steps during descriptive analysis. The full-service solution links all previously mentioned tools, such as *isiPanelManagement*, *isiBookingPlanner*, *isiTrainingTool* and *isiSensoryProfiling* with DIY questionnaires and automated reporting with other useful elements, such as the project request platform, similar to the Sensory Self Service mentioned in Chapter 8, Sensory self-service — digitalisation of sensory central location testing by Sven Henneberg, Annika Ipsen, Manuel Rost, Femke W.M. Damen and Torsten Koch (Fig. 4.5).

With the integration of a project request platform, clients or stakeholders can easily request or commission projects. Profound expertise of descriptive analysis is not required, as the request progress guides easily through the process and advises on predefined options, such as method, numbers of replicates per sample or number of panellists. In addition, the initiator can add project background, optionally select mandatory attributes, view and book available panel slots.

The project request automatically appears in the project overview and can be customised by the project leader. Via the booking planer, facilities are automatically booked and panellists are invited.

The panel leader can then set-up the isiTrainingTool or generate other training questionnaires or documents-based attribute template and adapt these during project progress. Templates for training protocols can be used to record training results and shared with stakeholders. Individual training or panel performance results are automatically saved in each panelist's profile. Performance parameters can easily be monitored to meet the need for action at an early stage. A reward system for panellists can be integrated based on project attendance, short notice cancellations and performance results.

The advantage of this tool is that it is customisable for each user according to their individual needs. Instead of purchasing a full-service programme with an overload of unneeded functions, specific needs are identified in advance and individually programmed and combined. In addition, the software can quickly be extended on demand at any time, when requirements change over time or new features are introduced. All layouts and language requirements can be adjusted. A company's standard settings for

isiSensorySuite

FIGURE 4.5 isiSensorySuite
Features of the isiSensorySuite.

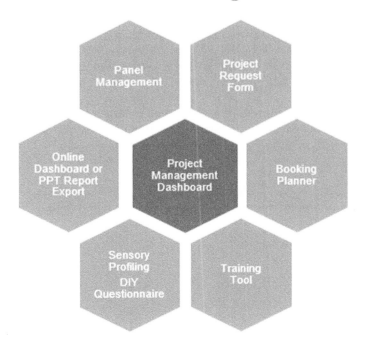

project set-up, questionnaire generation and analysis can be defined in collaboration with the relevant experts.

By linking the different programs to a multifunctional tool, efficiency is increased and errors or unnecessary duplicate efforts (as they are common when working with multiple programs to manage different project steps) are reduced.

4.4 Status Quo and a glimpse into the future

The degree of digitalisation of sensory and/or in-house testing processes still varies varies between industries and even within similar companies. While most companies are already using digital questionnaires for data collection, there are still gaps in the use of digital programs in training or panel management. Even when usage is routine, it is often a patchwork of multiple software solutions that focus on single aspects rather than on streamlining and linking the workflow as whole. Hence, although digitalisation has taken an enormous leap forward, there are still numerous ways to optimise it.

The vision is to have one software that digitises and connects all steps, hence reducing time-consuming and error-prone steps, making life easier for panel leaders and panellists. isi took a first step into this direction with the *isiSensorySuite*.

The *isiSensory Suite* has already become a multifunctional and highly appreciated software solution, but this not the end of the road. The future vision is that the project manager or sensory scientist just needs a few mouse clicks to handle a project comprehensively. This would allow panel leaders to finally focus on their primary expert function: consulting stakeholders and clients, delivering comprehensive insights to management and contributing with all their expertise to overcoming business challenges.

In addition, why should it not be possible for one sensory software to automatically create and send out requests for quotations, proposals or research briefs? Or to connect to already established software, such as SAP? And why should it not be possible to automatically place orders for products or palate cleaners online with grocery stores? Or to book and organise laboratory staff for upcoming panel sessions?

The technical capabilities to bring these visions to life are already there and we at isi predict that more and more companies will realise the benefits of digitalisation and bring this vision to life in the near future.

References

Compusense. 2022 2022 1 27 https://compusense.com/compusense-cloud/overview-2/.
EyeQuestion. 2022 2022 1 27 https://eyequestion.nl/.
Google Sheets. 2022 2022 1 27 https://www.google.com/sheets/about/.
isi Sensory Suite. 2022 2022 5 18 https://www.isi-insights.com/en/sensoryservices.
Miro. 2022 2022 1 27 https://miro.com/.

'Hybridisation' of physical and virtual environments in sensory design

Anne-Sophie Marcelino[1], Chantalle Groeneschild[2], Nathalie Boireau[2] and Victoire Dairou[1]

[1]Danone Global Research & Innovation Center, Paris, France [2]Danone Nutricia Research, Utrecht, The Netherlands

5.1 Introduction

5.1.1 Setting the scene — moving from physical, to a mix of both physical and digital, environments in sensory design

In sensory design, sensory acts from the start of the project articulating the 'ideal sensory experience' in the project brief. Thus sensory intent is anchored in the project brief, resulting from the translation of the ideal user experience into specific product criteria. The sensory brief is the unifying thread guiding the design of experience throughout the project in order to achieve the sensory intent. The brief is progressively materialised by prototypes, ranging from stimuli to mock-ups to prototypes to the final industrial product.

Consequently, with this intentional design approach, sensory evaluation plays a pivotal role throughout the project in supporting the team's informed decision taking. All along the design process, prototypes are evaluated according to sensory attributes that are matched against the sensory brief. Evaluation is performed by either trained sensory panels, some of which are conducted in-house, or by project team members. These so-called team tastings in which internal project teams evaluate a set of products from the project is a valuable tool used within companies including Danone to make faster and more agile decisions, mainly at early stages of design (Rogeaux et al., 2022). It is therefore used as a complementary tool to the traditional sensory profiling during projects, to ensure that, end-to-end, the sensory intent, defined at the beginning of the project, will be embodied in the final industrial product.

49

To run a proper project team tasting, a preliminary sensory upskilling training is delivered to Danone employees taking part in innovation and renovation projects.

Sensory trainings and evaluations used to take place physically, in specific locations where sensory materials and prototypes were available. However, increasingly they are operating in hybrid settings, in which there is a combination of both physical and digital environments.

There are multiple reasons for this including:

- The globalisation of projects, accelerated by digitalisation that enables the connection of teams across separate locations.
- The remote way of working, accelerated by the Covid-19 crisis.
- The increase in awareness of sustainable behaviour with limitations on travel.

Dinnella et al. (2022) showed that remote at home product testings were appropriate for studies with trained panellists. However, they highlighted certain requirements including that samples should be handled and shipped without causing any safety hazard for participants, that samples have a high stability (physicochemical, microbiological, and sensory), and, finally, that samples do not require specific conditions of evaluations (such as controlled serving, storage conditions, etc.).

In addition, Dinnella et al. (2022) pointed out the requirement of specific equipment and environmental conditions (such as internet connection), for remote evaluation at assessor's home, as well as all additional panel leaders' activities upfront for a remote session (such as evaluation box preparation and delivery, instructions to assessors to samples handling and session connection, internet connections check with each assessor).

5.1.2 The phygital concept and learnings for sensory design

Combining physical and digital environments has already been previously leveraged, especially in the retail industry. Along the shopper journey, the consumer might experience both online shopping followed by in-store shopping (or the other way around). Developing this further, the phygital term was created in 2013 by an Australian marketing agency, contracting 'physical' and 'digital'. The aim was to digitalise the store, evolving from a pure physical environment to a hybrid physical and digital environment. One example (Johnson & Barlow, 2021) is the Amazon Go store which offer a 'grab and go' cashierless shopping experience, enabled through a specific digital app download on shoppers' phone and stores equipment with cameras and sensors on shelves (e.g., for weight)

The phygital concept relies on merging both physical and digital settings to provide an enhanced shopper's experience, thus potentially impacting shopper's behaviour and increasing sales (Johnson & Barlow, 2021). Hybridising physical and digital environments can provide the best of both worlds for people: for instance, going back to the the Amazon Go store example, the ease and seamlessness of digital payment with the tangibility and the immediacy of physical product experiences. Considering UBER, the seamless payment was in the top three reasons for success.

The phygital concept is extending to other sectors beyond retail. During Covid, for hygienic reason, restaurants placed QR codes on tables instead of paper menu to limit the risk of exposure to the virus. It thus started to create a new habit with other benefits, such as flexibility to add other information to the menu of the day. Batat (2021) investigated the impact of augmented reality in the foodservice field thus opening the possibility of bringing the benefits of such a hybrid environment to a restaurant for the consumers. Batat leveraged 3D video animation projected onto restaurant customers' tables, presenting a 'little chef' character who cooks in front of diners' eyes, with the aim to immerse customers' in the kitchen. Batat found that consumers reported having a more fulfilling experience, influenced particularly by an increased multisensorial stimulation (such as visualising food from the menu, its preparation, the ingredients, etc.). It also facilitated social interactions and enhanced food cognitive skills.

Finally, phygital is also gaining in interest in the field of education. Fromm et al. (2021), examined the potential of virtual reality to afford a holistic experiential learning cycle for students, highlighting the potential to consider flipped classrooms, that integrate virtual reality experiences at relevant moments of the curriculum, for an increased learning efficiency. Incorporating the use of virtual reality in the classroom may contribute to experiential learning as it supports learners to apply first-hand knowledge and to experience consequences directly. For instance, in educational studies, students can have access to a classroom simulator, interacting with an intelligent agent proposing multiple-choice options of reaction to a critical teaching situation (e.g., a student insulting another student), illustrated by a 360-degree video from a database and providing the relevant video extract figuring out the outcome of the selected reaction to the situation. Furthermore, students can debrief together about the situations they experienced with the simulator and about their learnings.

Phygital literature brings a valuable perspective, inviting us to consider how merging of both physical and digital environments might bring additional value versus a pure physical or digital environment.

5.2 Case studies — exemplification of hybridisation of physical and digital environments in concrete situations of sensory design at Danone

In the following, we will exemplify how we have evolved our sensory design practices to better adapt to hybrid ways of working, leveraging both physical and digital environments for internal sensory training, internal team tastings, as well as for in-house sensory panels.

5.2.1 Designing a digital sensory training with physical stimuli

Team tastings are one of the sensory evaluation tools used by Danone to achieve sensory design. But although Danone employees are not selected with the same level of demand as a trained panel are, it is important to make them aware of their own sensory capabilities and assure team tasting sessions are set-up under the right conditions, for example blind coded, randomised order, same temperature, etc.

Therefore within Danone a sensory awareness training is developed by the User eXperience team (UX team), that is rolled out globally within R&I (Research & Innovation). The objective of this training is to make colleagues aware of what sensory science is about, their personal sensory acuity, what team tasting is and the basic rules of taking part. In addition, a training is available on the key sensory attributes depending on the category of products they are working on. The training consists of a theoretical part but also includes many exercises to let them practically experience how the senses work. Only persons that have participated in this training are allowed to participate in a team tasting. Therefore, this course is mandatory for any newcomer in R&I as all project members are encouraged to take part in team tastings.

As result of the hybridisation a digital version of the course was developed, which is available through the digital learning system within Danone. The course can be followed at home or at the office and a dedicated toolkit with tangible physical stimuli delivered to the participant with all the materials needed to check their sensory acuity. The participant can start the course at any time after having received the package. As it concerns the senses, it is important that this awareness of the senses is highlighted by real time exercises in which all of the senses are triggered. The package consists of a basic taste test, ranking test on intensity, flavour recognition test, colour blindness test, etc. For basic tastes and flavours recognition and ranking, tests are performed thanks to papers 'impregnated' strips.

The digital application of the training records the trainee's answers to tests such as recognition and ranking, indicates the correct answer and calculates the trainee's score. In addition, the digital version permits the introduction of questions on theoretical knowledge with a test and learn loop. As a result, the training is experienced by participants as engaging through the combination of both theoretical learning and practical exercises. Leveraging educational triggers (such as gamification, immediate reward, etc.), contributes to increasing the learning efficiency of such a hybrid training. At the end of the module, trainees receive a certificate for passing the sensory course and are officially allowed to participate in any team tasting.

5.2.2 Projects' team tasting: engaging all senses in a remote meeting

Team tasting is used by project teams throughout the development process as they have the need to characterise products and match them against the product brief. Starting in the early phases of a project with a need to understand the current product category and the differences between the products on the market (1), but also during development to monitor and evaluate, on a regular basis, product improvements and changes (2) or, in the case of a renovation, to understand the size and type of differences between old versus the replacing recipe (3). To answer these questions, team tasting take place using more agile methods that are quicker and less complicated in design and analysis compared to the traditional sensory evaluation with screened and trained panels. Examples of such methods that are used during team tasting are Check All That

Apply method (CATA), comparative team tasting, ranking and the discrimination test: A-Not A with sureness (Rogeaux et al., 2022). To manage team tasting evaluation it is important to have a standardised and formalised way of performing team tasting. Therefore an additional training course has been developed by the UX team to make sure the coordinator of the team tastings knows and uses the right methods, can organise and monitor the actual team tasting, can analyse and interpret the results.

Just like for the basic sensory awareness training, the team tasting coordination training is also available digitally and can be followed online to learn the basic principles to organising and conducting a team tasting. For these, the program EyeQuestion has been made available to perform the basic team tasting methods such as CATA, A-Not-A with sureness judgements, comparative team tasting, ranking, etc. That gives the advantage that within whole Danone the same way of working is used across team tastings and that also all data are digitally stored and available, making it possible to set-up studies across regions and track studies for a project.

This use of EyeQuestion makes it also easier to perform hybrid team tastings. In addition, in team tasting it is also possible to combine the physical and digital environments in which a part of the team is evaluating the product at a specific tasting location while some team members are testing from another location around the globe or from home. In this hybrid team tasting products are sent to separate locations and project team participants can give their feedback using EyeQuestion software or through a shared file by the use of Microsoft OneDrive. In case the hybrid team tasting is performed with a low number of team members it is possible to have a 'live' tasting all at the same time using video conferencing tools (Microsoft teams, Webex etc.), to directly meet, evaluate and discuss the outcome. However, when working with a large group of team members we advise that all participants perform the evaluation in their own time after which a joint session can be organised to go through the results together.

This hybrid type of team tasting can be used for the more descriptive methods such as comparative team tasting and CATA, but is not advised in the case of discrimination testing (e.g., A-not A with sureness judgement), or Ranking as here it is important to make sure that all participants receive products blind, randomised, having similar preparation and temperature, which are all better organised using a central location.

5.2.3 Running remote sensory descriptive in-house panels

Sensory descriptive in-house panels can also undergo this hybridisation, mainly as a result of the Covid-19 situation in which panellists were not allowed to be on-site with too many persons in the same room and the need of social distancing. This hybrid version also combined the physical interaction of panellists with the products while being in a digital environment. One option for hybridisation is the Out Of The Lab (OOTL) testing in which panellists are fully equipped at home to perform the evaluation out of the lab and during evaluation/training being connected in EyeQuestion and/or video conferencing to ask questions and have discussions around attributes and product. Fully equipping means sending all products anonymised, preparation instructions and equipment.

The advantage of this OOTL testing is that it retains panel contact and interaction, keeps panellists trained so they do not lose their finely tuned sensory skills and is a concrete solution to continue with sensory evaluations. But there are also some limitations such as increased product volume and time needed to prepare and conduct a study. Also, it is not possible for all product types and methods (e.g., discrimination tests), due to limitations in re-packing, fully blind testing, temperature control or other environmental conditions that need to be controlled.

This hybrid way of working of combining the physical and digital environments with a sensory panel can also be used at a central location especially in cases when social distancing may be required. This can be done by using separate rooms within a building in which panellists are divided into groups, or even having individual booths, which still facilitates interactions between each other via video conferencing tools and shared files in Microsoft OneDrive. The advantage of having this hybrid way of working within a central location is that the product preparation can still be done at one location and every panellist can receive the same product quality and at the right temperature which means fewer individual differences between the product. In the end this is more time efficient from a preparation perspective than making individual packages per panellists.

With a hybrid way of working at a central location or at home (OOTL), the main advantage, even though panellists are not all in one room, is that every panellist can still contribute to the discussion and training by filling in their individual sheet in OneDrive. Panellists have indicated they also felt 'heard' during the discussion as they all had their individual sheet and everyone's feedback could be seen which is not always the case during training session when all panellists are physically together in one room and not everyone speaks up or comments. This also has the additional advantage that all panellists feedback was captured digitally during the training which was easy to use for the panelleader afterwards for administration reasons.

Using the hybrid mode of testing with a sensory panel, in which a study is partly performed physically in the lab and partially at home, has also demonstrated advantages when measuring certain attributes over time. For example, when performing a time intensity measurement project on the lingering of certain attributes like refreshing it was decided to perform the training on the products and methods centrally on site in the sensory facilities with all panellists to ensuring that they understood the tested attributes, scale, and way of evaluating the products. The test itself was performed by the panellists at home as there was a maximum amount of product to take in on a day and the lingering was so intense that only one product a day could be evaluated. This resulted in a more time and cost efficient set-up in which the panellists received detailed instruction on how and when to test product and receive a package per week at home until the end of the study, when performing the study, they could online give their feedback using EyeQuestion. To determine if there was an effect of the test location the same time measurement of one of the products was performed at home and at a central location resulting in no significant difference, indicating that the at home testing for over time measurement can be reliable done at home.

In general, hybrid way of working with a sensory panel, being at home or at the central location has the advantage of being able to continue activities when social distancing is required. In addition, it has also demonstrated to be an efficient way of testing in cases where the products and preparation allows for it, such as in the case of evaluating attributes that need to be measured over time or with a restricted intake of panel work. However, while working in this hybrid mode for several months during the pandemic, the panellists indicated that they missed the 'social interactions' of being all together in one physical room. In this online environment, it is more difficult to have individual conversations and align on attributes' interpretation as lively discussions are more difficult to have within a group and it is difficult to see each other's facial expression and body language.

5.3 Conclusions

Hybridising physical with digital environments will be part of the future of sensory evaluation and has already demonstrated many opportunities. In our previous examples, hybrid sensory awareness training demonstrated added value in accelerating employee upskilling on needs and learning effectiveness. This hybrid e-learning is currently the preferred choice over the full in-person physical learning for upskilling Danoners on sensory awareness.

Regarding team tastings and sensory in-house panels, hybrid mode, meaning remote digitally enabled evaluation with physical products, led to better team engagement for members based in different locations and the systematisation of digital data collection enabling further data valorisation (on top of using less paper). In addition, we also noticed the value of a more balanced share of voice as only one person can talk at the time with the use of video conferencing.

On the other hand, this hybrid way of work also has some limitations as it is difficult to perform for all product types and test methods. For example, performing a discrimination test can be quite difficult to perform as there can be a lot of variation in preparation but also blind coding or using different lighting conditions for instance are a challenge.

As reflected in other chapters of the book, hybridisation of physical and digital environments is also leveraged as an opportunity to better take into account the impact of usage context on sensory experience: allowing for instance to deconstruct the various sensory touchpoints of a user experience (see Chapter 6, Sensory evolution: deconstructing the user experience of products by Phiala Mehring, Thibault Delafontaine and Huizi Yu), or to assess consumer's overall emotional response towards product experience in more ecological conditions (see Chapter 16, Multisensory immersive rooms: a mixed reality solution to overcome the limits of contexts studies by Adriana Galiñanes-Plaza, Agnés Giboreau and Jacques-Henry Pinhas).

Combining physical with digital environments for sensory design activities, should it be with users, panellists or employees, will expand along with digital technologies. Today digital technologies allow us to easily display visual and auditive cues, making digital environment relevant while designing a pack experience for instance, although tactile and ergonomic sensations require further investigations with physical environments. Tech developments are ongoing to trigger other senses, like a lickable display recently prototyped at Meiji University and that enables to taste what is on a digital screen.

Finally, thinking futuristic in the context of the Metaverse, there might be a future where digital experiences will inspire the design of physical sensory experiences.

References

Batat, W. (2021). How augmented reality (AR) is transforming the restaurant sector: Investigating the impact of "Le Petit Chef" on customers' dining experiences. *Technological Forecasting & Social Change, 172,* 121013.

Dinnella, C., Pierguidi, L., Spinelli, S., Borgogno, M., Toschi, T. G., Predieri, S., Lavezzi, G., Trapani, F., Tura, M., Magli, M., Bendini, A., & Monteleone, E. (2022). Remote testing; sensory test during Covid-19 pandemic and beyond. *Food Quality and Preference, 96,* 104437.

Fromm, J., Radianti, J., Wehking, C., Stieglitz, S., Majchrzak, T. A., & Vom Brocke, J. (2021). More than experience? On the unique opportunities of virtual reality to afford a holistic learning cycle. *The Internet and Higher Education, 50,* 100804.

Johnson, M., & Barlow, R. (2021). Defining the phygital marketing advantage. *Journal of Theoretical and Applied Electronic Commerce Research, 16,* 2365−238.

Rogeaux, M., Van Bommel, R., & Lawlor, J. B. (2022). Technical team tasting in the food industry. *Rapid sensory methods,* 2nd ed.

6

Sensory evolution: deconstructing the user experience of products

Phiala Mehring[1], Thibault Delafontaine[1] and Huizi Yu[2]

[1]MMR Sensory Science Centre, Wokingham, Berkshire, United Kingdom [2]MMR Sensory Science Centre, Pleasantville, NY, United States

6.1 What kept us in the laboratory?

Since its inception, sensory science has been driven through a very academic framing with the goal of eliciting robust, repeatable and discriminatory sensory data. In this chapter, we want to challenge this framing. We put forward a more commercial approach to sensory methods and techniques. We want to demonstrate that products are more than physical objects; they are experiences. And sensory can be used to understand and quantify these experiences.

Within this chapter, we take you on a journey, starting with reviewing why 'classic' sensory was established, the galvanising impact of Covid before rapidly moving onto how sensory is evolving to meet the demands of the world of fast-moving consumer goods, the challenges of understanding the user experience of products and the rapidly evolving digital world that we live in.

6.1.1 Sensory laboratories and booths

Sensory evaluation has become synonymous with sensory booths; placing sensory panellists into sensorially sterile environments to remove bias and external influence. This followed the academic requirement for scientific rigour in the approach of food and drink evaluation. These workstations or booths allow for a seemingly complete and unbiased evaluation of products.

This is achieved by carefully managing key factors when evaluating products. Isolation is one, each panellist is separated from each other to avoid distraction but also prevent

57

them influencing each other and facilitating a full focus on the product, and only the product, in front of them. Sensory booths provide a stimuli seal avoiding any odours, light or noise which may influence the panellist's evaluation. Using devices such as air extraction enable the testing of challenging products (such as coffee, fragrances, etc.).

The above underscores the need for standardisation. Sensory booths have remained unchanged over the years giving us reassurance that evaluations will be carried out through the exact same conditions time after time avoiding many biases along the way apart.

The principle applied to the booths also applies to the sensory laboratory. Standardising the preparation and protocols are seen as important as the standardisation of the booths. Our focus on removing bias and noise from our data has led to the design of the working sensory environment and as we argue in this chapter, whilst these are important elements of sensory as a scientific tool, this rigid approach stymies commercial development and innovation.

That is not to say there has not been change within the world of sensory science, with new methods, new approaches but these are designed to push the boundaries of academic sensory science rather than utilise sensory through a commercial framing.

And yet, within the commercial world, there is a strong desire to be more and more consumer centric, to integrate the consumer voice into sensory science. From our experience, we have learnt that excluding the consumers' voice from sensory projects can create artefacts and bias within sensory data. The need for standardisation can sometimes result in consumer behaviour being overlooked. For example, whilst working on a cereal project, we initially followed the consumption instructions of the pack as advised by our client and set precise assessment protocol, which, it transpired, negated regional behaviours. By listening to consumers, in this case in Poland, we discovered that they consumed the cereal in different ways, therefore necessitating the need to flex our protocols and develop more dynamic and adaptive ones which were not standard by any means.

We are not arguing for the abolishment of old practices here, we are challenging the rigid and, in many ways, unchanging ways of working of classic sensory science. Sensory needs to think outside the standardisation box, become more commercially facing and move to more consumer-centric ways of working. May we be so bold as to say that sensory needs to commercially evolve?

6.1.2 Social

One core element of a sensory laboratory is obviously the sensory panel and the individuals that comprise it. The drive for standardisation, removing bias and noise from our data has often led us to treat these individuals as 'sensory machines', like a weighing scale, that you must install, test and monitor. Yet every sensory professional has experienced the complexity of human behaviour in a sensory trained panel and will have spent many hours trying to take account of these 'human' characteristics in a bid to manage panellist performance.

Humans are social creatures by nature, panellists are recruited for (amongst other things), their ability to articulate sensations. One of their strengths is also to be able to brainstorm and reach consensus on the exact nature of a flavour, on the feeling in the mouth, or the specific definition of a sensation.

We recruit sensory panellists, in part, for their human abilities and yet manage them like machines.

6.2 What is a product?

The consumers' experience of using a product is set within a complex world of time, place, space, past experiences, emotion etc. All products are made up of a bundle of sensory touchpoints, shape, size, colour, feel, smell, flavour, ease of use, ease of application, after use experience as well as brand, brand promise, pack etc. The elements, such as brand, packaging, physical product and how/when it is used, combine to create the product experience and this is something of a user journey made up of many touchpoints.

Many of these touchpoints are temporal, for example how long odours or sensations linger or a whisky opening up and you slowly sip it. These touchpoints communicate with consumers, cueing up perceptions about the products. If you consider a toilet cleaner the strength, nature and duration of its aroma will have a strong impact on the consumers' views on efficacy and it is only through connecting consumer research with sensory research that you can really unpack that. Using the panel to understand what sensory touchpoint that influences the consumer perception of a product and in the experience of using it, in this case, makes a determination of the efficacy of the toilet cleaner.

In summary, we are using sensory panellists to deconstruct a more holistic user experience. If you start to reframe sensory science in this manner, then you quickly start to ponder why we constrained panellists in sterile sensory booths which standardises the evaluation process to such a degree that it is sometimes stepping away from normal consumer behaviour. This can abstract some of the more important elements of the user experience. Taking panellists out of the laboratory enables us to deconstruct the entire user experience of products.

We have already acknowledged that products don't work in isolation: pack, product, brand, marketing communication, product promise, values etc., must all work together. Creating 'Hero' products requires each of the sensory touchpoints in the user experience to be optimised. In truth, 'Hero' products are all about 'Hero' user sensory experiences and sensory panels can deconstruct those in a clear and consistent manner.

Hence, products and product experiences are a bundle of sensory touchpoints, interactions and sensations starting from the very first experience of the product. This, of course, could start at a supermarket, online, in an advert or through chatting to a friend. Nonetheless, these are sensory touchpoints with consumers interacting with whole sensory experiences. Evaluating these does not fit neatly into a purely classical academic approach to sensory science. There is no standard northern day lighting online, no way of removing stimuli in a supermarket or controlling outside odours in the street. And yet, getting a clear objective understanding of these sensory touchpoints is critical when innovating, optimising or renovating products.

At MMR, for quite a few years already, we were moving away from 'classic' sensory profiling to a more deconstruction approach to understand the sensory experiences of products.

What our clients need is to unpack the entire consumer experience through sensory, emulating how consumers actually use products from the first touchpoint through to the last post-consumption/use touchpoint and possibly beyond. Not only is it not possible to conduct all of this in sensory booths, but it also requires a complete turnaround on how you use technology. Hence, over the last 4 or 5 years MMR's commercial take on sensory has created much more digital sensory tools and methodologies.

As soon as you spin sensory around a commercial axis, you start to realise there are other ways of optimising outputs. By putting consumers, people, at the heart of sensory you quickly appreciate that tackling the optimisation of panel processes and the robustness of data in a very numerical and process orientated way, belies a fundamental truth, that panellists are people just like consumers. Why are we not tapping into that? Probably because of the sensory science view that panellists are part of the machinery of robust data generation.

6.2.1 Putting the 'human' back into sensory research

There is something very liberating in terms of what you can do with a sensory panel once you put 'THE human' at the heart of the process, make it people centric. There is also something exceedingly daunting about stepping away from the robust standardised approaches to managing data; blinded samples, triplicate data collection, 95% level of confidence and 90%−100% discrimination, consensus, and repeatability; this is a known, and a comfortable space to work in.

Qualitative deconstruction of a supermarket experience is scary and yet, the outcome you get from having a panel enter a supermarket and deconstruct is amazing, in fact this is how one of our clients described it as 'our little bit of magic research gold dust'. Here, we are taking the panellists' trained skills at deconstructing all sensory touchpoints of a product experience and simply expanding what we mean by product experience.

Coming from the commercial world of sensory, we, the authors of this chapter, would (strongly) argue whilst academic rigour is important, it doesn't always fit with the commercial imperatives of our clients nor the need to ensure that sensory research is consumer orientated and we are most certainly missing tricks by treating panellists as machines!

For the sensory science team at MMR, the recent industry shift only reinforced our previous belief that balancing up academic standards with commercial needs and imperatives was ultimately important to delivering better insights.

This is indeed fundamentally all about agile sensory, sensory tools and approaches that exactly meet our clients' requirements, taking long-standing, clearly defined sensory ways of working and building on them. However, this can only work if you have a sound scientific foundation.

6.3 Kicked out of the laboratory

6.3.1 The elephant in the (sensory) room

It is unavoidable not to acknowledge that the Covid-19 pandemic had a significant impact on the food industry from manufacturing to supply chain. But it also had an important impact on R&D entities as well and, here, we focus on how it impacted sensory facilities and functions.

Due to various local and national lockdowns around the world, sensory teams were not able to meet anymore for informal tasting to decide the next steps on innovation and renovation projects, but more importantly sensory panellists were not able to sit in those discussion rooms and work together. All our academic protocols came under intense scrutiny, how could we adapt to ensure that sensory research could continue and continue to create meaningful data?

6.3.2 Elevating the sensory outputs

Whilst for many research companies, Covid brought new challenges to researchers, at MMR, we were more fortunate as those 'new' challenges were not so new to us. Challenged by the need for sensory science to move closer to the consumer, we had already started to modify our ways of working, moving away from the former academic framing of sensory to a more commercial model, whilst still holding firm onto the scientific principles that are the foundations of sensory science.

For all the right reasons, sensory science has been driven by the scientific rules which govern what constitutes good sensory practice. However, being a commercial sensory scientist, endeavouring to stick religiously to these high academic standards can greatly limit the flexibility we need to be able to work in a commercial environment. The commercial world is becoming increasingly agile and time and cost sensitive. The industry has commercial imperatives that won't always fit within the academic approach or time scales of traditional sensory science.

Traditional academic ways of working see the consumer through the lens of preference, liking, purchase behaviour, branding etc., and sensory panels designed around precision, the absence of hedonics and quantitative data. The 'human' is abstracted as far as possible so as not to bias data or create noise in our results. Here, we witness two clear and distinct paradigms when conducting research set in contrast to a much more blurred, even blended, approach in the consumer world.

What was particularly troubling for us at MMR was, on the one hand, there is a perceived split between sensory and consumer, whilst on the other, for many of our clients, the two are synonymous, albeit that they have different 'departments' managing each research dynamic.

Our clients need to understand the entire sensory experience of products which led us to develop approaches which deconstruct packaging, understand how a product was cooked in the home and all the sensory touchpoints of this process reveal the sensory touchpoints of 'food on the go'. There was an ever-growing need of matching consumer behaviour in order to understand the true consumer experience better.

Think about a recent trend; plant-based food and let's take the example of burgers. Imagine having a sensory assessment of the burgers in a laboratory with products being prepared in the kitchen in a standardised way. This would give very robust data on how each burger performs against other competitor burgers, potentially with meat benchmarks and unpacking key elements of plant-based burgers such as the textural journey of eating

2. How digital is creating new opportunities for sensory science

the patty. Panellists will generate a vocabulary and score the attributes to give a finger-print of the products and highlight the differences.

However, we know that consumers would be influenced by more than just how products look and taste from the perspective of 'on the plate' evaluations. These are all important elements, but let's be clear that they are only part of the puzzle. For example, the smells of the burger cooking will cue up perceptions of how the burger will taste. And the noises of cooking, the sizzling in the pan? Surely, it is important to deconstruct and understand that. What about the packaging? This will equally cue up perceptions of how the product will taste or the texture in the mouth. And more fundamentally, all these sensory touchpoints need to work together, to be aligned and create a compelling product proposition.

6.3.3 Beyond classic sensory

The beauty and hair care categories also stretched our boundaries and forced us to be more creative in our approach of sensory. Due to the nature of the products, evaluating them in classic sensory booths will not enable you to fully deconstruct the experience of using them. Here, home evaluations are an asset as opposed to a risk or a Covid work around.

In addition, the logistics required in a sensory laboratory to accommodate, for example, a shampoo evaluation, are fairly extensive when this can be done relatively easily at home provided that you have the right protocol and the right tech to support it. Voice surveys, for instance, are now becoming an important tool in at-home data capture, where panellists would give their score by voice command. This can be used for other categories where preparation or application requires full use of hands as shown in Fig. 6.1 in the case of evaluation of tomato sauce.

And now with the advent of AI, chatbots can also be used to guide and capture the information that panellists generate as they deconstruct products in the home, in the conditions that products are designed to be used in.

Having addressed the challenges of home, on the go or personal care evaluation, we started to think about how this could be further developed. If you consider, for example,

FIGURE 6.1 Voice survey application to capture cooking experience. This enables to gather sensory data on the moment whilst acting as a consumer.

skin care products, then how they perform will vary from skin type to skin type. At first, this appears to be a challenge too far for sensory panels.

However, with our newly honed out of the laboratory sensory tools and techniques, we came up with a solution to train the actual target and user of the product for them to become panellists creating digital-trained user panel — more on that later.

6.4 The paradigm shift: using digital sensory to meet the new industry challenges

When reflecting back to how we tackled these adjustments and how we moved to more agile ways of working, we can summarise these changes in three broad categories:

1. tech adjustment
2. protocol adjustment — environment
3. the human touch

6.4.1 Tech adjustment

Because we had been using the panel at home for a number of years, we had already implemented ways of capturing data, both quantitative and qualitative from panellist's homes, cars and in other venues. The challenge for us, like many other sensory panels, was how to run discussion and training sessions online. The theory is, of course, very simple, use Teams or Zoom. The practice turned out to be a little more challenging.

We started off with some training sessions with the panel, teaching them the etiquette of working online, being on mute if the environment you are working in is noisy, raising your hand to ask a question, not talking over other people when having discussion sessions etc. We also quickly learnt that, for example you need two screens to work effectively as a panellist, one for the Teams/Zoom and one for the data capture software and both ideally need to be full size. Sharing and discussion training data via a phone just doesn't work!

When sensory works in this digital manner, it becomes much more inclusive. We had clients dialling into sessions from around the world, trainee panel leaders from around the world learning their trade quickly through joining other panels online that included client panel leaders.

We then began to understand that this digital approach to running panel training and data collection had huge potential beyond the simple things:

- As discussed earlier, voice-activated technology enables panellists to discuss or score a product when they are actually using it, in the shower, in the kitchen, vacuuming the floor, washing their car, serving dinner etc.
- Recoding product experiences — looking back to the supermarket example, it is not just the visual elements of a supermarket that will inform on a consumer's experience but also the sounds and the space itself.

In essence, digital sensory enables us to track the entire product deconstruction process, to unpack all the sensory touchpoints that will, ultimately, influence consumer preference.

6.4.2 Protocol adjustment — environment

Again, at MMR, we had already been thinking about how to assess products in a manner that emulates consumer behaviour. If your project is to deconstruct kettles, then the panellists need to be using the kettle in a conventional way to enable them to unpack all the sensory touchpoints which could impact the consumer perception of the kettle. If a panellist is sat in a booth, then the sounds that a kettle makes when being used won't be heard correctly — water going into the kettle, the click of the lid, the sounds of the kettle fitting into its base etc., all heard from a small booth rather than in the context of a kitchen environment.

This all stood us in good stead for working from home. But it didn't cover everything. We still had to run 'conventional' sensory research which meant thinking through how to standardise lighting as much as possible, by having the panellists working directly in front of a window evaluating products at exactly the same time so the outside light was as consistent as possible, or how to you control the impact of different cookers. By having a standard product which the panel will score so we can understand how each cooker impacts the sensory profile of this standard product. And the latter could then become an approach for evaluating the cookers themselves, deconstructing their use to understand all the sensory touchpoints and consumer interaction points.

Protocols had to become very precise. With a panel leader no longer in full control of their preparation, we had to think hard about managing protocol noise in the data. There are practical things we could do to control things — sending the panel home with a single weighed/measured portion of a product, but technology also came into it, in a very simple and obvious way. Protocols became digital, embedded into the data capture system, standard videos detailing exactly how to stir your porridge, while it's cooking, how to apply a hair product or dispense shower gel etc.

Digital protocols open up the world of global reach sensory projects, where panels around the globe profiling a section of the entire sample set which we then combined to enable our clients to understand the complete sensory space that their products work within globally. This has been a real blessing for digital sensory profiling.

6.4.3 The human touch

This was probably our biggest challenge. As scientists the above two points were a matter of thinking things through in a systematic and logical way to solve the problem and often applying simple digital solutions. We, as sensory scientists, are good at that.

However, panel engagement and motivation are more in the realms of social science rather than sensory science. If you are not careful, you can readily create a 'them and us' culture, exacerbated by being unable to have face-to-face chats.

Thinking and working our way around this problem has probably yielded the greatest amount of sensory innovation for MMR. The moment you step away from thinking of a panel as a machine, sensory tools and methodologies start to reveal themselves. If agile sensory is about using or developing the correct tools to meet our clients' research needs, then thinking more about the human element of research and challenging the sensory norms leads you to engage with your panel differently. For many of our projects we are no longer seeking robust triplicate data because that level of robustness is simply not required to meet our clients' research needs, timeframes, or budgets. If we are seeking deep understanding about all the sensory touchpoints of a product and a product experience then you need tools that enable panellists to use products 'as consumers' but report back 'as panellists', for example, describing the process of picking up and opening a pack and comparing this to other packs and if it is, for example, a baby product, doing this in a home kitchen whilst simulating holding a baby. Classically you'd think of this during vocabulary development, however, turning it into a qualitative deconstruction method generates a vast array of rich, detail driven information about product and the user experience.

And here digital sensory plays a big role. Data doesn't need to be numbers. Sometimes words, video or online streaming are more powerful. They will still be analysed with the same rigour providing our clients with very agile ways of generating the critical information they need to develop products or optimise them and to optimise the user experience of those products.

6.4.4 Digital innovation

What this digital shift made us realise is that panels are very agile and can add huge power to research by deconstructing the sensory touchpoints of products and experiences, and that this can happen remotely away from the sensory booths.

With panel leaders who were now exceedingly skilled at running panels online and getting robust, repeatable data from virtual panels, the logical step was to extend this further, conducting sensory deconstruction with end users: digital-trained user panels.

We took the learnings from moving panels to working at home, what science-based thinking was critical to retain and where a dose of healthy pragmatism worked and extended our construction of sensory panels to consumers, the end user.

Digital-trained user panels are recruited to be the users of the products being researched. And this is where consumer and sensory are indistinguishable. The objective of the exercise is to train the user panel to evaluate key sensory touchpoints in the context of the people who actually buy and use the products. A trained user panel can not only deconstruct products but also talk meaningfully about what it means to them as a consumer. This provides our clients with the ability to conduct early-stage or rapid product assessment evaluating them from a human point of view rather than a technical one. A crucial advantage of this approach is its quickness, typically the project can take only few weeks from recruitment to getting actionable data.

6.5 Future for sensory scientists

All the experiences accumulated on the path of moving away from 'classical' sensory profiling to a more deconstructive approach, and the skills of collecting robust data from a virtual panel naturally contributed to the fast development of moving sensory panels towards consumer behaviours and building digital-trained user panel at MMR. In the following sections, we address the opportunities and the challenges in the sensory evolution.

2. How digital is creating new opportunities for sensory science

6.5.1 Overall challenges and opportunities

The recent digital shift has rapidly sped up digital marketing. You have probably experienced this personally through the increased number of promotional emails, SMS and advertisements from the social media and live streaming we daily receive, whereas previously most of the product communication we were getting was from the product packaging.

A classic example is the chocolate from the brand Meiji in Japan which used to have a flavour spider profile on the back of the product pack. This is designed to inform consumers and guide choice through clear sensory cues whilst setting up expectations. We have seen more and more companies starting to build on it. Many times, when we talk about sensory, we are talking about its contribution to product innovation and renovation; however, moving into a digital world, its fundamental essence of promoting product communication cannot be ignored. And the opportunity for us to continue this sensory evolution is how to convert our sensory profile outputs into a compelling narrative around product experience and the expectation that sensory touchpoints can create in the digital world.

With a more crowded and competitive marketplace, a lot of pressure has been put on brands for the perfect innovation. Whether it being a disruptive innovation redefining the category with a dramatically better product experience or a co-creation innovation using sensory and consumers as a focus to identify need-based innovation, the level of alignment of an intended brand promise and the actual delivery of a compelling product experience are going to make a difference.

At MMR, we focus on offering rapid and efficient solutions to land real-added value outputs to our clients. In terms of sensory, we aimed at defining and quantifying the touchpoints that matter to consumers and provide impactful insight to drive product experience excellence. The digital-trained user panel approach provides us a unique opportunity to push beyond the boundaries of sensory evaluation by leveraging consumer-centric innovation resources thanks to the collaboration with our engaged product users.

6.5.2 Challenges and opportunities moving towards digital-trained user panel approach

6.5.2.1 *How to put the human at the heart of the process*

Understanding the consumer experience of products is fraught with complexity. Whilst they and only they can tell us about what they like and dislike, unfortunately extracting precise and actionable diagnostics can be challenging. However, with the right approach, users can provide crucial user relevant insight including sensory experience.

With digital-trained user panel, planning is everything to ensure a high degree of reliability and repeatability of the data. Somewhat reminiscent of classic sensory! Starting from the protocol design to incorporating input from the panellists efficiently, the application and product evaluation is reflecting the real-life product use experience. Then comes the panel performance monitoring. This new approach to sensory challenges the usual concept of panellist performance, do the traditional approaches of analysing this still work? Could there be any other ways we could use to measure data and performance?

Naturally, the following question to ask would be: how much noise in the data can we tolerate because using trained consumers rather than 'expert' sensory panellists is going to increase the amount of noise you get in the data? This does not render this new approach wrong or inaccurate; it simply requires a change in how we view panellist's performance and the metrics we use to assess it.

Through our exploration on developing digital-trained user panels, three interactions have been identified as very critical to make the process successful:

1. Select the most engaged and active product users to work with — screening. Based on the principles of 'classic' sensory.
2. Make effective engagement with the users in collecting quantitative and qualitative feedback — data collection.
3. Create the outlet for user experience sharing — motivation.

6.5.2.2 Remote screening

Panellists' screening is key for any traditional sensory panel, but it is even more for a digital-trained user panel as they will have less training time. As these projects are often quick turnaround, we also need to develop a rapid screening process.

Much like consumer research, the first step is about defining the product users and the demographics to recruit against. Akin to recruiting expert sensory panellists, the next stage is to access candidates' sensory acuity articulacy and ability to work in a group. Unlike a classic sensory panellist, you may also want to consider the consumer knowledge and engagement with the relevant product category.

In order to select such category-targeted demographics for each project, we had to move onto remote screening stages by using different platforms as shown in Fig. 6.2.

We also used chatbots where candidates are probed automatically depending on their previous answers, for example 'you said level of flavour is differentiating product A versus product B, which flavour is it in particular' which is generated because the candidate mentioned 'flavour' as main difference between two samples. Using all these tools also meant that we could have our clients take part to the entire process, they were able to see live progress and the results of the screening, making it a collaborative process.

6.5.2.3 Digital platforms for data collection

As mentioned in the previous section, optimising the interface of the digital platform for data collection and developing new techniques have contributed to a lot of innovation in conducting sensory studies at MMR. Working digitally can mean losing a part of the control that we have a in a typical sensory setting, let's say having a panel leader instructing when to start the evaluation and indicating each time point on a specific temporal method.

Therefore we invested time and resources for the platform with our partners to ensure that we had straightforward and simplified instructions, user-friendly environment and a highly tailorable platform to accommodate any type of tasks.

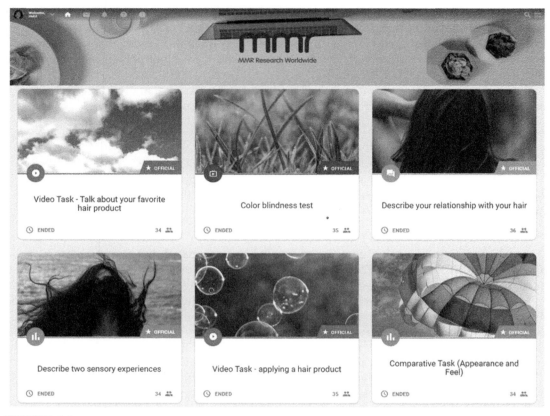

FIGURE 6.2   Remote screening digital platform. This enables to set up agile screenings globally.

6.5.2.4 Create the outlet for user experience sharing – motivation

Challenges around motivation are similar to those in a classic panel laboratory setting and online setting but at a much higher level. One key aspect is to make the panellists understand that they are contributing fully to the research and development of the next bestselling product. As users, we also want to understand their personal engagement with using the product which is why we introduced some qualitative tasks as part of the evaluation which makes them feel even more part of an improvement process. Flexibility is also key for panellists in a digital-trained user panel, typically sessions will be run in the evenings on weekdays and during the weekend.

6.5.2.5 Amplify the sensory power and beyond

We have seen the digital-trained user panel can not only provide our clients with the ability to conduct early-stage or rapid product assessment evaluating them from a consumer point of view rather than a technical one, but also an integrated approach to speed up new product development in terms of defining the target consumers' pain points and expectations, whilst providing unique perspectives on product experiences and insights for marketing the sensory experience digitally. The move to digital sensory is enabling if not empowering this change.

We used a digital-trained user panel with teenage makeup users to create quantitative sensory profiles of three makeup products at home using consumer language relevant to the category. Fig. 6.3 shows the outcome of the profile and how we could deliver clear insights on which key sensory touchpoints to convey and further product optimisation. In addition, qualitative feedbacks were shared on branding, advertising and price point for each of the products. These results were only achievable through the voice of real users. The teens also shared their preferred e-commerce platforms and social media influences, to provide further insights on how to better market the sensory experience of the products online, which can be used to maximise chances of sales success in e-commerce platforms.

Overtime we found that the digital-trained user panel approach is the perfect bridge between sensory research and consumer research, transforming personal user experience to objective quantitative product characterisation all through the lens of sensory science.

6.6 What does it mean for young sensory scientists about to enter the digital sensory era?

As we mentioned, the development of sensory science was inherently connected with the rapid growth of FMCG/CPG companies. It is also a science developed based on the foundations of multiple other sciences: physiology, psychology, statistics etc. As we are

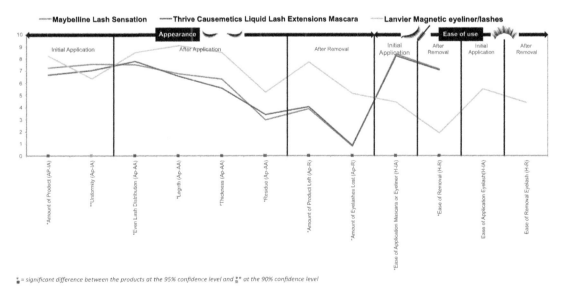

FIGURE 6.3 Profiling results of lash curlers using the digital-trained user panel approach. Results were showing clear significant differences on many major attributes giving clear directions for improvement. Please check the online version to view the colour image of the figure.

heading into a digital world, it is important that we, as sensory scientists, see the bigger picture that products don't work in isolation, they exist as holistic product experiences.

A lot of the adjustments talked about previously heading into digital sensory are from collaborative teamwork in seeking continuous innovation not just between panel leader and project manager but with experts in statistics, new technology etc. We believe that in blurring the lines between sensory and consumer, you can generate insights to provide comprehensive product feedback to our clients to unpack the users' experience of products throughout the user journey.

We can also foresee that data correlation will be even more important for the future, for example connecting technical data to explore the impact of how adding different ingredients may impact the sensory experience (e.g., how skin care products with SPF may deliver a different sensory experience) or adding product sales data to consumer and sensory data to understand not only which product is liked and how it should sensorially be designed but also to highlight business priorities. We also can't ignore recently the development of AI and its potential in correlating data or its potential to predict data.

6.6.1 Deconstruct beyond the obvious

With the internet, our sensory panel can deconstruct beyond food and personal care products out of the booth to a world of experience. We can connect the experience of cooking and preparing food, defining the best aroma profile from the moment the plant-based burger patties are turned golden brown on the grill in the kitchen. We can design sequential profiling to capture their drinking experiences of drinks designed to be consumed 'on the go', for example sports drinks. With makeup users, we could explore the experience of hair products overtime, when applying the products, and wear experience at multiple time points.

The user experience of a product is a function of all its sensory touchpoints and all the sensory cues it delivers. Whilst many sensory characteristics and cues work at an almost subconscious level, they are extremely important to understand, evaluate and optimise and we believe that digital sensory can facilitate the move from a classic sensory laboratory to explore a world of experience. Bring it on!

Using digital tools to understand individual differences in flavour perception which impact on food preferences

Frances Jack

The Scotch Whisky Research Institute, Research Avenue North, Edinburgh, United Kingdom

7.1 Introduction

Sensory scientists often refer to their panel as being the equivalent of an instrument for measuring flavour. Traditionally, the output of this 'instrument' has been the consensus opinion of the panel, which they use as a guide to how their products will be perceived in the market. Popular methods, such as Quantitative Descriptive Analysis (Stone et al., 1974) and the Texture Profile Method (Brandt et al., 1963) involve panellists scoring the intensities of predefined attributes. Panellists require extensive training. This often involves the presentation of reference samples, formulated to illustrate specific intensities of flavour attributes. For example, training for sweetness might entail tasting different concentrations of sucrose solutions formulated to represent 0, 2.5, 5, 7.5 and 10 on a 10-point scale. The aim is to train panellists to score samples in a similar way. In these tests, any residual inconsistencies are evened out by averaging scores across the panel. Similarly, sensory difference testing, using methods such as the Triangle Test (ISO, 2021), are centred around the overall judgement of the panel. Results are interpreted based on the proportion of panellists selecting the 'different' sample in these tests. Other sensory approaches, such as Free Choice Profiling and Napping (Delarue et al., 2015) allow panellists to evaluate samples using their own constructs and descriptions of characteristics. Even with these techniques, the default is to apply statistical analysis to obtain an overview of the data.

In Chapter 9, Predicting sensory properties from chemical profiles, the ultimate flavour puzzle: a tale of interactions, receptors, mathematics and artificial intelligence, Andy Taylor describes application of technologies such as Machine Learning may be a route to achieving this. He touches on the additional complicating factor that the panel is not homogeneous, with individuals perceiving sensory characteristics differently. Consumers will vary in their perceptions. This chapter considers whether the data that we hold on our sensory panellists can be used to our advantage to give better insights into the range of consumer perceptions. With this approach, our panel is more akin to a suite of instruments, than a single instrument, with individual members of the panel each providing a different view of how the product/products will be perceived. Before going on to consider the opportunities that this can provide, we will look at the differences that exist in flavour perception and the implications of this.

7.2 Differences in flavour perception

Aroma is inarguably the most complex of our perceptions. It is detected in the olfactory epithelium located at the top of the nasal cavity, with humans having around 400 odour receptors. Each odour receptor can be stimulated by multiple compounds, and odour compounds can stimulate more than one receptor. Consequently, aroma perception is a pattern recognition process, often arising from the complex stimulation of multiple receptors. It is estimated that humans can smell over 1 trillion different aromas (Bushdid et al., 2014). To add to this complexity, each person has a unique set of genetic variations that influence the function of individual olfactory receptors. This has a significant impact on perception, as described by Trimmer et al. (2019). Typically, reduced receptor function can be linked to a reduction in aroma perception. All in, this results in a high degree of person-to-person variability in the way that aromas are perceived.

In addition to genetic differences, physiological and cultural factors can also impact aroma perception. Our age, for example, has an influence, with olfactory acuity decreasing as we get older (Boyce & Shone, 2006). Familiar aromas are often perceived differently from novel aromas. This is influenced by our diet and food experiences, which are intrinsically linked to our cultural background and upbringing. Long-term exposure to aromas can decrease sensitivity, while short-term exposure (training) can enhance it. Lifestyle factors, for example being a regular smoker, can also have an impact on our sensory acuity. Furthermore, our perceptions can be different at different times of the day, and we can experience day-to-day variation. Finally, the Covid-19 pandemic has highlighted how our sense of smell can be diminished or distorted by infections, either on a short-term basis or for an extended period. Similarly, alterations in perception are well-recognised in other circumstances, such as in cancer patients or during pregnancy.

Similar differences between people can be observed for other sensory characteristics: gustation, chemesthesis and texture sensations. The most extensively researched of these is our ability to taste bitterness. The first cited evidence of differences in bitter taste is attributed to an observation made by Fox in the 1930s (Blakeslee & Fox, 1932). As the result of an accidental release of phenylthiocarbamide (PTC) in his laboratory, he noted that some individuals found this compound extremely bitter while others were unable to, or could barely, detect it. This led to a multitude of other studies on PTC and a similar bitter compound 6-n-propylthiouracil (PROP). Research resulted in the identification of a polymorphism in the TAS2R38 gene (Kim et al., 2003), which is now well established as one of the key genes influencing our sensitivity to bitterness.

Although it is unlikely that the sensory panel leader will have access to genetic information on their panellists, other data relating to differences in their perceptions are likely to be collected. Regular proficiency testing is good practice for any trained sensory panel. Application of digital technology, such as an interactive data visualisation platform, will help ensure any issues are identified and trends in good and poor performance highlighted. The data collected while carrying out proficiency tests can be useful not only to include or exclude panellists based on performance but also as an aid to interpreting panel results. There is also information to be gained from the day-to-day performance of each panellist in regular sensory tests, that is the types of flavours an individual tends to pick up on and score highly, and conversely flavours they omit. Panellists may also have participated in other tests, such as threshold testing or gas-chromatography olfactometry, which will again show their ability to detect individual flavour compounds. Other chapters of this book provide more information on diversity of other types of sensory data that can be collected from panellists. Consolidation of information from each data source and centralisation using a digital dashboard contributes to more visible and actionable results. This can inform day-to-day decisions and contributes to more powerful business insights.

Finally, it is important to highlight that it is not just our perceptions that vary. Another important element in sensory evaluation is the ability to communicate our perceptions. Describing flavour characteristics relies on having the language skills required to verbalise what we detect. Some people are naturally better at this than others, with flavour recognition being closely linked to memory. Unlike our fundamental perceptions, this can be enhanced by training, namely giving panellists the language and vocabulary they required to describe what they perceive. However, even after training, there may be flavours that some panellists find difficult to verbalise or descriptors that they confuse. Knowledge of these issues and integration into a data visualisation platform will also help when it comes to interpreting the data.

7.3 Implications of differences in flavour perception

Differences in sensory perception are important determinants of food preferences and consumption. Hence, there are clear benefits to be gained if we can take this into account in our sensory panel tests. Feeney et al. (2021) produced a detailed review of research on the links between genetic variation in flavour perception and food liking and intake, which provides a useful resource for those interested in exploring this topic in detail. For the purposes of this chapter, some examples have been chosen to highlight the implications of differences in perception.

The first is the polarised response that individuals have to the herb *Coriandrum sativum*, commonly known as coriander or cilantro. Many people love its flavour, but others express an extreme dislike of its 'intense soapy' characteristics. A study of over 14,000 European participants by Eriksson et al. (2012) demonstrated that these preferences are linked to the function of the olfactory receptor (OR6A2) that contributes to the ability to detect a range of aldehydes which are particularly prevalent in this herb. Individuals that have the genetic variant of this receptor that results in greater sensitivity to these compounds tend to be the people who dislike coriander.

Similarly, genetic differences in taste sensitivity can be key drivers of food preference. Since there are a relatively small number of taste characteristics which are common to many different foods, this can have more of an impact on a person's nutritional intake, particularly if this results in dislike of key food groups. Again, comprehensive review of the literature in this area is available for those that would like to learn more (Diószegi et al., 2019).

Other interesting studies on the implications of differences in flavour perception are those that look at variation among larger groups. A good example of this is the Italian Taste Project. This study brought together researchers from organisations with common interests in exploring the influences on food choice in Italian consumers. It involved 1225 individuals, with PROP status and sensitivities to key aroma and taste compounds being central to the evaluation (Monteleone et al., 2017). This broad study was one of the first of its kind, revealing both the diversity in sensitivities and the impact of this on food preferences.

In this chapter, two case studies have been selected to demonstrate how knowledge of individual panellists' strengths and weaknesses in perception can provide the sensory scientist with useful insights which would not have been apparent when viewing overall panel data. Both examples are drawn from the author's experience in sensory research at The Scotch Whisky Research Institute (SWRI), and each considers how digital tools could be used to leverage the insights from the data more efficiently.

Scotch Whisky flavour arises from the complex interaction of a diverse range of naturally generated flavour compounds. These compounds originate from the raw materials used in whisky production, their processing in the distillery and subsequent maturation in oak casks. The relationship between these compounds and the sensory characteristics of a whisky is complex, with additive, suppressive and synergistic effects of different compounds all playing a role in overall flavour (Harrison, 2022). The SWRI Flavour Wheel (Fig. 7.1) further highlights the complexity of whisky flavour. Developed back in the 1970s, this flavour wheel was established to aid communication across the industry as whiskies were frequently traded between companies (Shortreed et al., 1979). Individual whiskies typically comprise a combination of characteristics from the wheel, with intensity and balance varying between brands.

Sensory evaluation is used in the whisky industry as the primary method for measuring and controlling flavour character. Typically, tests are carried out by trained or expert panellists. In recent years, the industry has embraced digital technology for the collection of sensory data, with companies using specialised sensory software such as Compusense or Red Jade, or developing their own bespoke data collection and sensory information management systems. The case studies give two very different examples of the advantages to be gained from applying information held on panellists when executing or interpreting sensory tests. The first case study describes a quality control application where the aim of the test is to identify and control potential off-notes. The second describes a flavour profiling approach that might be used during new product development or in similar applications.

7.3.1 Case study 1: detection and control of off-notes in whisky samples

Sensory evaluation is used in the whisky industry to detect and eliminate undesirable, off-note, characteristics. These attributes can be found on the flavour wheel (Fig. 7.1), the

FIGURE 7.1 The Scotch Whisky Research Institute Flavour Wheel. The flavour wheel is a comprehensive whisky vocabulary covering both desirable and off-note aromas and tastes. *Source: Reproduced with permission from The Scotch Whisky Research Institute.*

main categories of off-note characteristics being sour, sulphury and stale. Passing or failing of samples is often based on the consensus opinion of the panel. Such an approach runs the risk of passing samples with off-notes that will be detectable by sensitive consumers. The most obvious way around this is to only use highly sensitive panellists to carry out these tests. However, from our research at SWRI, we know that it is hard to find individuals that have good sensitivity across this full range of aromas. A panellist that can easily detect sour notes might not be so good at detecting sulphury characteristics. The best panel is one which has enough members with the range of sensitivities to cover all the off-notes of interest. Combining this with a good understanding of each panellist's strengths and weaknesses will further help in the interpretation of panel data. Digital technology makes this achievable, giving the company greater reassurance that off-notes will be detected.

A good example of an off-note that requires sensory evaluation for its detection is the musty taint caused by the presence of 2,4,6-trichloroanisole (TCA). TCA is familiar in the wine industry, being the compound responsible for 'corked' wines. It is formed from the microbial degradation of trichlorophenols. Although not hazardous to health, its presence can have a severe detrimental impact on product quality. At lower levels, although not perceived as characteristically musty, it can still have a negative impact, suppressing other, desirable flavour characteristics. TCA can occur in whisky, again due to defective corks. It also occasionally arises in the oak casks used for maturation. This is of greater concern due to the greater volume of product which will be tainted. TCA is formed in the cask when it is empty, so its presence can be detected before filling and the cask rejected. 2,4,6-trichloroanisole has an extremely low sensory threshold. Currently, the instrumental limits of detection for this compound are around 10 ppt, but it can be detected on the nose at much lower levels.

In addition to TCA having a low sensory threshold, our research has shown that there are large differences in sensitivity to this compound. Table 7.1 shows the aroma detection thresholds for TCA in 20% ethanol, the standard alcohol strength at which whiskies are typically assessed in quality control applications. The data presented are a summary of the distribution of thresholds for 217 industry assessors, all of whom were actively involved in sensory testing of whiskies and related spirits at the time of testing.

The first point to note from the data in Table 7.1 is that almost 70% of the people tested were able to detect TCA below, or around, the instrumental limit of detection (10 ppt), confirming the need for sensory testing in the control of this compound. The data highlight another challenge. Some assessors detect TCA at concentrations of 0.25 ppt or less. Although the percentage of individuals with this high degree of sensitivity was low (3.2%), these results indicate that there will be a small but important group of consumers that could potentially detect this taint-forming compound when present at these trace levels.

Consider the scenario where a customer has returned a bottle of whisky with the complaint that 'it doesn't taste right'. The company has a retained reference sample against which this customer complaint can be compared. Compositionally they look identical, so a follow up sensory comparison is carried out. Triangle Testing is used for this. The

TABLE 7.1 Distribution of aroma detection thresholds for 2,4,6-tricholoroanisole (TCA) in a group of 217 whisky industry assessors. This table shows the number and percentage of assessors able to detect TCA at levels ranging from 0.25 to 10 ppt.

	≤0.25 ppt	0.50 ppt	1 ppt	2 ppt	5 ppt	10 ppt	>10 ppt
Number of assessors	7	14	29	29	44	27	67
Percentage	3.2	6.4	13.4	13.4	20.3	12.4	30.9

company is in the fortunate position of having a relatively large group of 21 trained panellists. In this test, only seven of the panellists correctly select the odd sample, well below the 12 required to infer statistical significance at a P value of .05 (ISO, 2021). One of the seven panellists that correctly selected the odd sample described the customer complaint as musty. This is where understanding of individual panellist performance comes in to play. Data held on this panellist show that they are extremely sensitive to TCA. This could indicate that, although most of the panel could not detect any problem with the sample, a musty taint may still be detected by a small proportion of the population. This would potentially validate the customer complaint. Further blind testing with this individual panellist would be required to confirm this observation, but this may be the first step in pin-pointing an issue that would otherwise have been overlooked.

In the past, the panel leader may have picked up this anomaly, but only if they had built up a good knowledge of panellist performance and had enough time and attention to detail to pick up on the single 'musty' comment. Retained sensory threshold data for each of the panellists allows this process to be automated and integrated in a digital platform connected to the sensory software, triggering a notification when an off-note is picked up by a particularly sensitive individual. Equally important might be the triggering of a notification to alert the panel leader to overall weaknesses, for example if the test session did not contain any panellists with high sensitivities to certain off-notes. Overall, it is clear to see how this approach can give more confidence in the data obtained from quality control panels and enhance the sensory decision-making process for passing or failing samples. If we assume that the data in Table 7.1 are broadly representative of variations in sensitivity across the population, companies could make informed risked-based decisions on the percentage of the population likely to detect an issue.

7.3.2 Case study 2: flavour profiling of mature whiskies during new product development

A common means of creating new and interesting flavour profiles in whisky is to use different types of casks during the maturation process. According to its legal definition, The Scotch Whisky Regulations 2009 (UK Parliament, 2009), Scotch Whisky must be matured in oak casks for a minimum of three years. During maturation, the flavour of the whisky changes. The pungent aromas of the new distillate are transformed into the typical mellow characteristics of a mature whisky. There is addition of flavour compounds which are extracted from the wood, and at the same time, some of the most volatile components are lost through evaporation. Finally, reactions between the different spirit components alter the flavour over time (Conner, 2022). The Scotch Whisky industry does not typically use new casks, instead sourcing ones that have previously been used for other purposes, such as the maturation of bourbon or sherry. Recently the legislation has been modified to allow a wider range of cask types. The whisky industry is taking advantage of this to produce whiskies with different flavour profiles. Trained sensory panellists are used to understand the resulting sensory characteristics, and how the flavour profile of the whisky is likely to be perceived by the consumer. Again, we need to take into account that these perceptions might not be uniform.

This case study looks at the impact of variation in sensitivity to a single flavour compound, oak lactone. As the name suggests, this compound is extracted from the oak casks during maturation, imparting a nutty/coconut flavour. It is present in all types of oak, but levels and ratios of the different stereoisomers of this compound can vary, depending on the species of oak used to construct the cask (Masson et al., 1995). These differences have been shown to have an impact on the final character of the whisky (Noguchi et al., 2010). Research, again undertaken at SWRI in this instance using their expert sensory panel (n = 14), has shown that some individuals have a much higher sensory threshold, or are even anosmic, to oak lactones. The average aroma detection threshold in 20% ethanol across the panel was 0.11 ppm, but results showed that five panellists had a much lower sensitivity than the rest of the panel. Further work was carried out to determine if these lactone-insensitive panellists would perceive the mature characteristics of a whisky differently from the other (lactone-sensitive) panellists. Since these panellists were all trained and experienced in the quantitative descriptive profiling of whiskies, it was a straightforward task to get them all to evaluate the same whisky. The average scores across the full panel and for the lactone sensitive and insensitive subgroups are shown in Fig. 7.2.

Sensitivity to oak lactone had a clear influence on how the panellists perceived the flavour characteristics of the whisky (Fig. 7.2). Splitting the panel in this way provides useful insights for whisky producers, which would not otherwise have been apparent. As anticipated, the lactone-insensitive group gave lower scores for a nutty character, the attribute most associated with oak lactone. Interestingly, this group also gave lower scores for other oak-derived flavours, such as woody and spicy aromas. Further work would be required to determine if these panellists also had low sensitivity to other mature flavour compounds. But, since these are chemically quite different from oak lactone, they are not likely to be linked. Another interesting observation from the data is the higher scores for other attributes, such as cereal and soapy, from the lactone-insensitive panellists. These attributes are generally associated with immature spirit, with intensity of these flavours reducing as the whisky matures. One hypothesis is that oak lactone plays a role in masking

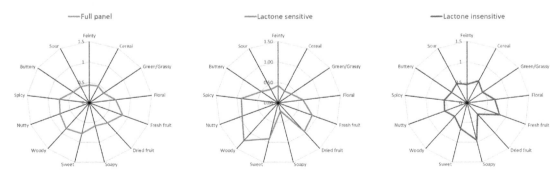

FIGURE 7.2 Mean sensory scores across the full panel (n-14) and for lactone-sensitive (n = 9) and lactone-insensitive (n = 5) panellists from the quantitative descriptive profiling of a mature Scotch Whisky. This figure shows three flavour profiles of the same whisky which are based on data from the full panel and the two subsets of assessors, lactone sensitive and insensitive groups. It demonstrates the differences in perception that can exist depending on sensitivity to key flavour compounds, in this case, oak lactone.

these aromas. There would be less of this masking in the lactone-insensitive group, so these attributes are perceived as being stronger.

One point that is clear from the data in Fig. 7.2 is that the consensus opinion of the panel is not a good representation of how the product is likely to be perceived by all consumers. The data from the two subgroups may not be a good representation either, as there are likely to be further influences on the flavour profile caused by sensitivity to other compounds. The case could be argued for looking at the individual profiles produced by each panellist. This understanding of the variability would potentially be more useful than the consensus. The application of digital technology could then allow any differences to be related back to differences in sensitivity to maturation related flavour compounds. It would allow the producer to understand how different consumers might view the flavour profile of their products before they are released on the market. This helps the whisky producer make more informed decisions on cask selection and aid in the development of marketing strategies around flavour.

These two case studies are examples from the whisky industry, but the linking of information on panellist sensitivity with panel outputs is equally applicable in a full spectrum of other consumer products. Detection of off-notes or undesirable flavours during production and flavour profiling to understand consumer perception during new product development are standard approaches used by most sensory scientists.

7.4 Conclusions and future perspectives

This chapter has demonstrated the benefits to be gained from understanding differences in perception among our sensory panellists and applying this information in the interpretation of panel data. Digital technology makes the exploration of these links possible in a more accessible and efficient way. Developments in this area are only in their early stages. It is relatively easy to start with the implementation of simple systems, such as the one described in the first case study for the identification of off-notes. Sensory scientists need to work with software suppliers to incorporate the computational requirements into existing systems, or alternatively develop their own bespoke software to address these needs. These benefits go beyond the work at the SWRI, applying to quality control, and in particular, the identification of off-notes, at both large and small distilleries. Digital tools could help minimise products contaminated with off-notes reaching the consumer, increasing their enjoyment and ultimately profitability. Producers of other consumer goods have similar challenges to identify off-notes as part of quality control procedures. As digital tools become cheaper and more user-friendly, many more companies can benefit from the opportunities offered.

The second case study showed that even a simple understanding of differences in sensitivity to a single compound can have a significant impact on how products are perceived. Again, this type of knowledge is easy to build into existing software systems. However, as this chapter describes, the influences on overall perception are complex and extensive research is still required for a more complete picture. Application of technologies such as Machine Learning, described in detail in Chapter 9 of this book, may be a route to achieving this. The challenge described in this chapter was to predict sensory character from compositional data. There is potential to expand on this to predict individual perceptions, again based on measured concentrations of flavour compounds but also taking data on sensitivities to these compounds into account. This would require significant computational power. As with any statistical prediction, good quality data will be central to success. In the future, advances in genotyping may provide useful information on our panellists, though the ethics of this would need to be fully considered. Currently, threshold testing or performance in sensory proficiency schemes are good options. These data need to be regularly updated as perceptions change over time. Development of digital tools that allow the rapid collection of information on compound sensitivity would be useful.

Finally, as technologies continue to advance, the scope of this approach magnifies. In the future, we may be able to collect information directly from consumers more readily. Who knows, the time may even come when digital technology allows the creation of products specifically tailored to our own flavour sensitivities and preferences.

References

Blakeslee, A. F., & Fox, A. L. (1932). Our different taste worlds: P. t. c. as a demonstration of genetic differences in taste. *Journal of Heredity, 23*(3), 97–107. Available from https://doi.org/10.1093/oxfordjournals.jhered.a103585.

Boyce, J. M., & Shone, G. R. (2006). Effects of ageing on smell and taste. *Postgraduate Medical Journal, 82*(966), 239–241. Available from https://doi.org/10.1136/pgmj.2005.039453.

Brandt, M. A., Skinner, E. Z., & Coleman, J. A. (1963). Texture profile method. *Journal of Food Science, 4*, 404–409. Available from https://doi.org/10.1111/j.1365-2621.1963.tb00218.x.

Bushdid, C., Magnasco, M. O., Vosshall, L. B., & Kellar, A. (2014). Humans can discriminate more than 1 trillion olfactory stimuli. *Science, 21*, 343. Available from https://www.science.org/doi/10.1126/science.1249168.

Conner, J. (2022). Maturation. In I. Russell, G. Stewart, & J. Kellershohn (Eds.), *Whisky and other spirits technology, production and marketing* (pp. 291–311). Academic Press. Available from https://doi.org/10.1016/C2019-0-03286-4.

Delarue, J., Lawlor, J. B., & Rogeaux, M. (2015). *Rapid sensory profiling techniques: Applications in new product development and consumer research.* Woodhead Publishing. Available from https://doi.org/10.1016/C2013-0-16502-6.

Diószegi, J., Llanaj, E., & Ádáy, R. (2019). Genetic background of taste perception, taste preferences and its nutritional implications: A systematic review. *Frontiers in Genetics, 10*, 1272. Available from https://doi.org/10.3389/fgene.2019.01272.

Eriksson, N., Wu, S., Do, C. B., Kiefer, A. K., Tung, J. Y., Mountain, J. L., Hinds, D. A., & Francke, U. (2012). A genetic variant near olfactory receptor genes influences cilantro preference. *Flavour, 1*(1). Available from https://doi.org/10.1186/2044-7248-1-22.

Feeney, E. L., McGuinness, L., Hayes, J. E., & Nolden, A. A. (2021). Genetic variation in sensation affects food liking and intake. *Current Opinion in Food Science, 42*(2021), 203–214. Available from https://doi.org/10.1016/j.cofs.2021.07.001; http://www.journals.elsevier.com/current-opinion-in-food-science/0.

Harrison, B. (2022). Volatile compounds formation in whisky. In F. R. Reis, & C. M. E. dos Santos (Eds.), *Volatile compounds formation in specialty beverages* (pp. 79–108). CRC Press. Available from https://doi.org/10.1201/9781003129462.

ISO. (2021). ISO 4120: Sensory analysis – Methodology – Triangle test Unpublished content ISO 4120: Sensory analysis – Methodology – Triangle test.

Kim, U. K., Jorgenson, E., Coon, H., Leppert, M., Risch, N., & Drayna, D. (2003). Positional cloning of the human quantitative trait locus underlying taste sensitivity to phenylthiocarbamide. *Science, 299*(5610), 1221–1225. Available from https://doi.org/10.1126/science.1080190.

Masson, G., Guichard, E., Fournier, N., & Puech, J. L. (1995). Stereoisomers of beta-methyl-gamma-octalactone. II. Contents in the wood of French (*Quercus robur* and *Quercus petraea*) and American (*Quercus alba*) oaks. *American Journal of Enology and Viticulture, 46*(4), 424–428.

Monteleone, E., Spinelli, S., Dinnella, C., Endrizzi, I., Laureati, M., Pagliarini, E., Sinesio, F., Gasperi, F., Torri, L., Aprea, E., Bailetti, L. I., Bendini, A., Braghieri, A., Cattaneo, C., Cliceri, D., Condelli, N., Cravero, M. C., Del

Caro, A., Di Monaco, R., ... Tesini, F. (2017). Exploring influences on food choice in a large population sample: The Italian Taste Project. *Food Quality and Preference*, *59*(2017), 123–140. Available from https://doi.org/10.1016/j.foodqual.2017.02.013.

Noguchi, Y., Hughes, P. S., Priest, F. G., Conner, J. M., & Jack, F. R. (2010). *Distilled Spirits — New horizons: Energy, environment and enlightenment. The influence of wood species of cask on matured whiskey aroma — The identification of a unique character imparted by casks of Japanese oak.* Nottingham: Nottingham University Press.

Shortreed, G. W., Rickards, P., Swan, J., & Burtles, S. (1979). The flavour terminology of Scotch whisky. *Brewers Guardian*, *1979*, 2–6.

Stone, H., Sidel, J., Oliver, S., Woolsey, A., & Singleton, R. C. (1974). Sensory evaluation by quantitative descriptive analysis. *Food Technology*, *28*(11), 24–34.

Trimmer, C., Kellar, A., Murphy, N. R., Snyder, L. L., Willer, J. R., Nagai, M. H., Katsanis, N., Vosshall, L. B., Matsunami, H., & Mainland, J. D. (2019). Genetic variation across the human olfactory receptor repertoire alters odor perception. *Proceedings of the National Academy of Sciences of The United States of America*, *19*, 116. Available from https://doi.org/10.1073/pnas.1804106115.

UK Parliament. (2009). The Scotch Whisky Regulations 2009. Retrieved from https://www.legislation.gov.uk/uksi/2009/2890/contents. Accessed March 9, 2022.

CHAPTER

8

Sensory self-service — digitalisation of sensory central location testing

Sven Henneberg, Annika Ipsen, Manuel Rost, Femke W.M. Damen and Torsten Koch

isi GmbH, Goettingen, Germany

8.1 Introduction

For many years, there has been an increasing demand in the food and beverage as well as in the nonfood and cosmetic industries to increase the speed of innovation (Horvat et al., 2019; Santeramo et al., 2018). Products need to be developed faster and more efficiently than ever.

Trends like sustainability, the plant-based food revolution, naturalness of products and reduction of 'unwanted' ingredients that result in a clean label policy represent an increasing challenge. Consumers have more diverse needs (Damen et al., 2019; Lesschaeve & Bruwer, 2010; Linnemann et al., 2006), and the industry wants to offer product innovations as quickly as possible (Guiné et al., 2020) to meet these needs. At the same time, product innovation failures remain a costly risk.

To anticipate and satisfy consumer needs, new or optimised products need to be developed and evaluated by consumers in a sensory consumer test to get their opinions on the products.

To reach this goal, simple and fast sensory product test methods are needed that deliver validated results and reflect consumers' preferences. To achieve faster research procedures, a standardisation of methods and processes is required. A predefined process and the preselection of the methodology for standard research projects lead to faster research results. Digitally implemented, even nonexperts in sensory consumer tests will be able to carry out sensory consumer research studies in a quick and smart way. Here, a digital platform that takes into account all of the user's needs and requirements is the obvious choice. The platform can overcome problems and difficulties in key areas of sensory consumer research, for example saving time from design to data collection to results. Of course, it is also associated with budget savings compared to nondigital approaches. As a result, users will get faster and high-quality implementation with valid results. Besides that, the workload of sensory consumer research specialists is reduced as well.

But how can digitising processes make sensory product tests simpler and faster? Let us first take a look at the typical kinds of people in the food and beverage industry whom we at isi Sensory Marketing Research (Germany) encounter again and again over 25 years as a partner. In particular, we will look at their needs and requirements.

The product manager: his bottleneck is a lack of expertise in sensory testing and a limited budget.

Let us imagine a product manager in the food industry responsible for a portfolio of products under one brand umbrella. Every year, several new products are developed for this brand portfolio. Before these products are launched, the product manager has to conduct a sensory product test with consumers to check the performance of the concepts and the associated recipes. In addition, all market products in his portfolio must be tested against competitors at regular intervals, also in a sensory consumer test. All these tests have to be paid out of the product manager's limited budget. He might prefer to use the budget for project innovation rather than for mandatory standard tests. Actually, the product manager only needs to have a good overview of his product portfolio without overspending resources like time and money. Moreover, he is most likely not a trained market researcher, which makes it difficult for him to choose the right methodological approach for the required consumer tests and to evaluate the resulting data. For this reason, he enlists the help of a consumer research specialist for all market research projects.

The consumer research specialist: his bottleneck is staff capacity, staff time and turnaround time of stakeholder-friendly insights.

The consumer research specialist's role could be to support all product managers in conducting market research studies, including methodological selection and data evaluation. However, a common problem for the specialist is the large number of tests he has to accompany. If these are standard tests, a large part of his time ends up being spent on the same type of projects over and over again. This specialist could use his experience much more efficiently in more complex research projects. Another task of the consumer research specialist is to provide the research results in simple, understandable overviews for different groups of stakeholders. For this purpose, the most important results are compiled for each research report produced, with different focal points as required, such as marketing, product development and product management. In addition, the different hierarchical levels, such as product managers or marketing directors, must be taken into account.

The marketing director: his bottleneck is obtaining relevant insights in a simple manner.

For example, the role of the marketing director is managing many different information resources. He needs to get information, including context and benchmarks, very quickly and concisely. He normally has little time to digest the information and therefore needs concise information directly relevant for decisions.

All these people are similar in their limited resources such as time and budget. They also partly lack the knowledge and experience to be able to carry out research projects independently from start to finish; they are dependent on support. A platform that addresses all the requirements, supports the handling of research projects with recurring questions, processes data and presents results to all participants in a way that satisfies their needs is the solution for all of their bottlenecks.

However, the usage experience of the platform should be intuitive. A focus on getting basic necessary insights is to be favoured over complicated methods, test designs and statistics. That way, everyone can concentrate on the essentials of their work.

8.2 Digital project

8.2.1 Project planning

A well-documented research project usually starts with a briefing document that summarises all available information and gives a rough description of the project. A product-test briefing often consists of two parts. The first part includes the requirements for the planned sensory product test. It covers the rationale for this project and all relevant objectives and facts from the business perspective (see Table 8.1). This part of the briefing is often created by a product manager.

In the second part of the briefing, the methodological parts are defined: the key objectives of the product test and the relevant questions to be covered. At this point, action standards can also be added. Furthermore, the methodology of the sensory consumer test is determined, and a precise definition of the consumer target group is chosen. This part of the briefing shows clearly the challenges of a product manager. If he is not a market research specialist, he has difficulties deciding about possible methodological options. Therefore this part of the briefing is created by a sensory consumer research specialist. At this point, the consumer research specialist's bottleneck becomes apparent: due to decision date and launch date deadlines, the number of projects from different stakeholders often pile up in similar periods of the year for the sensory consumer research team.

This bottleneck can be circumvented. Rather than forcing the research team to select methods, it is more efficient to predefine test standards and guidelines on a digital platform. The platform then enables other users to access this knowledge when planning their research, be it product managers, new colleagues, trainees or interns. Users who execute projects are guided to the right methods and standards via the digital interface/creation wizard. In other words, the role of the consumer research specialist becomes more of a designing, predefining and supervising role — he becomes an enabler for less knowledgeable colleagues and team members. In this role, specialists can set company-wide standards and support business units in standard projects rather than having to manage all

TABLE 8.1   Needs for project briefing.

Briefing part 1 responsibility: product manager	Briefing part 2 responsibility: sensory consumer research specialist
• Business objectives	• Key objectives
• Rationale why this project shall happen	• Key questions
• Known key fact	• Action standards if available
• Test stimuli	• Test method
• Test market (country or region)	• Target group
• Overall time line	• Consumer sample size

the studies on their own. This allows them to allocate more resources to nonstandard strategic projects that actually require their full expertise without quality concerns.

Such a platform can do even more. If standards are defined for the test method or the target group to be selected, the entire project planning can be carried out by the digital study-management platform. So, once the appropriate method has been determined by the system, further project-relevant parameters can be integrated as options – again defined in advance as framework conditions.

Via the platform, users can be asked, for example whether they want to receive results for the total consumer group or for subgroups. The latter then leads to an automatic predefined increase in the sample size to be able to evaluate the desired subgroups separately. The user receives a series of selection decisions for the methodological procedure, like easy or complex product preparation, number of samples, etc. At the end of this guided project planning management, a research briefing is automatically generated. In addition, the recruitment questionnaire as well as the product questionnaire can be generated as output by just one click. This is because all the characteristics have already been predefined and embedded in the platform.

Product questionnaire content often depends on product category and flavour variant – this can be predefined and selected interactively by the study design wizard. If necessary, optional questions can be predefined that users select upon demand. The role of the consumer research specialist here is merely to define which key attributes are mandatory and which are optional. It seems sensible to specify the total number of questions in advance to keep the survey for the test persons within reasonable limits. This results in a standardised yet flexible test design. In the same manner, scale levels, anchoring of scales and similar aspects can be set to being fixed or flexible, as seems appropriate to the consumer research specialist with its research knowledge for meeting quality standards

Sensory analysis usually takes place on site, that is in the market, in the region of distribution, with consumers tasting products. If the product test is conducted as a Central Location Test, the environmental conditions should be controlled. Alternatively, products can be brought to the consumers. In that case, they test at home (Boutrolle et al., 2007). However, all this leads to the necessity that the right consumers have to be recruited, the products and the test persons should come together, and the products must be prepared if necessary. Essentially, the fieldwork has to be planned and set up. Usually, the services of specialised field agencies are used (Horvat et al., 2019; Symmank, 2019), who can provide locations in the markets or regions.

Digitalisation can help simplify the selection and preparation process. All field partners in question can be embedded in the platform, listed and categorised by relevant characteristics like region, method and possibly even recommendations based on experiences with the partners. For this purpose, a whole set of detailed information can be embedded, such as the possibilities of carrying out quantitative or qualitative tests, whether or not a sensory lab is available, existing certifications, available IT technology, storage and preparation possibilities and meta-learning based on previous studies. A search platform can then be used to map the selection of a suitable field partner. The selection can be narrowed based on multiple filtering of those embedded characteristics so that the user is shown only such test studios that actually fit the individual project requirements. This way, even a product manager without extensive market research experience can select the right field partners without knowing them directly.

8.2.2 Fieldwork

The digitalisation of the questionnaires is already advanced. For many years already, questionnaires have been displayed on PCs and the answers collected digitally. Now, questionnaires can also be accessed on tablets or mobile phones. Mobile devices and apps enable the researcher to measure the sensory experience at the right place in the right moment. There are numerous software programs for generating questionnaires available. These offer many possibilities for questionnaire design, test design and partly also for evaluation. However, the software products are sometimes so complex that the researchers need to know what they want to do or achieve. Thus these software tools serve mainly specialists with deep methods knowledge. However, an integration of questionnaire programming into a holistic study-executing platform as mentioned above will also allow easy questionnaire generation following the very same principles as in briefing, methods setup and planning: A user has to choose the questions of interest and the questionnaire will be generated automatically but is limited by standards, key attributes, scales and input that the researcher has predefined. Hence, increasing speed does not mean lowering quality but happens instead within the framework of predefined best practices.

8.2.3 Analysis and reporting

After the execution of the fieldwork, data need to be analysed and reported. This can be done by a reporting tool, especially when the questionnaires are always the same or very similar as described above for a project with a standard questionnaire. As the type of questions and scales can be kept the same, requirements for analyses and charts in a report or dashboard remain similar. Therefore analysis and report generation can be predefined and automatically run, requiring the expertise of the responsible consumer research specialist predominantly for interpretation and support in decision-making based on the obtained findings. Results output by statistics software or most questionnaire generators, layout/graphics/representation can be customised and predefined exactly the way the addressed marketing director would like it, without delay and to the point. By using corporate design, standardised graphics, presentation styles for post hoc tests, penalty analyses, etc., it becomes possible to then use the results directly.

8.3 Business case: a digital study-management platform

8.3.1 Background

A global food company conducts product tests for its various brands in local markets worldwide. Local product managers are supported in the planning and execution of these product tests by sensory consumer specialists from the Consumer Technical Insight (CTI) team in the central innovation centre.

The researchers help product managers with project conception. Together, the test method is determined, the sample size is determined, and the target group is defined. The researchers assist in the design of recruitment and product questionnaire. In addition, they support the specification of statistical methods and processing of the results into a meaningful report.

A recent development in the company was that product managers were increasingly carrying out product tests. For this reason, the researcher team was also increasingly required to tailor and fine-tune approaches, which in turn led to an increased workload for this team. As a result, the team decided to automate some of the projects. This appeared possible because many projects have the same general objective and, from the experts' point of view, have a fairly low level of methodological complexity. Up to this point, researchers' involvement had always been essential in those projects to avoid deviations from required specifications even for relatively simple projects.

8.3.2 Objectives

It was therefore decided to create a digital study-management platform with the following objectives. Research projects should continue to be methodologically designed and controlled centrally by researchers but should be able to be planned and set up independently by local product managers for standard tests.

The tool should take into account internal company standards, which vary depending on the project objective, for example specification of the relevant test method, defined case numbers, generally applicable target group criteria for recruitment, statistical evaluation routines or also graphic representations of the results. The questionnaire should also contain standard questions per product category, for example overall liking, taste liking, etc. But it should also allow to select further questions from a given list. This will result in a platform that is based on standard elements. At the end of the test, data should be automatically analysed and the results should be integrated into a predefined report.

8.3.3 Setting up the platform

isi was asked to assist in the entire process and set up the platform together with the researcher team. In several workshops, all relevant aspects were defined. First, the standard product test questionnaires were developed and translated into all relevant languages. Questions were separated into questionnaire blocks and components that can be selected and deselected, for example liking questions, JAR questions, open-ended questions, etc. Furthermore, target group specifications were listed and further screening criteria as well as exclusion criteria were defined. All this was the basis for the platform and enables the platform user to easily select from predefined items.

For all types of questions, statistical analyses were specified (e.g. ANOVA models, post hoc tests, significance level hierarchical structure for the analysis of JAR questions, etc.). After that, all aspects of the report were defined with variables that refer to the data base (e.g. structure, customised layout, charts per type of question, etc.). Finally, the booking process was determined and the user experience discussed. The platform also contains an administration centre that links to all relevant study steps, from the field partner briefing to the questionnaires, data and the report. This administration centre also contains business-relevant content like the purchase order and it handles the individual assignment of access rights for all team members.

8.3.4 Hosting the platform

Then the digital study-management platform was hosted. It allows a user to set up a new project easily with the guidance of the tool (see Fig. 8.1). After setting up a project by means of the booking tool, the user receives a price according to the chosen options and books the study within a matter of minutes.

Shortly after, the CTI team and the relevant field partner receive the specifications of the study. This automatically commissions the study. The questionnaire is automatically configured by the tool, based on the questionnaire library available in relevant country languages, and can be opened on a PC or any mobile device (see Fig. 8.2).

The field partner is setting up and running the fieldwork according to all specifications. The collected data flow directly back into the tool. The user and the CTI team are informed when fieldwork ends. Right after that, the dashboard and the report are available. Raw data and a report can be downloaded and used outside the platform if necessary.

8.3.5 Summary

Simple recurring product tests can now be handled directly by the local project manager via the digital study-management platform. With standardised yet flexible test designs and automated reporting, the platform significantly accelerates the testing process and ensures consistent testing standards across markets and product categories. The experts are informed, but they need invest less time in these projects in the future. As a result, they can devote more of their expertise to more complex research projects.

8.3.6 Outlook

It is planned to integrate further background information on the platform, for example additional external reports on relevant topics. Sensory profiles in the relevant categories

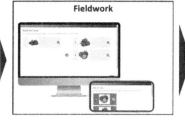
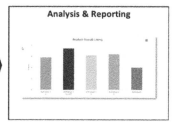

- The user sets up a project with the guidance of the booking tool
- By booking the study the field partner receives all specifications of the study

- Questionnaire is automatically configured
- The field partner sets up and runs the fieldwork

- Data feeds directly back to tool
- After fieldwork ends, dashboards and automatic report are available

FIGURE 8.1   General process flow: From research booking over fieldwork to analysis and reporting.

DiY Product Test Suite

General

Research Type: productTasting   **Product Category:** food

Project Setup

Project Title: Test123   **Sample Size:** 150

Product Interests   ## Related Questions

Overall Liking   ☑ How much do you like the product OVERALL?[3]

Open Ended Question   ☑ You indicated that you LIKE or DISLIKE this sample.

Product Appearance   ☑ How much do you like the APPEARANCE?[3]

Product Texture   ☐ How do you perceive the CONSISTENCY?[2]

Taste Profile   ☑ How do you like the TASTE?[6]

FIGURE 8.2   isi Product Test Suite: Questions can be easily included or excluded.

will be published as well. After the platform is running for a while, internal benchmarks can be established based on the surveys and made available on the dashboard. As a future next step, the booking of the field location could be integrated. By selecting the field partner in the booking platform and a final confirmation at the end of the booking process, the partner is directly commissioned. Such an integration of fieldwork partners for the fieldwork phase of sensory consumer tests will then lead to a comprehensive digital study-management tool. Users just have to decide what they want to do and will get a full result report within a specific time.

8.4 Status quo and a look into the future

Digitalisation, currently one of the top topics in the business world, is advancing in sensory consumer research. Since personal computers became affordable, they have been used to generate market research questionnaires and conduct the surveys. The researcher now has full control over data collection and data quality.

Today, digital tools can support sensory product testing much more comprehensively than just during field execution. Even the planning of research projects can be carried out using a booking platform. With recurring objectives and the associated questions, methodological standards can be specified. The user who wants to carry out a project is guided through the platform in such a way that a meaningful project briefing is available at the end. With the help of a digital test studio finder, the user can then select the right partner in the desired market. The project briefing created is helpful in providing the partner with comprehensive information.

Digital tools have also been supporting researchers in the postfield phase for many years, especially in evaluation and reporting. Also here, the processes for recurring standard projects can be made simpler and faster. Statistical evaluation routines could be defined and reports could be generated automatically.

This automation of processes in all phases of the project means that even nonexperts of consumer tests will be able to carry out sensory consumer research studies in a quick and smart way. As a result, users receive a fast, high-quality project implementation with valid results. Due to the clearly defined guard rails in the planning wizard and the predefined evaluation routines that are part of the available test methods, the user cannot do much wrong. Nevertheless, the specialist should always be informed about the processes, which can be done automatically by the platform, in order to be able to intervene if necessary. Of course, the nonexpert may have objectives that go beyond the standard tests and are not covered by the platform. A clear boundary must be drawn here and the user must be provided with an option in the platform to address these issues outside the platform. Further, the automation process reduces the workload on sensory consumer research specialists because automation processes change their role. Instead of assisting users in the choice of the right methods and with the analyses and reporting of the data, the role of the sensory researcher becomes more and more of a designing, predefining and supervising role. In this role, the researcher could set company-wide standards and support business units in standard projects rather than having to manage all studies on its own. This allows researchers to allocate more resources to other projects like nonstandard strategic projects.

Ideally, all digital tools are implemented in a single tool: a digital study-management platform. The user is guided through the planning phase — including field partner selection — and triggers the project at the very end with a click on 'confirmation'. This directly informs the field partner and hands over all instructions fully automatically. Even the sample sourcing is done automatically by the project confirmation. Then, the field partner takes over the fieldwork and data control and signals the end of the fieldwork when done. The user now generates the automatic evaluation and report at the push of a button. It seems we are not far from this situation, at least for standard sensory product tests.

When such a digital study-management platform has been running for a while, it collects more and more data. Benchmarks can then be derived from this database and successively adjusted automatically. In this way, the system can continue to develop automatically.

References

Boutrolle, I., Delarue, J., Arranz, D., Rogeaux, M., & Köster, E. P. (2007). Central location test vs. home use test: Contrasting results depending on product type. *Food Quality and Preference, 18*(3), 490–499. Available from https://doi.org/10.1016/j.foodqual.2006.06.003.

Damen, F. W. M., Luning, P. A., Fogliano, V., & Steenbekkers, B. L. P. A. (2019). What influences mothers' snack choices for their children aged 2–7? *Food Quality and Preference*, *74*, 10–20. Available from https://doi.org/10.1016/j.foodqual.2018.12.012.

Guiné, R. P. F., Florença, S. G., Barroca, M. J., & Anjos, O. (2020). The link between the consumer and the innovations in food product development. *Foods*, *9*(9). Available from https://doi.org/10.3390/foods9091317, https://www.mdpi.com/2304-8158/9/9/1317.

Horvat, A., Granato, G., Fogliano, V., & Luning, P. A. (2019). Understanding consumer data use in new product development and the product life cycle in European food firms — An empirical study. *Food Quality and Preference*, *76*, 20–32. Available from https://doi.org/10.1016/j.foodqual.2019.03.008.

Lesschaeve, I., & Bruwer, J. (2010). *The importance of consumer involvement and implications for new product development. Consumer-driven innovation in food and personal care products* (pp. 386–423). Canada: Elsevier Ltd. Available from http://www.sciencedirect.com/science/book/9781845695675, https://doi.org/10.1533/9781845699970.3.386.

Linnemann, A. R., Benner, M., Verkerk, R., & Van Boekel, M. A. J. S. (2006). Consumer-driven food product development. *Trends in Food Science and Technology*, *17*(4), 184–190. Available from https://doi.org/10.1016/j.tifs.2005.11.015, http://www.elsevier.com/wps/find/journaldescription.cws_home/601278/description#description.

Santeramo, F. G., Carlucci, D., De Devitiis, B., Seccia, A., Stasi, A., Viscecchia, R., & Nardone, G. (2018). Emerging trends in European food, diets and food industry. *Food Research International*, *104*, 39–47. Available from https://doi.org/10.1016/j.foodres.2017.10.039, http://www.elsevier.com/inca/publications/store/4/2/2/9/7/0.

Symmank, C. (2019). Extrinsic and intrinsic food product attributes in consumer and sensory research: Literature review and quantification of the findings. *Management Review Quarterly*, *69*(1), 39–74. Available from https://doi.org/10.1007/s11301-018-0146-6.

Digitalization in instrumental, neurological, psychological and behavioural methods: Current applications and opportunities

CHAPTER

9

Predicting sensory properties from chemical profiles, the ultimate flavour puzzle: a tale of interactions, receptors, mathematics and artificial intelligence

Andrew J. Taylor

Flavometrix Limited, Loughborough, United Kingdom

9.1 Introduction

Until the advent of commercial gas chromatography (GC) equipment around the 1960s, the analysis of food flavour was mostly driven by sensory methods. The ability to quickly identify and quantify the volatile components of a flavour by GC coupled to mass spectrometry (MS), led to attempts to correlate flavour composition with sensory flavour profiles. Despite many attempts over the last 60 years, the link between chemical composition (whether measured by GC-MS, liquid chromatography (LC-MS) e-tongues or e-noses) and sensory descriptors remains somewhat of a mystery. This chapter explores some of the reasons behind this impasse. Specifically, the complexities of correlating abstract, sensory perception data with highly structured chemical data are called out as well as the limitations of chemical analysis, plus the issue of data transformation prior to digitisation. The use of Artificial Intelligence (AI) and Machine Learning (ML) to address the connection between chemical compounds and sensory properties will be surveyed to indicate future opportunities in this area.

The ability of computers to mimic human behaviour caught the public imagination in 1997 when the IBM computer system (Deep Blue) beat the then World Chess Champion Gary Kasparov by $3\frac{1}{2}$ to $2\frac{1}{2}$. More recently (2021), protein folding was modelled by Google's Deep Mind computer and, in the same year, an advert from Google sought a

DOI: https://doi.org/10.1016/B978-0-323-95225-5.00009-2

researcher to join their team to develop sensors to give their computers a 'sense of smell'. These activities have brought the terms AI and ML into common usage amongst scientists and lay people. Although the terms are often used interchangeably, a definition from javatpoint.com (Anon, 2021) seems to highlight the key differences between the two methods:

> 'Artificial intelligence is a technology which enables a machine to simulate human behaviour. Machine learning is a subset of AI which allows a machine to automatically learn from past data without programming explicitly. The goal of AI is to make a smart computer system like humans to solve complex problems'.

Given the success of computers in mastering complex games such as chess and GO (a well known, Chinese board game), it was inevitable that the task of predicting the sensory flavour from any chemical formulation would be proposed as a future goal. The purpose of this chapter is to assess the feasibility and readiness of this proposal and point out the differences between current successes in ML and AI and the processes involved in flavour perception. Specifically, issues like data quality, data transformation and variability in many parts of the flavour perception process (genetic, cultural and physiological) will be outlined to give an idea of the scale of the task. To make the comparison simpler, scenarios that mimic sensory perception of flavour will be used to evaluate the feasibility and readiness parameters, as well as a third success factor, namely, how useful such models might be in commercial practice.

9.2 Computing strategies

In the examples of chess and GO given above, the computing strategies involved three stages: firstly, some input data, secondly, a collection of rules (that described the behaviour of the input data) and thirdly, the goal of winning the game. In the chess example, the board position of the chess pieces at the beginning of the game was the starting input data, the rules were the way that each piece was allowed to move around the board and the end goal was to surround the opponent's king so that it could not escape. In this case, the rules are absolute (there can be no deviation or variance from them), so algorithms can be written to predict the outcome and opportunities after each move. The computer analyses each and every move of both players to choose the best route to a win. For ease of understanding in this chapter, this type of strategy will be referred to as 'rule-based'. In fact, the successful programs for the other well-publicised computer successes (GO and protein folding) are also 'rule-based', although in protein folding, there are some unknowns, and the strategy overcomes this lack of knowledge by 'learning' from a wider range of samples. From the definition of AI and ML in the previous paragraph, it seems these programs largely use the ML approach.

If the same strategic analysis is applied to flavour perception, then the input data will be the chemical composition of the food or flavouring, the rules will have to include all the stages as the flavour compounds (e.g. tastants and odourants) are released from food during eating and travel to the appropriate receptors in the mouth and nose, and the goal will be sensory descriptors and intensities developed in the human brain and delivered in visual format in the typical spider-web diagram. The difficult part of the computing strategy is how to deal with the complexity of the flavour perception process; Fig. 9.1 shows a schematic that breaks down the process into discrete stages.

Fig. 9.1 demonstrates that correlating flavour composition with sensory attributes requires consideration of a variety of subprocesses and the task of predicting sensory properties from the flavour composition of a food will be complicated. When confronted by an overly complex situation, the usual scientific approach is to simplify it and study individual parts of the process to gain some understanding. Taking odour perception as an example, we know that the volatile compounds released from the food can be detected orthonasally (direct introduction into the nose) or retronasally (air from the mouth enters the nose from the direction of the throat due to swallowing and breathing). Orthonasal odour detection involves release of odours into the air above the sample (the 'headspace' profile), a process that has been well-researched, and odour release is governed by physicochemical rules such as partition coefficient, vapour pressure and hydrophobicity (Taylor, 2002). In the orthonasal case, the headspace and the in-nose odour profiles (the 'nosespace' profile) are very similar, thus simplifying the overall process. The transfer of odours from the nasal air to the odour receptors, through the so-called peri-receptor region, is not so well-documented with differing views on the role of odour binding protein (OBP) in the process (Taylor et al., 2008). Binding of single odours to odour receptors has been partially studied and there is some information on the specificity and binding conditions that produce a signal when the receptor is activated by an odour compound: see, for example Cong et al. (2022). Neural processing has been more difficult to study and the effect of odours on neural signalling has mainly been examined in laboratory rodent experiments (Xu et al., 2020). With this descriptive analysis of the flavour perception process, it is now possible to map the subprocesses onto a hypothetical plan as shown in Table 9.1.

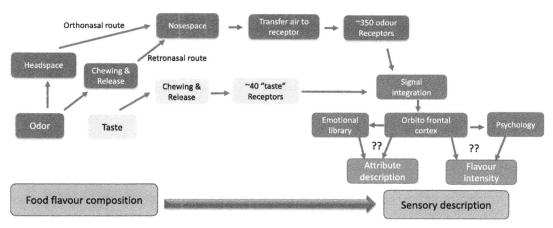

FIGURE 9.1 Stages of flavour perception. A simplified scheme illustrating some key stages of flavour perception from chemical composition to sensory descriptors. Colours denote the stages of odour and tastant transport to the receptors, neural processing and the creation of sensory data in panellists. Question marks indicate uncertainty about the connections in these areas.

TABLE 9.1 Processes involved in odour release from aqueous solutions. Knowledge available on the individual subprocesses for sensory analysis of a single odour in water.

Subprocess	State of knowledge	Factors
Odour transfer from water to the air phase (headspace)	Very good	Vapour pressure, partition coefficient, hydrophobicity
Transfer of headspace odour to nosespace	Good	Assume no change in profile
Transfer from nosespace to receptors	Basic — role of odour binding protein	Must depend on solubility hydrophobicity
Activation of odour receptor(s)	Improving — receptors for certain odours identified	Activation involves binding, so hydrophobicity and solubility involved
Neural processing of receptor signal	Poor but for a single odour the process is greatly simplified due to lack of neural interactions	Emotional library to identify odour of the test compound and quantitative analysis to determine its intensity

The simplicity of the single odour-in-water scenario has been recognised by several research teams who have used the physicochemical parameters listed in Table 9.1 to produce predictive models. Bill Cain's laboratory published a series of papers describing the use of the quantitative structure-property relationship (QSPR) technique to link compounds with their odours (Abraham et al., 2002, 1998), while Joel Mainland's group at Monell (Mayhew et al., 2021) used volatility and hydrophobicity criteria to define the 'borders of olfactory space', that is whether any given compound has an odour or not.

If the single odour-in-water scenario is adjusted to consider mixtures of odours, then other subprocesses, such as odour interactions, need to be added to the list in Table 9.1. Interactions can occur in several subprocesses, especially when odours bind to proteins, when there can be competition for binding sites that affects the perception process. OBP, found in the mucus surrounding the odour receptors, is one such example where odours with higher affinity can block odours with lower affinity depending on their relative concentrations as shown by in vitro experiments (Nespoulous et al., 2004). Some odour receptors bind several odorous compounds, whereas others are specific for a single odour compound (Kurian et al., 2021); in the former case, competition for the binding site can change the odour perception in terms of intensity or odour quality (Belhassan et al., 2019). The presence of even subthreshold levels of odour can interact with other odours or tastants to significantly change the sensory perception of the mixture (Adhikari et al., 2006; Miyazawa et al., 2008), another factor that needs to be included in the computing strategy. Since these subprocesses are poorly understood, it is difficult to propose general rules and, therefore, the computing strategy is to use ML or AI to address these unknowns. The lack of rules also means that data sets need to be larger to provide enough information for ML and AI systems to mimic what is happening in human flavour perception.

Building models to predict the perceived odour of a mixture of volatile compounds in water would undoubtedly be a major step forward, but extending the scenario to predict the sensory properties when a flavouring is added to a real food will require a significant step change in strategy (Regueiro et al., 2017; Spence et al., 2017). Many more subprocesses will be involved, as will a recognition that the properties of the food matrix play a major role in the rate and extent of flavour release, compared to water. For example, it is well-established that fat content affects flavour release in foods (Shiota et al., 2011; Taylor et al., 2008), as can other food ingredients (Guichard, 2002). When proteins are present in the food, then binding of aldehydes, ketones and sulphur-containing odour compounds can occur and change the odour profile considerably (Adams et al., 2001; Viry et al., 2018). With real foods, physiological factors such as chewing as well as air and saliva flows play a major role in the flavour release process. This not only introduces complexity into the release mechanisms but also brings in the human variation factor, which then needs additional computing solutions. Studies on model aqueous systems have identified the key physicochemical processes involved in aroma release (de Roos & Graf, 1995) and have recently been expanded to consider the effect of viscosity (Weterings et al., 2019). Theoretical models of aroma release from simple foods during eating have been proposed (Harrison & Hills, 1996; Harrison et al., 1997, 1998) but few have been tested, although experimental data from gelatin gels did not tally with the predicted release from the Harrison model (Taylor, unpublished data).

One of the most difficult subprocesses to express in mathematical terms is the release of tastants and odourants from food during eating and their transport to the taste and odour receptors. There are many factors that control release, such as mastication, surface area, diffusion, hydration rate of the food, air and saliva flow, plus physicochemical properties of the odours and tastants themselves (Taylor, 2002). While some factors have fixed values and behaviours (and are therefore easy to model), the physiological processes (mastication, saliva and air flow) vary considerably from person to person and are not amenable to a rule-based strategic approach. Again, simplification could be helpful, for example by initially studying 'homogenous' foods (beverages, milk, yoghurt) to simplify the 'mastication' phase of the overall process.

In conclusion, the multistage nature of the flavour perception process, plus human variability and those subprocesses where the mechanisms are not clear, mean that a wholly rule-based approach is not the best computing strategy and a hybrid model with a combination of rules (where appropriate) and some AI or ML modules seem to offer the best chance of success in the simple odour-in-water scenario. By applying neural networks to a two-odour mixture, some success was achieved in predicting the perceived odour (Debnath et al., 2021). The Deep Sniffer project used a fabricated odour sensor and algorithms to identify the composition of aromatic oils (Liu et al., 2021), showing that some complex systems can be resolved if a single sensor system is used and potential human variability is ignored. Predicting the sensory profile when a flavour is added to a real food will require a very different approach. AI seems to be the best choice for this task as it tries to mimic human behaviour; however, the wide variation in human physiology, culture and genetics will be a stern test even for AI.

9.3 Importance of data quality

Like all computer modelling programs, the success of AI and ML depends on the quality of the data used to establish the model. The old saying 'rubbish in equals rubbish out' summarises the fact that reliable and robust computer models need high-quality data to be successful and useful. In the case of modelling flavour perception, the key input data sets are the chemical composition of the flavour and the sensory descriptors and values provided by the sensory panel (Chambers & Koppel, 2013). The following sections examine factors that could affect data quality for both chemical and sensory analyses, along with some alternative ways to obtain or present data.

9.3.1 Chemical composition of flavours

At first sight, defining the chemical composition of a flavour seems an easy task. If the flavour was made by a flavourist, surely, all that is needed is the formulation information. While this is largely true, changes in chemical composition after formulation do occur. One well-known reaction occurs between aldehydes or ketones with compounds containing hydroxyl groups (-OH), to form acetals or ketols, respectively. An assessment of the potential for aldol/ketol formation from common food odourants showed that many reacted with common solvents used for flavourings (ethanol and propylene glycol) under weak acid conditions (Petka & Leitner, 2020). Reports of chemical changes in formulated flavourings involving sulphur compounds, and vanillin can also be found (Chu et al., 2015; Weerawatanakorn et al., 2015). Therefore formulation data are not always ideal for inclusion in ML or AI processes.

When flavour data on common food ingredients are required, the natural variation in commodities such as herbs and spices or soy sauce can be considerable. These differences are due to the different plant varieties used, the growing conditions and the treatment of the materials after harvest, or, in the case of soy sauce, the fermentation and manufacturing process. Then, the difficulty is which chemical composition to use and how to define flavours like tomato when there are so many varieties and multiple formats (fresh tomato, canned tomatoes, sun-dried, etc.).

There are well-established methods for analysing the flavour composition of foods and the Sensomics method (Meyer et al., 2016) is considered the gold standard by the flavour community, although, like all methods, it has its limitations. Briefly, Sensomics is designed to identify and then quantify key odourants that, when recombined, match the orthonasal sensory profile (Granvogl, 2022). Not all foods (especially high-fat products) are amenable to the extraction process and taste compounds need to be analysed separately (Glabasnia et al., 2018). Therefore analysing multiple plant samples to understand natural variation in aroma profile is a time-consuming process using Sensomics. An alternative method to measure both taste and smell compounds simultaneously has been proposed (Hofstetter et al., 2019) and could replace the current odour-based Sensomics analysis in the future. Two-dimensional GC analysis provides a greater resolution of odour compounds by taking compounds that are not separated on one type of GC column and subjecting them to separation on a different column, therefore, providing a greater depth of analysis compared to single column GC analysis. Although more complex than classical GC-Ms, the ability to resolve and identify/quantify more compounds makes it attractive from the point of view of providing high-quality data for computer modelling. The technique has recently been reviewed (Cozzolino, 2022), with a focus on flavour applications.

Other scientific disciplines have experienced similar analytical problems, and the metabolomic community switched their strategy from a targeted analysis of compounds to an untargeted approach some years ago (Alonso et al., 2015). The aim was to generate large data sets with comprehensive coverage of all analytes but without identification of each compound. In metabolomics, compounds in biological pathways are linked by well-defined chemical reactions like oxidation or hydrolysis where the molecular weight of each metabolite changes by a known amount (e.g. loss of 18 Da for a dehydration reaction), and thus it is possible to allocate the unknown compounds to specific biochemical pathways. Applying this concept to flavour analyses, the untargeted data are fed into the computer model, and compounds that show no relationships or correlation with sensory properties are not subjected to further identification, thus allowing high-throughput chemical analyses and reducing the significant burden of identifying each and every compound. The untargeted approach to food flavour analysis has been reported in several recent publications (Reyrolle et al., 2022; Ronningen & Peterson, 2015; Sherman et al., 2020). Fourier Transform-Ion Cyclotron Resonance-Mass Spectrometry (FT-ICR-Ms) is a sophisticated analysis that has been adapted to analyse food and drink samples using a simple and rapid extraction technique and with an analysis time of around 10 min (Marshall et al., 2018). It can measure both volatile and nonvolatile compounds, so it achieves both taste and smell analysis in one step, although there are some molecular weight and solubility cut-off restrictions. A pet food sample analysed by methanol extraction and FT-ICR-Ms yielded several thousand molecular formulae (Marshall et al., 2018). With this kind of capability, it is theoretically possible to measure hundreds or thousands of samples to build a flavour database that covers the variation seen in typical food flavour compositions (Garcia et al., 2013; Marshall et al., 2018; Roullier-Gall et al., 2014). With the availability of both high-throughput and high-quality flavour data, it should be possible to populate the input data adequately; how modellers deal with the variation in the flavour composition of food ingredients is not so clear.

9.3.2 Alternative flavour profile measurements

So far, chemical data have been discussed purely as a list of flavour compounds and their concentration, as shown by the simple strawberry composition in Table 9.2.

Table 9.2 contains very little information about the properties of the odour compounds, although Table 9.1 showed that physicochemical properties were involved in many of the subprocesses of flavour perception. This fact was identified many years ago and led to the concept of QSPR. In this approach, the compound name is enhanced by values for its vapour pressure, hydrophobicity and solubility in water and oil, as well as more detailed molecular features such as polarisability and hydrogen bonding potential (Abraham et al., 2011). The attraction of QSPR is that estimated values of the physicochemical properties can be obtained by calculation, therefore it is relatively easy to set up a database containing the values for any compound.

So far, the focus for chemical analysis has been on the flavour profile in the product or sample. However, measuring the aroma profile in the headspace can be advantageous for odour modelling as the process of odour release from the sample/product can be ignored.

TABLE 9.2 Chemical composition of a model strawberry odour. Odour compounds and their concentration in a model strawberry flavouring.

Compound	Concentration/kg
2,3-Butandione	0.005
Ethyl butanoate	5
4-Hydroxy-2,5-dimethyl-3-furanone	10.725
Cis-3-hexen-1-ol	10.8
Ethyl hexanoate	3.36
Butanoic acid	0.92
Methyl(E)-3-phenylprop-2-enoate	2.7
Gamma-decalactone	1.33
Methyl dihydrojasmonate	0.003
Propylene glycol	965

Similarly, measuring the odour profile sensed by panellists during eating (the so-called nosespace (Beauchamp, 2021; Taylor & Linforth, 2010)) gives information on the odour profile that is delivered to the mucus layers around the odour receptors and again allows the complexities of flavour release to be ignored. However, nosespace profiling is very time-consuming and probably impractical if large numbers of panellists need to be screened to obtain sufficient data for the AI and ML processes. An alternative approach (He & Chung, 2019) selected just 20 key odourants of soybean curd and transformed the data by taking ratios between each compound to reflect potential interactions in the flavour perception process. They also included other stimuli (protein, moisture and salt content) in the 'formulation' data, thus developing a more holistic approach to predict the sensory flavour of a real food product. A critique of the issues around chemical and sensory analyses in studying dairy product quality (Andrewes et al., 2021) concluded that 'what chemical analysis has yet to achieve to even begin to challenge human sensory analysis as an objective measurement tool (needs to be) considered', that is chemical analysis alone does not contain sufficient information to describe the complexity of sensory attributes.

9.3.3 Data from sensory analysis

Over the last decade, developments and refinements in sensory methods have provided a range of tools to assess the perceived flavour of foods. Methods range from the use of highly trained panels with the ability to rate sensory attributes accurately and reproducibly, to larger-scale methods, designed to capture the responses of consumers, for example the use of Rapid Profiling (RP) techniques (Delarue et al., 2015). From the point of view of selecting data to use as the 'target' in computer modelling, the choice depends on the purpose of the model. If the aim of the computer model is to describe flavours in terms of their attributes (floral, sweet, sour, etc.) then the use of quantitative descriptive analysis

(QDA) with a trained panel is one option, that will deliver the classical 'spider diagram' showing the perceived intensity of each attribute. RP methods were originally designed to measure the difference between samples but can be adapted to descriptive analysis (see, e.g. Dehlholm et al. (2012)), although a comparison between QDA and RP found that QDA showed less variance than the Flash Profile method (He & Chung, 2019). The choice then is to either use a data set that represents a very small proportion of the population but with high-quality data, or to use a wide sampling of the population but in the knowledge that variance will be higher and the model may not be very accurate.

Another impediment to using sensory data is that each sensory panel has their own vocabulary to define the sensory descriptors, so harmonising data from several panels to build a larger data set is difficult. Studies to standardise descriptors have been reported for single odour compounds that used a technique called 'natural-language semantic representations' (Gutierrez et al., 2018; Thieme et al., 2022). Data from four online databases containing the odour descriptors for single odour compounds were compared to determine which descriptors were associated with each odour compound; this resulted in a standardised vocabulary comprising 27 odour descriptors. This type of harmonisation could be useful in connecting data sets to provide more wide-ranging information; however, the technique is based on single odour compounds, not odour mixtures as found in real foods or flavourings, which are likely to be significantly more complicated.

Coupled with the variance between trained panellists and naive consumers, genetic variance between individuals and cultures also needs consideration. A study on several hundred subjects demonstrated that genetic differences in the odour receptors had a knock-on effect on the pleasantness and intensity of 68 odours (Trimmer et al., 2019). A similar study on 49 participants showed that genetic differences in taste receptors caused a change in odour threshold and preference (Chamoun et al., 2021). These findings suggest that different computer models may be needed for different genetic groups and that the genetics of panellists' taste and smell receptors should ideally be recorded. Alternatively, is it possible to select a sensory panel that represents the global genetic variation and use this data as a global estimate?

Another factor in selecting sensory data is whether the odours are sensed by the orthonasal or retronasal routes. Experiments using an olfactometer to deliver odours via the two different routes showed that the perception of the odours was different (Diaz, 2004; Heilmann & Hummel, 2004; Negoias et al., 2008). This emphasises the need to define the route that odour compounds take to reach the olfactory receptors and appreciate that using orthonasal perception data in computer modelling of a sample that is sensed retronasally is not valid.

Since overall flavour perception is the result of the brain combining (and processing) signals from the olfactory, gustatory and chemesthetic receptors, another potential method to measure the sensory response is to record brain signals during smelling or tasting of samples. Measuring signals in the brain can record not only the signals at the primary locations in the brain but also record interactions in areas such as the orbitofrontal cortex (Calvert, 2001; Rolls, 1997, 2010). Electroencephalography was originally used to detect brain signals through electrodes attached to the skull, but functional magnetic resonance spectroscopy (fMRI) and positron emission tomography produce more detailed information and have been used to monitor the areas of the brain that are stimulated when humans experience flavour compounds (De Araujo et al., 2003). fMRI analysis records signal with time so the data are three-dimensional and data processing to align the signals from different participants into a normalised data set is a time-consuming task. Data acquisition is relatively slow with only one participant being screened at a time, so collecting significant amounts of data is both time- and cost-limited. fMRI has been extremely useful in researching interactions in the brain to support the hypothesis that flavour is a multimodal construct (Eldeghaidy et al., 2016; McCabe & Rolls, 2007; Small & Prescott, 2005) but is not currently regarded as a method to record sensory data.

9.4 Case histories

Correlating flavour composition with sensory properties is a long-standing quest and has gone through several stages of development. The sections below outline some of the methods proposed for predicting sensory properties from flavour composition. The underpinning hypotheses and the relative success of each method are described along with commentary on future needs.

9.4.1 Tea

Early attempts used statistical methods such as multiple linear regression (MLR), partial least squares (PLS) and principal component analysis (PCA) to identify relationships between 77 tea volatiles and 16 sensory attributes (Togari et al., 1995). The MLR method delivered equations (e.g. Eq. (9.1)) that expressed the flavour intensity of a sensory attribute (in this case, fresh floral) in terms of the relative contributions of individual odour compounds. The equations were obtained from three tea types: oolong, black and green tea, with the aim of building a general predictive model for tea taste. GC data were normalised by dividing by the internal standard response of each sample and then converted to a logarithmic scale.

$$\text{Fresh floral} = -0.482 \times \text{methylpyrazine} + 0.931 \times \text{linalool} + 0.329$$

$$\times (E)\text{-2-hexenyl hexanoate} - 0.217 \times 5, 6\text{-epoxy-ß-ionone} + 0.615 \times \text{jasmine lactone} - 0.958$$

$$(9.1)$$

Eq. (9.1) *Equation relating the sensory attribute 'Fresh floral' with weighted concentrations of selected compounds.*

Application of this approach to other tea samples was unsuccessful (Taylor, unpublished data), suggesting that the models were highly product-specific and difficult to transfer to other tea blends. This may be because the approach was entirely based on statistical analysis; no attempt was made to incorporate factors that are known to be important in flavour perception. For example, the GC response was the sole measurement used to represent the amount of volatile compound present. It did not take into account the GC response factor (to calculate the exact amount of compound present) nor the sensory contribution of each compound. Although published odour thresholds (OT) are notoriously variable, calculating the Dose over Threshold value (odour concentration/OT; DoT) will transform the profile of the odour data and could deliver a model with wider applicability.

9.4.2 Trends in current research publications

A review of recent publications on flavour composition and sensory properties indicates that many researchers use PCA or PLS to analyse their compositional and sensory data and present the results in a visual format. Fig. 9.2 shows a typical result with individual flavour compounds marked by triangles and overlaid with sensory attribute data marked by circles (Owusu et al., 2013).

Both PCA and PLS methods are based on 'drawing' different axes (called principal components (PC)) through the data set to minimise variation and then plotting compounds and sensory data on the new PC axes. Compounds close to a sensory attribute are recognised as being associated with that attribute. Drawing random axes to minimise variation between the two data sets is a sound statistical principle but does not

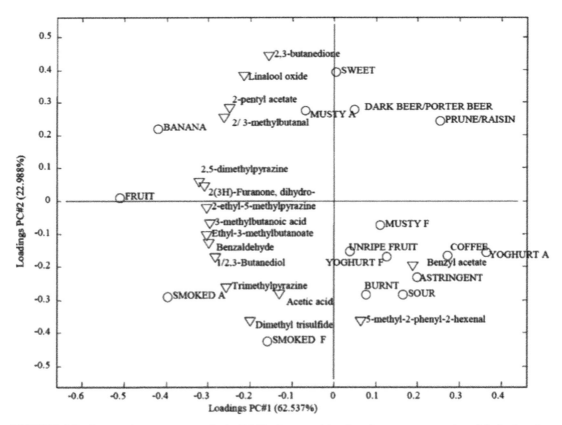

FIGURE 9.2 Principal component analysis (PCA) of compositional and sensory properties of dark chocolate samples. The PCA software plots sensory properties and chemical compounds on the grid and proximity of a compound and a sensory attribute suggests there may be an association between the two. *Source: From Owusu, M., Petersen, M. A., & Heimdal, H. (2013). Relationship of sensory and instrumental aroma measurements of dark chocolate as influenced by fermentation method, roasting and conching conditions. Journal of Food Science and Technology, 50(5), 909–917. Available from https://doi.org/10.1007/s13197-011-0420-2.*

consider the processes occurring in flavour perception, so the results are difficult to interpret and, in the author's opinion, should be taken as being qualitative indicators at best.

9.4.3 DREAM project

Using data from Andreas Keller and Leslie Vosshall, Pablo Meyer organised the DREAM olfactory challenge project, supported by an unusual financial resource – crowd funding (Keller et al., 2017). The concept was to use carefully selected sensory and chemical data and establish an odour data set that could be used to explore different computing and modelling strategies, with the aim of predicting sensory attributes, overall pleasantness and odour intensity. In total, the data set comprised the sensory properties of 476 compounds, assessed by 49 subjects, as well as data on 4884 physicochemical features of each of the 476 compounds. The challenge was carried out in a highly structured way. Initially, a data set of 338 compounds was released to the teams to build some initial models. Perceptual data from another 69 compounds were used to compare performance and establish a 'leader board' amongst the teams. Towards the end of the process, more data were released to the teams to allow them to refine and optimise their models. Eighteen teams delivered models to predict individual perception and nineteen teams delivered a population model. Absorbing the considerable amount of data produced by the challenge requires detailed reading of the report (Keller et al., 2017) but the top line message is that success was achieved in certain parts of the programme. The project team reported that the models predicted some odour intensity and pleasantness ratings and successfully predicted 8 (out of the 19) sensory attributes in the data set. An example is given in Fig. 9.3 where attributes like garlic, fishy and pleasantness are the best predicted. The approach taken by the teams in building models varied in the statistical techniques and methods used, but by incorporating such a wide range of physicochemical properties, a strong element of QSPR was also evident in the models.

9.4.4 Odorify

A predictive model (Odorify) to relate odours to one (or more) of the 400 human odour receptors used deep neural networks and AI learning from existing knowledge of odours and their receptors (Gupta et al., 2021). The model was able to identify which molecules bind to human odour receptors and could predict potential interactions between odourants. Although the model did not include sensory data, the relationships between odours and their receptors could be a valuable part of future, more extensive predictive models.

9.4.5 Commercial platforms

Several companies offer services based on correlating sensory properties with chemical composition. Aromyx use high-throughput screening of human olfactory receptors to measure binding of odours. According to their website (http://www.aromyx.com), the data obtained are expressed quantitatively and in terms of flavour quality. The patterns of activated receptors are paired with sensory ratings and chemical composition for further analysis. There is no readily available information on whether aroma release and food matrix effects are

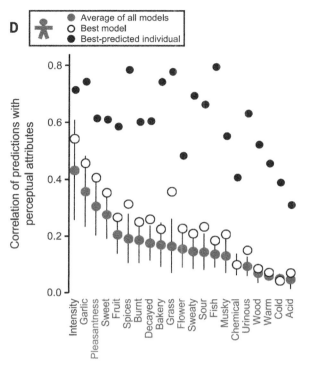

FIGURE 9.3 Sensory predictions from the DREAM project. Plot showing the correlation of predictions for the sensory attributes in the DREAM challenge. Correlation scores for the average correlation, the best model and the best predicted individual are shown. *Source: From Keller, A., Gerkin, R. C., Guan, Y., Dhurandhar, A., Turu, G., Szalai, B., Mainland, J. D., Ihara, Y., Yu, C. W., Wolfinger, R., Vens, C., Schietgat, L., De Grave, K., Norel, R., DREAM Olfaction Prediction Consortium, †, Stolovitzky, G., Cecchi, G. A., Vosshall, L. B., & Meyer, P. (2017). Predicting human olfactory perception from chemical features of odor molecules.* Science, 355, 820−826.

considered in the model. Aryballe also use biosensors to measure odour selectivity but the sensors are incorporated into a silicon photonic system that produces a pattern for each odour sample. The software can not only record the patterns but can also identify odours from previously recorded patterns. iSense takes a different approach by taking commercially available formulations and measuring their sensory profile using a set of standard attributes. An example is the sensory analysis of 132 vanilla flavours in water followed by an analysis of variance, hierarchical cluster analysis and PCA to visualise the location of each sample either by spider diagram or on a PCA plot. The outcome is the ability to describe and compare different vanilla formulations in sensory terms to suit a particular application. Vivanda have analysed hundreds of thousands of food products, menus, etc. and mapped the data onto 16,000 aroma chemicals, then used ML to establish a FlavourPrint profile for each food. The suggested uses of the model are focused on consumer behaviour, for example identifying opportunities for new flavours to attract new customers. In conclusion, each of the companies offers a tool based solely on odour prediction and in very well-defined niche areas. While these tools provide useful information, they are not focused on the ultimate goal of predicting the sensory properties of flavourings in different food matrices.

9.4.6 Compositional-sensory relationships in a real food

The sensory attributes of 12 samples of soybean curd were measured using QDA and a Flash Profile method (He & Chung, 2019). Using multifactor analysis, these values were correlated with chemical data, based on 20 targeted, volatile compounds and some physico-chemical properties of the curd itself. Although at first sight, this looks like a conventional approach to establishing correlations, there were some key differences. One was to include some nonodour factors in the data set (moisture, salt and protein content, textural profile, colour) that were related to the curd textural and visual properties (an important factor in product acceptability). The other difference was to use the headspace odour profiles of 20 targeted volatile compounds to represent the orthonasal signal, instead of the full chemical composition profile. The odour data were also transformed prior to statistical analysis to develop values for potential odour interactions by dividing the headspace intensity of each compound by the intensity of the other 19 compounds to produce intensity ratios. PCA plots indicated that both the concentration and the ratios of the volatile compounds influenced the perceived flavour profile and discriminated between the 12 different curd samples. This publication also delivered information on the difference between rapid and trained-panel sensory methods in assessing the samples. Sensory attributes of the 12 curd samples were better discriminated by QDA compared to RP (Fig. 9.4). The data in

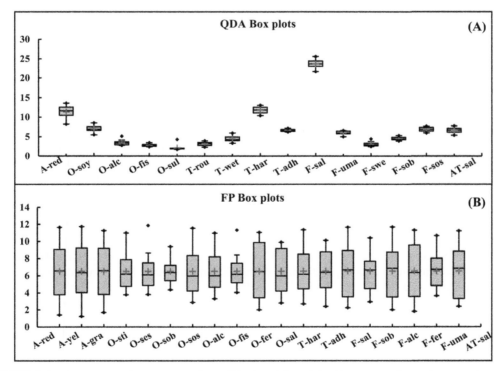

FIGURE 9.4 Quantitative descriptive analysis (QDA) and Rapid Profiling scores for soybean curd samples. Box plots showing the QDA intensity scores (y-axis, A) and the Rapid Profiling ranking scores (B) for the 19 attributes (x-axis) in the 12 curd samples. *Source: From He, W. M., & Chung, H. Y. (2019). Multivariate relationships among sensory, physicochemical parameters, and targeted volatile compounds in commercial red sufus (Chinese fermented soybean curd): Comparison of QDA (R) and Flash Profile methods.* Food Research International, 125. *Available from https://doi.org/10.1016/j.foodres.2019.108548.*

Fig. 9.4 support the general feeling that there are pros and cons for QDA and RP methods. In summary, the data from QDA are more detailed and robust but the relatively small number of people limit the applicability of the data to the general population. In contrast, RP methods can access data from hundreds of people but it is not so well-detailed (e.g. fewer attributes are monitored) and the quantitative measures are not so accurate.

Similar studies have been reported on matcha (Wu et al., 2022) and grape quality (Ferrero-Del-Teso et al., 2022). None of these publications achieves the full goal of predicting sensory flavour from chemical composition but they do show that it is possible to match certain sensory properties to physicochemical properties, especially if the properties are well-chosen and relevant to that particular food product.

9.5 Conclusions and further perspectives

Previous sections of this chapter have focused on the three main areas that are hindering the development of predictive flavour models, namely, the complexity of the flavour perception mechanisms and the need for high-quality data in both chemical and sensory analyses. Current approaches to relate chemical composition with flavour perception mainly use the physicochemical properties of the flavour compounds in the model because many of the subprocesses are governed by these factors. However, a fully rule-based model of flavour perception seems very far away. It is clear from the assessment of existing models that high-quality data are essential for success both in AI and ML approaches. High quality refers to data sets that contain accurate chemical and sensory analyses and have ideally been obtained from a set of samples where the flavour composition has been changed in a controlled manner, so the effects of subtle compositional changes on interactions can be related to sensory differences. For example, the DREAM project used the best available sensory data, with a wide range of modelling techniques, to determine which modelling approach was best for that particular application (Keller et al., 2017).

Such high-quality sensory data are relatively scarce and the development of new data resources, focused on different areas of flavour perception, could be extremely useful. To provide high-quality data for the simplest scenario (odourants in aqueous systems), where chemical and sensory data are relatively easy to obtain, it may be necessary to regulate the sensory protocols in order to minimise variation between panellists. This might involve asking panellists to sip a known volume of a sample, roll it around the mouth and either swallow it after a certain time period or expectorate the sample. Alternatively, gustometers could be used to mix taste and odour solutions accurately and deliver a constant liquid stream to the mouths of panellists (Hort & Hollowood, 2004). This technique can be useful when testing the sensory impact of a single component in a flavour formulation as the full profile can be delivered, then single components can be omitted or reduced in concentration to determine their contribution to the sensory profile of the sample. An example of this approach is a study on the role of mixtures of acids and sugars on the perception of fruity flavours using a statistically designed experiment (Pfeiffer et al., 2006). The study demonstrated that the ratio of acid and sugar not only changed the perception of a strawberry flavour but that different ratios of acid and sugar could give the same sensory intensity of strawberry despite the amount of strawberry volatiles being the same. Fig. 9.5

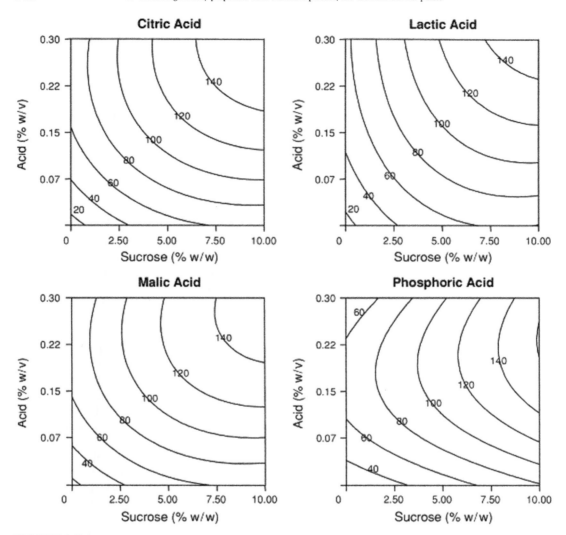

FIGURE 9.5 Sensory contour plots showing perceived strawberry intensity. Intensity is expressed by the numbers on the contours for different acid/sugar ratios but constant strawberry aroma concentrations. *Source: From Pfeiffer, J., Hort, J., Hollowood, T. A., & Taylor, A. J. (2006). Taste-aroma interactions in a ternary system: A model of fruitiness perception in sucrose/acid solutions.* Perception & Psychophysics, 68, 216−227.

shows this effect in the form of contour plots with four different food-grade acids and the same concentration of strawberry aroma in all experiments. The numbers on the contours represent the perceived intensity of the strawberry aroma and show that different ratios of acid and sugar can give the same sensory intensity of strawberry. Although such studies deliver in-depth data on tastant−aroma interactions, they are complex and very time-consuming to design and carry out.

Another approach (Desforges et al., 2015) was to adapt the conventional omission test-ing protocol where one component is removed from a mixture and the 'n-1' mixture reas-sessed by the sensory panel to determine whether that component alters the sensory profile and therefore its sensory importance in the formulation. By removing 25%, 50% or 75% of a single flavour component, and assessing the mixtures by sensory analysis, it is possible to get a better understanding of how that single component interacts with the rest of the flavour formulation. These different ways of assessing sensory properties of the sample provide more detailed information on interactions and could prove useful when building predictive models.

The previous paragraphs refer to aqueous solutions of flavours, but if modelling of real foods is to be attempted, then the contribution of the food matrix should definitely be included in the data set. Some examples have been given already, but the well-known effects of colour, viscosity and textures on flavour perception should also be represented in the data set.

Obtaining sufficient sensory data to be representative of population groups is another fac-tor that limits the usefulness of any predictive model. Genetic and anatomical variations have been shown to affect perception (Feeney & Hayes, 2014; Feeney et al., 2021; Running & Hayes, 2016), and although the well-known response to the bitter compounds PROP and PTC is often tested when recruiting sensory panels, other genetic variations, such as those in taste receptors (Chamoun et al., 2021) and in the odour receptors (Trimmer et al., 2019), are present in the human population and known to change flavour perception. Currently, there is insufficient information to be able to identify clusters of humans with specific patterns of flavour receptors, which might help in developing predictive models that are tailor-made for the key human population groups. An alternative proposal is to use a mimic of the human receptors (e.g. a bioelectronic nose) to produce standard responses for human olfac-tion and based on an agreed set of human receptors (Son et al., 2017). Although an interest-ing idea, it does not address the question of which variants of the odour receptors should be used to obtain a representative view of the human population.

The issue of how to develop large-scale sensory databases presents a dilemma. The choice is between highly detailed and accurate data from small-scale cohorts or less detailed and more variable data sets from much larger populations. Therefore difficult decisions have to be made around quality of data or quantity of panellists. It is difficult to see how this issue can be resolved in the near future. What is clear, however, is that knowledge and expertise from flavour and modelling experts will be of great help when setting the objective of any model and choosing which data sets, transformations and algo-rithms to use.

References

Abraham, M. H., Gola, J. M. R., Cometto-Muniz, J. E., & Cain, W. S. (2002). A model for odour thresholds. *Chemical Senses*, 27(2), 95–104. Available from: http:// <Go to ISI>://000174037400001.

Abraham, M. H., Kumarsingh, R., ComettoMuniz, J. E., & Cain, W. S. (1998). An algorithm for nasal pungency thresholds in man. *Archives of Toxicology*, 72(4), 227–232.

Abraham, M. H., Sánchez-Moreno, R., Cometto-Muñiz, J. E., & Cain, W. S. (2011). An algorithm for 353 odor detec-tion thresholds in humans. *Chemical Senses*, 37(3), 207–218. Available from https://doi.org/10.1093/chemse/bjr094.

Adams, R. L., Mottram, D. S., Parker, J. K., & Brown, H. M. (2001). Flavor-protein binding: Disulfide interchange reactions between ovalbumin and volatile disulfides. *Journal of Agricultural and Food Chemistry*, *49*(9), 4333–4336. Available from https://doi.org/10.1021/jf0100797.

Adhikari, K., Hein, K. A., Elmore, J. R., Heymann, H., & Willott, A. M. (2006). Flavor threshold as affected by interaction among three dairy-related flavor compounds. *Journal of Sensory Studies*, *21*(6), 626–643.

Alonso, A., Marsal, S., & Julià, A. (2015). Analytical methods in untargeted metabolomics: State of the art in 2015. *Frontiers in Bioengineering and Biotechnology*, *3*(23). Available from https://doi.org/10.3389/fbioe.2015.00023.

Andrewes, P., Bullock, S., Turnbull, R., & Coolbear, T. (2021). Chemical instrumental analysis versus human evaluation to measure sensory properties of dairy products: What is fit for purpose? *International Dairy Journal*, *121*, 105098. Available from https://doi.org/10.1016/j.idairyj.2021.105098.

Anon. (2021). Difference between Artificial intelligence and Machine learning. https://www.javatpoint.com/difference-between-artificial-intelligence-and-machine-learning. (Accessed April 10, 2023).

Beauchamp, J. D. (2021). Dynamic flavor: Capturing aroma using real-time mass spectrometry. *ACS Symposium Series*, *1402*, ACS, Washington D.C.

Belhassan, A., Zaki, H., Aouidate, A., Benlyas, M., Lakhlifi, T., & Bouachrine, M. (2019). Interactions between (4Z)-hex-4-en-1-ol and 2-methylbutyl 2-methylbutanoate with olfactory receptors using computational methods. *Moroccan Journal of Chemistry*, *7*(1), 28–35.

Calvert, G. A. (2001). Crossmodal processing in the human brain: Insights from functional neuroimaging studies. *Cerebral Cortex*, *11*(12), 1110–1123.

Chambers, E., IV, & Koppel, K. (2013). Associations of volatile compounds with sensory aroma and flavor: The complex nature of flavor. *Molecules*, *18*, 4887–4905. Available from https://doi.org/10.3390/molecules 18054887.

Chamoun, E., Liu, A. S., Duizer, L. M., Feng, Z., Darlington, G., Duncan, A. M., Haines, J., & Ma, D. W. L. (2021). Single nucleotide polymorphisms in sweet, fat, umami, salt, bitter and sour taste receptor genes are associated with gustatory function and taste preferences in young adults. *Nutrition Research*, *85*, 40–46. Available from https://doi.org/10.1016/j.nutres.2020.12.007.

Chu, K., Jones, L. L., & Delime, P. (2015). *A cause for concern – the stability of a savoury flavouring. Flavour science: Proceedings of the XIV Weurman flavour research symposium* (pp. 443–446). UK: Context Products Ltd, Packington.

Cong, X., Ren, W., Pacalon, J., Xu, R., Xu, L., Li, X., de March, C. A., Matsunami, H., Yu, H., Yu, Y., & Golebiowski, J. (2022). Large-scale G protein-coupled olfactory receptor–ligand pairing. *ACS Central Science*. Available from https://doi.org/10.1021/acscentsci.1c01495.

Cozzolino, D. (2022). *Characterization of odorant patterns by comprehensive two-dimensional gas chromatography comprehensive analytical chemistry*. In *Comprehensive Analytical Chemistry*. Elsevier, ISBN 9780323988810.

De Araujo, I. E. T., Rolls, E. T., Kringelbach, M. L., McGlone, F., & Phillips, N. (2003). Taste-olfactory convergence, and the representation of the pleasantness of flavour, in the human brain. *European Journal of Neuroscience*, *18*, 2059–2068.

Debnath, T., Prasetyawan, D., & Nakamoto, T. (2021). Predicting odor perception of mixed scent from mass spectrometry. *Journal of The Electrochemical Society*, *168*(11), 117505. Available from https://doi.org/10.1149/1945-7111/ac33e0.

Dehlholm, C., Brockhoff, P. B., Meinert, L., Aaslyng, M. D., & Bredie, W. L. P. (2012). Rapid descriptive sensory methods – Comparison of free multiple sorting, partial napping, napping, flash profiling and conventional profiling. *Food Quality and Preference*, *26*(2), 267–277. Available from https://doi.org/10.1016/j.foodqual. 2012.02.012.

Delarue, J., Lawlor, J. B., & Rogeaux, M. (2015). Rapid sensory profiling techniques. *Woodhead*, 555. Available from https://doi.org/10.1533/9781782422587.2.197.

de Roos, K. B., & Graf, E. (1995). Nonequilibrium model for predicting flavor retention in microwave and convection heated foods. *Journal of Agricultural and Food Chemistry*, *43*, 2204–2211.

Desforges, N., O'Mahony, K., Delime, P., Hort, J., & Taylor, A. J. (2015). Measuring flavor interactions using fractional omission testingthe chemical sensory informatics of food: Measurement, analysis, integration. *American Chemical Society*, 77–86. Available from https://doi.org/10.1021/bk-2015-1191.ch007.

Diaz, M. E. (2004). Comparison between orthonasal and retronasal flavour perception at different concentrations. *Flavour and Fragrance Journal*, *19*(6), 499–504. Available from https://doi.org/10.1002/ffj.1475.

Eldeghaidy, S., Marciani, L., Hort, J., Hollowood, T., Singh, G., Bush, D., Foster, T., Taylor, A. J., Busch, J., Spiller, R. C., Gowland, P. A., & Francis, S. T. (2016). Prior consumption of a fat meal in healthy adults modulates the brain's response to fat. *The Journal of Nutrition*, 146(11), 2187–2198. Available from https://doi.org/10.3945/jn.116.234104.

Feeney, E. L., & Hayes, J. E. (2014). Exploring associations between taste perception, oral anatomy and polymorphisms in the carbonic anhydrase (gustin) gene CA6. *Physiology & Behavior*, 128, 148–154. Available from https://doi.org/10.1016/j.physbeh.2014.02.013.

Feeney, E. L., McGuinness, L., Hayes, J. E., & Nolden, A. A. (2021). Genetic variation in sensation affects food liking and intake. *Current Opinion in Food Science*, 42, 203–214. Available from https://doi.org/10.1016/j.cofs.2021.07.001.

Ferrero-Del-Teso, S., Suarez, A., Ferreira, C., Perenzoni, D., Arapitsas, P., Mattivi, F., Ferreira, V., Fernandez-Zurbano, P., & Saenz-Navajas, M. P. (2022). Modeling grape taste and mouthfeel from chemical composition. *Food Chemistry*, 371, 131168. Available from https://doi.org/10.1016/j.foodchem.2021.131168.

Garcia, J. S., Vaz, B. G., Corilo, Y. E., Ramires, C. F., Saraiva, S. A., Sanvido, G. B., Schmidt, E. M., Maia, D. R. J., Cosso, R. G., Zacca, J. J., & Eberlin, M. N. (2013). Whisky analysis by electrospray ionization-Fourier transform mass spectrometry. *Food Research International*, 51(1), 98–106. Available from https://doi.org/10.1016/j.foodres.2012.11.027.

Glabasnia, A., Dunkel, A., Frank, O., & Hofmann, T. (2018). Decoding the nonvolatile sensometabolome of orange juice (*Citrus sinensis*). *Journal of Agricultural and Food Chemistry*.

Granvogl, M. (2022). Sensomics principles. *Characterization of odorant patterns by comprehensive two-dimensional gas chromatography*. In *Comprehensive Analytical Chemistry*. Elsevier, ISBN 9780323988810.

Guichard, E. (2002). Interactions between flavor compounds and food ingredients and their influence on flavor perception. *Food Reviews International*, 18(1), 49–70.

Gupta, R., Mittal, A., Agrawal, V., Gupta, S., Gupta, K., Jain, R. R., Garg, P., Mohanty, S. K., Sogani, R., Chhabra, H. S., Gautam, V., Mishra, T., Sengupta, D., & Ahuja, G. (2021). OdoriFy: A conglomerate of artificial intelligence-driven prediction engines for olfactory decoding. *Journal of Biological Chemistry*, 297(2). Available from https://doi.org/10.1016/j.jbc.2021.100956.

Gutierrez, E. D., Dhurandhar, A., Keller, A., Meyer, P., & Cecchi, G. A. (2018). Predicting natural language descriptions of mono-molecular odorants. *Nature Communications*, 9. Available from https://doi.org/10.1038/s41467-018-07439-9.

Harrison, M., Campbell, S., & Hills, B. P. (1998). Computer simulation of flavor release from solid foods in the mouth. *Journal of Agricultural and Food Chemistry*, 46(7), 2736–2743.

Harrison, M., & Hills, B. P. (1996). A mathematical model to describe flavour release from gelatine gels. *International Journal of Food Science and Technology*, 31(2), 167–176.

Harrison, M., Hills, B. P., Bakker, J., & Clothier, T. (1997). Mathematical models of flavor release from liquid emulsions. *Journal of Food Science*, 62(4), 653–659.

He, W. M., & Chung, H. Y. (2019). Multivariate relationships among sensory, physicochemical parameters, and targeted volatile compounds in commercial red sufus (Chinese fermented soybean curd): Comparison of QDA (R) and Flash Profile methods. *Food Research International*, 125. Available from https://doi.org/10.1016/j.foodres.2019.108548.

Heilmann, S., & Hummel, T. (2004). A new method for comparing orthonasal and retronasal olfaction. *Behavioral Neuroscience*, 118(2), 412–419. Available from https://doi.org/10.1177/0735-7044.118.2.412.

Hofstetter, C. K., Dunkel, A., & Hofmann, T. (2019). Unified flavor quantitation: Toward high-throughput analysis of key food odorants and tastants by means of ultra-high-performance liquid chromatography tandem mass spectrometry. *Journal of Agricultural and Food Chemistry*, 67(31), 8599–8608. Available from https://doi.org/10.1021/acs.jafc.9b03466.

Hort, J., & Hollowood, T. A. (2004). Controlled continuous flow delivery system for investigating taste-aroma interactions. *Journal of Agricultural and Food Chemistry*, 52, 4834–4843.

Keller, A., Gerkin, R. C., Guan, Y., Dhurandhar, A., Turu, G., Szalai, B., Mainland, J. D., Ihara, Y., Yu, C. W., Wolfinger, R., Vens, C., Schietgat, L., De Grave, K., Norel, R., DREAM Olfaction Prediction Consortium., Stolovitzky, G., Cecchi, G. A., Vosshall, L. B., & Meyer, P. (2017). Predicting human olfactory perception from chemical features of odor molecules. *Science*, 355, 820–826.

3. Digitalization in instrumental, neurological, psychological and behavioural methods: Current applications and opportunities

Kurian, S. M., Naressi, R. G., Manoel, D., Barwich, A. S., Malnic, B., & Saraiva, L. R. (2021). Odor coding in the mammalian olfactory epithelium. *Cell and Tissue Research*, 383(1), 445–456. Available from https://doi.org/10.1007/s00441-020-03327-1.

Liu, C., Miyauchi, H., & Hayashi, K. (2021). DeepSniffer: A meta-learning-based chemiresistive odor sensor for recognition and classification of aroma oils. *Sensors and Actuators B: Chemical*, 130960. Available from https://doi.org/10.1016/j.snb.2021.130960.

Marshall, J. W., Schmitt-Kopplin, P., Schuetz, N., Moritz, F., Roullier-Gall, C., Uhl, J., Colyer, A., Jones, L. L., Rychlik, M., & Taylor, A. J. (2018). Monitoring chemical changes during food sterilisation using ultrahigh resolution mass spectrometry. *Food Chemistry*, 242, 316–322. Available from https://doi.org/10.1016/j.foodchem.2017.09.074.

Mayhew, E. J., Arayata, C. J., Gerkin, R. C., Lee, B. K., Magill, J. M., Snyder, L. L., Little, K. A., Yu, C. W., & Mainland, J. D. (2021). Drawing the borders of olfactory space. *bioRxiv*. Available from https://doi.org/10.1101/2020.12.04.412254.

McCabe, C., & Rolls, E. T. (2007). Umami: A delicious flavor formed by convergence of taste and olfactory pathways in the human brain. *European Journal of Neuroscience*, 25(6), 1855–1864. Available from https://doi.org/10.1111/j.1460-9568.2007.05445.x.

Meyer, S., Dunkel, A., & Hofmann, T. (2016). Sensomics-assisted elucidation of the tastant code of cooked crustaceans and taste reconstruction experiments. *Journal of Agricultural and Food Chemistry*, 64(5), 1164–1175. Available from https://doi.org/10.1021/acs.jafc.5b06069.

Miyazawa, T., Gallagher, M., Preti, G., & Wise, P. M. (2008). The impact of subthreshold carboxylic acids on the odor intensity of suprathreshold flavor compounds. *Chemosensory Perception*, 1(3), 163–167. Available from https://doi.org/10.1007/s12078-008-9019-z.

Negoias, S., Visschers, R., Boelrijk, A., & Hummel, T. (2008). New ways to understand aroma perception. *Food Chemistry*, 108(4), 1247–1254. Available from http://www.sciencedirect.com/science/article/B6T6R-4PFDDV3-4/2/890961845c340250c0b3a72a3cb16956.

Nespoulous, C., Briand, L., Delage, M. M., Tran, V., & Pernollet, J. C. (2004). Odorant binding and conformational changes of a rat odorant-binding protein. *Chemical Senses*, 29(3), 189–198.

Owusu, M., Petersen, M. A., & Heimdal, H. (2013). Relationship of sensory and instrumental aroma measurements of dark chocolate as influenced by fermentation method, roasting and conching conditions. *Journal of Food Science and Technology*, 50(5), 909–917. Available from https://doi.org/10.1007/s13197-011-0420-2.

Petka, J., & Leitner, J. (2020). Supplemental study on acetals in food flavourings. Proceedings 16th Weurman Flavour Research Symposium. E. Guichard, & J.-L. Le Quéré (Eds.). Available from https://doi.org/10.5281/zenodo.5752281.

Pfeiffer, J., Hort, J., Hollowood, T. A., & Taylor, A. J. (2006). Taste-aroma interactions in a ternary system: A model of fruitiness perception in sucrose/acid solutions. *Perception & Psychophysics*, 68, 216–227.

Regueiro, J., Negreira, N., & Simal-Gandara, J. (2017). Challenges in relating concentrations of aromas and tastes with flavor features of foods. *Critical Reviews in Food Science and Nutrition*, 57(10), 2112–2127. Available from https://doi.org/10.1080/10408398.2015.1048775.

Reyrolle, M., Ghislain, M., Bru, N., Vallverdu, G., Pigot, T., Desauziers, V., & Le Bechec, M. (2022). Volatile fingerprint of food products with untargeted SIFT-MS data coupled with mixOmics methods for profile discrimination: Application case on cheese. *Food Chemistry*, 369, 130801. Available from https://doi.org/10.1016/j.foodchem.2021.130801.

Rolls, E. T. (1997). Taste and olfactory processing in the brain and its relation to the control of eating. *Critical Reviews in Neurobiology*, 11(4), 263–287.

Rolls, E. T. (2010). Taste, olfactory and food-texture processing in the brain and the control of appetite. *Obesity Prevention: The Role of Brain and Society on Individual Behavior*, 41–56. Available from https://doi.org/10.1016/b978-0-12-374387-9.00004-0.

Ronningen, I. G., & Peterson, D. G. (2015). Application of untargeted LC/MS techniques (flavoromics) to identify changes related to freshness of food. *Chemical Sensory Informatics of Food: Measurement, Analysis, Integration*, 269–277.

Roullier-Gall, C., Lucio, M., Noret, L., Schmitt-Kopplin, P., & Gougeon, R. D. (2014). How subtle is the "Terroir" effect? Chemistry-related signatures of two "Climats de Bourgogne". *Plos One*, 9(5), 11. Available from https://doi.org/10.1371/journal.pone.0097615.

3. Digitalization in instrumental, neurological, psychological and behavioural methods:
Current applications and opportunities

Running, C. A., & Hayes, J. E. (2016). Individual differences in multisensory flavor perception. *Multisensory Flavor Perception: From Fundamental Neuroscience Through to the Marketplace*, 185−210. Available from https://doi.org/10.1016/b978-0-08-100350-3.00010-9.

Sherman, E., Coe, M., Grose, C., Martin, D., & Greenwood, D. R. (2020). Metabolomics approach to assess the relative contributions of the volatile and non-volatile composition to expert quality ratings of Pinot Noir wine quality. *Journal of Agricultural and Food Chemistry*. Available from https://doi.org/10.1021/acs.jafc.0c04095. 10.1021/acs.jafc.0c04095.

Shiota, M., Isogai, T., Iwasawa, A., & Kotera, M. (2011). Model studies on volatile release from different semisolid fat blends correlated with changes in sensory perception. *Journal of Agricultural and Food Chemistry*, 59(9), 4904−4912. Available from https://doi.org/10.1021/jf104649y.

Small, D. M., & Prescott, J. (2005). Odor/taste integration and the perception of flavour. *Experimental Brain Research*, 166, 345−357.

Son, M., Lee, J. Y., Ko, H. J., & Park, T. H. (2017). Bioelectronic nose: An emerging tool for odor standardization. *Trends in Biotechnology*, 35(4), 301−307. Available from https://doi.org/10.1016/j.tibtech.2016.12.007.

Spence, C., Obrist, M., Velasco, C., & Ranasinghe, N. (2017). Digitizing the chemical senses: Possibilities and pitfalls. *International Journal of Human-Computer Studies*, 107, 62−74. Available from https://doi.org/10.1016/j.ijhcs.2017.06.003.

Taylor, A. J. (2002). Release and transport of flavours in vivo: Physico chemical, physiological and perceptual considerations. *Comprehensive Reviews in Food Safety and Food Science*, 1, 45−57.

Taylor, A. J., Cook, D. J., & Scott, D. J. (2008). Role of odor binding protein: Comparing hypothetical mechanisms with experimental data. *Chemosensory Perception*, 1, 153−162.

Taylor, A. J., & Linforth, R. S. T. (2010). *On line monitoring of flavour processes. Food flavour technology* (2nd ed., pp. 266−295). Chichester: Wiley-Blackwell.

Taylor, A. J., Shojaei, Z. A., Bayarri, S., Hollowood, T. A., & Hort, J. (2008). Balancing flavour attributes in reduced fat foods. *Recent Highlights in Flavor Chemistry and Biology*, 53−58, Deutsche Forschungsanstalt fur Lebensmittelchemie, Garching.

Thieme, A., Korn, D., Alves, V., Muratov, E., & Tropsha, A. (2022). Novel classification of mono-molecular odorants using standardized semantic profiles. *ChemRxiv. Cambridge Open Engage*. Available from https://doi.org/10.26434/chemrxiv-2022-h64sb.

Togari, N., Kobayashi, A., & Aishima, T. (1995). Relating sensory properties of tea aroma to gas chromatographic data by chemometric calibration methods. *Food Research International*, 28(5), 485−493.

Trimmer, C., Keller, A., Murphy, N. R., Snyder, L. L., Willer, J. R., Nagai, M. H., Katsanis, N., Vosshall, L. B., Matsunami, H., & Mainland, J. D. (2019). Genetic variation across the human olfactory receptor repertoire alters odor perception. *Proceedings of the National Academy of Sciences*, 116(19), 9475. Available from https://doi.org/10.1073/pnas.1804106115; http://www.pnas.org/content/116/19/9475.abstract.

Viry, O., Boom, R., Avison, S., Pascu, M., & Bodnar, I. (2018). A predictive model for flavor partitioning and protein-flavor interactions in fat-free dairy protein solutions. *Food Research International*, 109, 52−58. Available from https://doi.org/10.1016/j.foodres.2018.04.013.

Weerawatanakorn, M., Wu, J.-C., Pan, M.-H., & Ho, C.-T. (2015). Reactivity and stability of selected flavor compounds. *Journal of Food and Drug Analysis*, 23(2), 176−190. Available from https://doi.org/10.1016/j.jfda.2015.02.001.

Weterings, M., Bodnár, I., Boom, R. M., & Beyrer, M. (2019). A classification scheme for interfacial mass transfer and the kinetics of aroma release. *Trends in Food Science & Technology*. Available from https://doi.org/10.1016/j.tifs.2019.04.012.

Wu, J. Z., Ouyang, Q., Park, B., Kang, R., Wang, Z., Wang, L., & Chen, Q. S. (2022). Physicochemical indicators coupled with multivariate analysis for comprehensive evaluation of matcha sensory quality. *Food Chemistry*, 371. Available from https://doi.org/10.1016/j.foodchem.2021.131100.

Xu, L., Li, W., Voleti, V., Zou, D.-J., Hillman, E. M. C., & Firestein, S. (2020). Widespread receptor-driven modulation in peripheral olfactory coding. *Science*, 368(6487), eaaz5390. Available from https://doi.org/10.1126/science.aaz5390; http://science.sciencemag.org/content/368/6487/eaaz5390.abstract.

3. Digitalization in instrumental, neurological, psychological and behavioural methods: Current applications and opportunities

CHAPTER

10

Electronic noses and tongues: current trends and future needs

Gianmarco Gabrieli, Michal Muszynski and Patrick Ruch

IBM Research Europe, Säumerstrasse 4, Rüschlikon, Switzerland

10.1 Introduction

Technology, in particular artificial intelligence (AI) and robotics, has often been developed based on nature-inspired insights reflecting capabilities of human counterparts in the context of specific applications. The synergy of human—machine collaboration will bring new cooperative tools to tackle problems while raising new questions on how to best put them into practice. We are witnessing a historical transition in which technology is involved in almost every aspect of everyday life (Varsani et al., 2018). In this context, olfactory and gustatory systems are extremely challenging to reproduce as their functionalities have not been completely unveiled yet. The complexity of such systems is what makes them so powerful and fascinating at the same time. In 2015 an experimental approach (Meister, 2015) demonstrated that humans are able to discriminate thousands of odours (a popular estimate is $\sim 10,000$), but the debate around how many odours and tastes we can perceive continues still. For this reason, we have not yet fully explored how to digitally mimic human multimodal information processing to support new applications serving humanity. Electronic noses (e-noses) first reported in the 1980s and electronic tongues (e-tongues) reported in the 1990s were the first digital analogues of human biological sensing functions (Gardner & Bartlett, 1994; Vlasov et al., 2005). These are of great interest for food analysis, in which sensory perception is extensively used to identify, describe and relate various products. E-noses and e-tongues comprise an array of sensors whose output is affected by the presence of chemical compounds, which can eventually be correlated to quantitative or qualitative analysis, intended to distinguish different odours, aromas and flavours. Even though such systems are inspired by human biological functions, they are not replicas of human physiology but may attempt to serve as their functional analogues.

Digital Sensory Science
DOI: https://doi.org/10.1016/B978-0-323-95225-5.00010-9

117

Copyright © 2023 Elsevier Ltd. All rights reserved.

IBM can use the Contribution for their internal purpose prior to or after publication of the Contribution. The members of the scientific community outside of IBM can also use the Contribution in response to specific requests for peer communication.

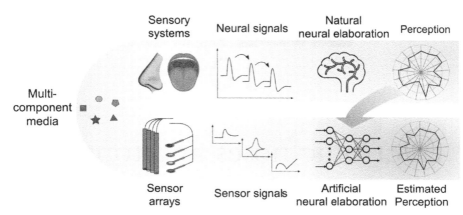

FIGURE 10.1   Analogy of physiological and artificial elaboration of sensory perception. Instrumental predictions are based on models informed by human descriptive analysis (red arrow).

The result of human sensory analysis is indeed the description of qualities derived from the interaction with a product, and e-noses and e-tongues may be devoted to the estimation of exactly that perception (Fig. 10.1).

Olfactory and gustatory responses to multicomponent media are complex because compounds activate many receptors simultaneously while the same receptors are able to bind multiple types of molecules. As a result, an ensemble of neural signals is activated and propagated to the brain for processing and, ultimately, for assessing aroma and taste perception. Artificial sensor arrays can be used to test the same media and to produce various types of signals depending on the corresponding sensing principle. A vast range of sensor types and sensing principles have been implemented for e-tongues and e-noses as reviewed previously (Ciosek & Wróblewski, 2007; Karakaya et al., 2020). Essentially, the design of sensor arrays for e-noses and e-tongues usually involves sensors that exhibit selectivity towards taste-eliciting compounds (Tahara & Toko, 2013; Toko, 1996), sensors that exhibit multiselectivity and cross-sensitivity towards a broader range of compounds (Vlasov et al., 2005), or a combination of both (Ciosek et al., 2004). In general, an array of sensors with imperfect selectivity generates combinatorial signals arising from the interaction with a range of analytes and interferents, and machine learning may be leveraged for effective data processing in order to map sensor measurements to sensory perception regardless of the selectivity of individual sensors (Ishihara et al., 2005; Vlasov & Legin, 1998). Eventually, e-noses and e-tongues could estimate the aroma and taste perception for liquids that have not yet been tested by humans. To achieve this, e-nose and e-tongue instruments may be trained with data from human sensory analysis in order to build discriminative and predictive models. The chemical sensors in these instruments may interact with a wide range of compounds, including tasteless and odourless species. Thus the information contained in the instrument response may be correlated to chemical parameters beyond the sensory space. For this reason, e-noses and e-tongues have also been used for various applications, such as car tuning, breath analysis, smoke detection, gas alarms, pollution detection, disease diagnostics and, importantly, for sensing materials that are hazardous or unpleasant for humans. Application of e-noses and e-tongues to a range of food products has been reported, including beverages, grains, cooking oils, eggs, dairy products, meat and fish, fruit and vegetables (Ali et al., 2020; Baldwin et al., 2011; Gallardo et al., 2005; Judal & Bhadania, 2008; Müller-Maatsch & van Ruth, 2021; Patel et al., 2019; Pérez-Ràfols et al., 2019; Podrażka et al., 2017; Sierra-Padilla et al., 2021; Tudor Kalit et al., 2014). These artificial systems could be disruptive for many food-related applications, including determination of freshness, quality, ripeness, shelf-life, acceptance and authenticity. Employing e-noses and e-tongues for food analysis bears the potential to accelerate routine sensory analysis of products in an objective manner, while at the same time supporting estimation of chemical parameters. Nevertheless, as consumer acceptance is the main objective of industrial food processes, e-noses and e-tongues are never expected to completely displace sensory panels, but could be effectively used to support and complement sensory panels in their mission by offering an objective platform that can be trained with sensory knowledge.

The rest of this chapter will first outline considerations pertaining to machine learning for e-noses and e-tongues in Section 10.2. Then, Section 10.3 will elaborate on applications of e-noses and e-tongues in two scenarios related to sensory evaluation, namely food safety and food innovation. Finally, future perspectives of the technologies will be contemplated in Section 10.4 and conclusions summarised in Section 10.5.

10.2 Machine learning methods for e-tongue and e-nose technologies

Today, advancements in machine learning support a large number of out-of-the-box methods to enable e-tongues and e-noses to accurately and precisely detect and/or quantify compounds. Most machine learning methods have recently been incorporated into feature extraction, qualitative and quantitative analysis, sensor calibration as well as drift compensation.

10.2.1 Feature extraction and learning

E-tongues and e-noses typically produce high-dimensional multivariate transient data that is contaminated with noise and includes redundant information. The main goal of the feature extraction process is to characterise patterns encoded in the raw response signals of sensors while reducing the volume of data, and removing redundancy and noise. The feature extraction process aims to distil bits of relevant information describing patterns in raw sensor signals and to create salient representations of the data that enable effective machine learning (Bengio et al., 2013). The discovery of appropriate representations may be derived from domain expertise and human knowledge about the characteristics of the sensor system and food safety and innovation application, or from unsupervised feature learning on unlabelled data (Jordan & Mitchell, 2015). In general, features in the time or frequency domain could be inspired and designed by prior domain knowledge using handcrafted features, or the features could be learned by a machine learning model from raw response signals to arrive at abstract features or high-level features (Yan et al., 2015; Ye et al., 2021). Those machine learning methods, such as Convolutional Neural Networks

(Shi et al., 2019, 2021), Restricted Boltzmann Machine (Längkvist & Loutfi, 2011), Deep Belief Networks (Luo et al., 2016) and Autoencoders (Yan & Zhang, 2016) involve the learning of high level representations of response signals in a supervised or unsupervised manner (Ye et al., 2021). Moreover, the feature extraction process could include parametric fitting with predefined functions (Cortina et al., 2006; Sharma et al., 2015; Su & Peng, 2014).

10.2.2 Qualitative analysis

Extracted features are used for qualitative and quantitative analysis assisted by appropriate machine learning modelling. Qualitative analysis is used to distinguish different odours, aromas, flavours and so on, while quantitative analysis is used to predict particular properties of targeted gas, solution, or solid material. Qualitative analysis has been extensively investigated for mapping responses of e-tongues and e-noses to targeted sensory predictions, such as odours, aromas, taste and flavour categories. Various machine learning classifiers have been deployed for this purpose, such as Linear Discriminant Analysis (Dai et al., 2015), K-Nearest Neighbour classifier (Nallon et al., 2016a), Multilayer Perceptron (Dong et al., 2017), Extreme Learning Machine (Shi et al., 2021), Support Vector Machine (Men et al., 2018), Random Forest (Nallon et al., 2016b), Convolutional Neural Networks (Wei et al., 2019) and Long Short-Term Memory Neural Networks (Zhang et al., 2018). To boost discriminative power, sensor fusion has been used in several studies by analysing combined responses from both e-tongues and e-noses (Han et al., 2014; Zhi et al., 2017). Two sets of features could be extracted from the responses of e-tongue and e-nose either to train two separate classifiers and then make a final decision by using decision-level fusion, or to train one classifier on concatenated features in feature-level fusion. Various classification performance metrics, such as overall accuracy, balanced accuracy, precision, recall, F1 score and receiver operating characteristic (ROC) curve have been used to assess the performance of classifiers regarding a specific task and class proportion (Ye et al., 2021).

10.2.3 Quantitative analysis

Quantitative analysis aims to estimate continuous properties of gases, odours, solutions and solid materials, and is more flexible in comparison to qualitative analysis that only indicates the presence of targeted categories or classes. Linear Regression, Partial Least Square Regression (Cui et al., 2015), Multilayer Perceptron Regressor (Viejo et al., 2020; Zhang & Tian, 2014), Random Forest Regressor (Du et al., 2019), Support Vector Regression (Han et al., 2014), Convolutional Neural Networks-based Regressor (Guo et al., 2021) and Long Short-Term Memory Neural Networks-based Regressor (Wijaya et al., 2021) are among the various machine learning methods that have been applied to predict targeted quantities such as chemical, physical or sensory parameters. Several evaluation metrics, including the t-value, Pearson correlation coefficient, coefficient of determination, mean squared error, root mean square error and mean absolute error have been applied to evaluate the performance of regressors (Ye et al., 2021).

10.2.4 Compensation of sensor variability

The use of nonlinear models to map multisensor responses to target properties might amplify the challenges encountered by any chemical sensor with respect to variability, drift and ageing. Sensor properties may change over time because of sensor ageing, environmental or conditioning changes (Vergara et al., 2012; Verma et al., 2015; Yan & Zhang, 2016). As a result, variability of measurements could diminish the performance of predictive models trained on previously collected data. E-nose and e-tongue measurements benefit from specific reference samples to provide consistent reference data for use in drift compensation, and for establishing precision and accuracy at different measurement times. Simple mathematical transformations can be applied to update training and calibrations by comparing the response of the array to the same reference samples over time. Univariate transformations, such as Single Wavelength Standardisation (SWS), allow each signal/feature to be corrected independently from each other. Instead, multivariate approaches, such as Direct Standardisation (DS) or Piece-wise Direct Standardisation (PDS), leverage existing correlations among features to transform them simultaneously (Rudnitskaya, 2018). Sensor drift can to some extent be compensated through machine learning techniques, for example domain adaptation instead of training a new predictive model on an aged sensor from scratch. Domain adaptation and ensemble learning are known for electronic sensor drift compensation (Ye et al., 2021). The main goal of domain adaptation techniques, such as Weighted Domain Transfer Extreme Learning Machine (Ma et al., 2018), Kernel Transformation (Tao et al., 2018) and Autoencoder-based Domain Transfer (Yan & Zhang, 2016), is to learn a new feature space in which the similarity between the source domain and target domain is maximised, preserving the discriminative power of features for a given prediction task. Furthermore, ensemble learning methods, including bagging, stacking and boosting to compensate sensor drift, are based on the training of several prediction models on sensor responses acquired over time. A weight is assigned to a single prediction model to produce a weighted average of all predictions while predicting scores for new instances (Liu et al., 2015; Vergara et al., 2012; Zhao et al., 2019). In addition to domain adaption and ensemble learning, prioritisation of drift-insensitive features during feature selection has been proposed (Rehman & Bermak, 2018).

10.2.5 Statistical evaluation

Cross-validation is typically applied to assess the validity of prediction models (Refaeilzadeh et al., 2009; Schaffer, 1993). An important challenge with the evaluation of prediction models is that the prediction capability may be adequate on the training data but fail to predict satisfactorily for unseen instances due to a lack of generalisation or model overfitting and sparse training data. The first main goal of cross-validation techniques is to estimate the model performance when applied to unseen data. The second main goal of cross-validation is to compare the performance of different prediction models in order to select the best model for an available dataset. For example, the average cross-validated performance metric on a dataset can be considered as an estimate of the accuracy and/or precision of a prediction model on unseen data. To compare the performance of two prediction models, predicted scores of the instance set through cross-validation might be used for two-sample statistical hypothesis tests. Various cross-validation procedures have been proposed to achieve those two main goals. One of the most common cross-validation procedures is k-fold cross-validation in which instances from a data set are first split into k almost equally sized folds (Refaeilzadeh et al., 2009). Subsequently, k iterations of training and validation folds are performed such that within each iteration a different fold of the data is held back for model validation, while the remaining $(k - 1)$ folds are used for model training. It is important to randomly rearrange the data in order to guarantee that each fold is representative of the entire population of instances before being partitioned into k folds. Leave-one-out cross-validation is a special case of k-fold cross-validation where k equals either the number of measurements or the number of samples in the data. Furthermore, in each iteration nearly all the dataset except for a single measurement or sample is used for model training, and then the model is tested on that single unseen measurement or sample. The advantage of leave-one-out cross-validation is that the performance estimate has very low bias while having high variance. This procedure is widely applied when only few training instances are collected. In all cases, it is essential that training sets and validation sets are completely independent in order to have a realistic assessment of prediction performance, and validation must only be carried out on samples unseen during training.

10.2.6 Reliability and agreement among sensory ratings

An e-nose or e-tongue experiment for sensory analysis might produce quantitative or qualitative results. To assess the validity of those results, the predicted scores need to be compared to the sensory scores provided by human sensory panels. There are disadvantages of involving human sensory panels, for example the time and cost needed for comprehensive training to provide consistent and reproducible ratings. One of the main factors that can influence the evaluation is the reliability and agreement among panellists. For example, a standard sensory panel should involve at least 10 trained panellists who are invited to participate in the establishment of a sensory profile (ISO 13299, 2016). The panellists evaluate the intensity of attributes covering smell and/or flavour taste on a measurement scale (e.g. 10 cm scale). To study the reliability of sensory panels, the agreement among ratings of sensory panellists needs to be investigated. Intraclass correlation coefficients, Cohen's Kappa agreement, log-linear association agreement model and clustering techniques can be used to measure agreement for each pair of panellists or the entire sensory panel (Newman et al., 2014; ISO 8586-2, 1994).

10.3 Case studies for e-tongue and e-nose applications in sensory evaluation

10.3.1 Case study 1: food safety

The ability of e-tongues and e-noses to discriminate a wide range of samples has been explored in the context of food safety as a means to authenticate the originality and verify the quality of various food products. One objective of food authentication is to validate the geographic origin of products in order to verify compliance with protected designation of origin labels and prevent dissemination of counterfeit products (Danezis et al., 2016). Another objective may be to identify food spoilage which may arise from microbial contamination during processing, storage or both (Adley, 2014). Finally, food authentication can be used to verify the conformity of a product to quality standards in terms of consumer acceptability (Astill et al., 2019).

The most commonly employed analytical tools to assess food authenticity notably include mass spectrometry (MS) in conjunction with gas chromatography (GC-MS) or liquid chromatography (LC-MS), nuclear magnetic resonance (NMR) and vibrational spectroscopic techniques such as near-infrared (NIR) or Raman spectroscopy for analysis of the food metabolome (Danezis et al., 2016). E-tongues and e-noses represent interesting complements to these tools for the evaluation of food authentication and quality due to their relative ease of operation, minimal to no sample preparation, affordability and applicability to a range of molecules and ions. Thus e-tongues and e-noses have been applied both separately and combined in food safety applications for the characterisation of various food matrices, including in dairy products, sweeteners, beverages, meat, fish, fruit, vegetables, fats and oils (Aouadi et al., 2020).

A common workflow for food safety applications typically follows a conventional scheme for supervised learning (Fig. 10.2). A series of labelled food samples are measured using a multisensor e-tongue or e-nose, or a combination of multiple analytical techniques. The sample labelling encompasses a balanced distribution of the properties required for the food safety application, which could take the form of binary labelling of acceptable versus nonacceptable samples, labelling of samples by provenance, or labelling of samples with numeric quantities associated with sensory attributes or physical or chemical properties. The availability of labelled samples is particularly important for data- and correlation-driven e-noses and e-tongues that lack built-in a priori sensor selectivity towards compounds that directly and uniquely affect the target parameters. In addition to the availability of balanced training data from labelled samples, the choice of representation of the measurement data is crucial to support effective machine learning, as discussed in Section 10.2. A variety of models may then be trained on such representations to learn

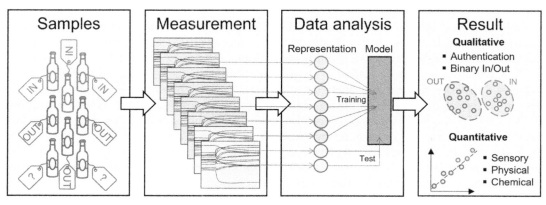

FIGURE 10.2   Supervised learning scheme for e-tongues and e-noses in food safety applications.

how to discriminate sample types, ultimately resulting in the ability to classify unknown food samples for a direct authentication or acceptance task, or to estimate quantitative parameters related to sensory, physical or chemical attributes that indirectly support a decision with respect to authenticity or acceptance.

Numerous applications of e-tongues and e-noses relying on the above scheme for various food safety scenarios have been explored (Rodríguez Méndez, 2016). The fusion of these instruments with other tools for food authentication and quality assessment has also been reviewed (Borràs et al., 2015; Di Rosa et al., 2017). In the wine industry, analysis of the quality of grapes, monitoring of fermentation and ageing, evaluation of organoleptic parameters and correlations with chemical parameters and human perception have been studied with e-noses and e-tongues (Rodríguez-Méndez et al., 2016). A study of Italian red wines using an e-tongue found that human sensory scores including general acceptance could be predicted with an accuracy between 4% and 27% depending on the taste and flavour parameter and the type of wine (Legin et al., 2003). The adulteration of fruit juices by dilution with water, sugar syrup or overripe fruit was shown to be clearly discernible by means of e-tongue and e-nose analysis, and correlation with sensory perception could be identified (Bleibaum et al., 2002; Hong & Wang, 2014; Rudnitskaya et al., 2001). Further, the characterization of provenance, floral origin and adulteration of honey was successfully demonstrated using e-tongues (Sobrino-Gregorio et al., 2018, 2019; Sousa et al., 2014).

In all of the reported cases, the sensitivity of e-tongues and e-noses to a broad range of volatile and nonvolatile food metabolites underlies the capability of these instruments to capture a wealth of chemical information during analysis of food samples. The final food safety outcome, whether it is related to authenticity or acceptance, is generally governed by a nontrivial superposition of the combined chemical information. Therefore besides and beyond the determination of individual chemical parameters, the training of surrogate models that attempt to directly connect measurement representations from e-tongues and e-noses to the final food safety result appears to be an attractive and viable path to simplify and accelerate the assessment of foods.

10.3.2 Case study 2: food innovation

Objective descriptions of the nature and intensity of sensory characteristics are the main outcomes of product descriptive analysis, whose recent advances have led to more rapid, customised and fast-forwarded profiling techniques (Marques et al., 2022; Delarue et al., 2015). Currently, such an approach requires recruiting a sensory panel, performing training based on both natural and artificial references, and establishing a sensory lexicon that provides the set of product qualities that need to be quantified for an effective profiling (ISO 13299, 2016). Nevertheless, sensory analysis was demonstrated to be time-consuming, expensive and strongly influenced by panellist fatigue and carryover effects (Ross, 2021). Such drawbacks are amplified when dealing with applications targeting food innovation processes, as panellists are required to test a large number of samples and provide continuous feedback to each step of the innovation process. In this context, cost-effective and fast analytical tools, such as e-tongues and e-noses, could represent a valuable complement to sensory analysis.

A study conducted by Waldrop and Ross (Waldrop & Ross, 2014) demonstrated the potential to leverage an e-tongue to predict the sweetness profile of unknown products. The increasing concern about the impact of sugar levels in food and beverage products on human health has forced industries to find alternatives such as natural sweeteners. Product optimisation requires finding the best sweetener blends that satisfy consumer acceptance thresholds with limited concentration levels. Sugar and sweetener blends were evaluated in aqueous solutions by a trained panel (i.e. eight panellists), 60 consumers and the e-tongue. The e-tongue was able to discriminate all mixtures under test and produced taste profiles that were well correlated with sensory panel data (i.e. coefficient of determination $R^2 > 0.79$) with consumer intensity ratings using partial least squares (PLS) regression analysis. The discrimination of samples based on the e-tongue response was correlated with consumer ratings, implying ability of the instrument to predict consumer acceptance and liking. These results suggest that such analytical tools could be used to determine sweetener substitutions to meet consumer demand or production constraints. Lorenz et al. (2009) used an e-tongue to estimate the bitterness of drug formulations in order to produce medication with tailored taste properties. The e-tongue was successfully employed to determine the effectiveness of artificial sweeteners as bitterness-masking agents by analysing the response to taste formulations including a bitter active pharmaceutical ingredient (API) in comparison to a placebo without API. Moreover, a high correlation (coefficient of determination $R^2 = 0.99$) was found between e-tongue bitterness scores with the bitterness assessed by a human sensory panel on the same drug formulations. Thus the possibility was considered to estimate the bitterness of new samples for which sensory data were not available in order to identify the matrix providing the highest level of bitterness masking. These findings suggest that e-tongue and e-nose systems have the potential to transform sensory analysis in the pharmaceutical industry as drug palatability significantly impacts compliance and commercial success of a given product.

Gabrieli et al. (2022) reported the training of an e-tongue on 33 coffee products through pattern recognition and machine learning models that unveiled interdependencies between sensory descriptors and provided a unique fingerprint of product sensory profiles. Due to the high number of sensory properties that are typically used to describe a product, dimension reduction techniques were applied to minimise the target parameters of regression algorithms. Inverse transformations were used to reconstruct the predicted sensory profile in the original sensory attribute space. The sensor array leveraged cross-sensitive materials that do not interact selectively with compounds associated with specific aroma or taste descriptors, thereby demonstrating how a hypothesis-free instrument can be applied to product sensory analysis based solely on training examples. The measurement data and trained models were used to reconstruct differences between coffee products, and every new test could be mapped and compared to the existing product base.

Since e-tongues and e-noses do not replicate human physiology, it is expected that the instruments will not completely replace human sensory analysis, but rather support the exploration of new product formulations within a sensory space used to calibrate these instruments. Compared to the changes in consumer perception and liking during food consumption, these instruments produce a unique and static objective fingerprint of a product under test. Some efforts have been reported to make measurements more similar to human processes, for instance, by use of artificial saliva as a background solution for measurements of samples with different drug formulations (Khaydukova et al., 2017). Nevertheless, the possibility to perform fast analysis of a greater number of samples is what makes these complementary tools suitable for accelerated evaluation of new products. A general scheme describing the integration of e-tongues and e-noses in the design cycle of new formulations is described in Fig. 10.3.

After defining the requirements for a new product formulation, the innovation cycle may start by collecting sensory knowledge about existing formulations and characterising those formulations using the e-tongue or e-nose systems. After verifying the correlation of human sensory perception with e-tongue and e-nose responses, prediction models are built based on the measurements recorded for a given set of reference samples. Cross-validation (Section 10.2) may be used to determine whether model performance is satisfactory compared to sensory panel evaluation, which enables new product formulations, within the sensory space used to develop the algorithm, to be evaluated by the e-tongues or e-noses to predict corresponding sensory properties in an automated and accelerated procedure. The new formulations with the most desirable predicted sensory characteristics may be downselected for validation by a targeted panel session and the results used to inform new product formulations. Data for new validated formulations are added to the sensory knowledge pool, thereby further growing the sensory space and potentially innovation capability over time. Such a pipeline to support product development is not only limited to the discovery of formulations, but can also aid in decision-making processes for development of optimal food packaging materials by detecting off-flavours and predicting quality and shelf-life (Ali et al., 2020).

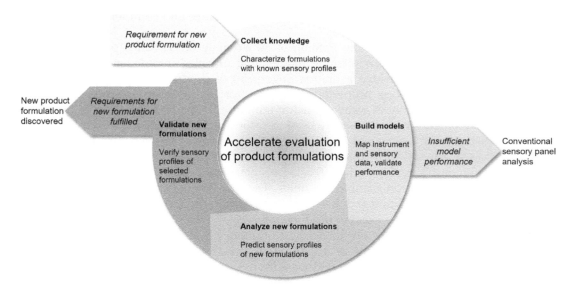

FIGURE 10.3 Principle of accelerated food product innovation cycle leveraging e-tongues and e-noses to help screen new formulations.

10.4 Future perspectives

Commercial and prototype e-tongues and e-noses have shown the potential to revolutionise sensory analysis through digitisation and automation of taste and aroma analysis. Advancements in deploying AI and data pipelines for analytical infrastructure, device connectivity through internet of things (IoT) and sensor miniaturisation are major driving factors that can benefit and accelerate the dissemination and application of these technologies. There are a number of perspectives related to technology governance that will help establish e-tongues and e-noses as reliable tools for sensory analysis, as well as several technological advances that promise to improve their functionality and performance:

1. Interoperability of data recorded with different instruments.
2. Standardised methodology for training, validation and performance benchmarking.
3. Shared datasets defining reference data and common data exchange formats.
4. Data-driven, sensor-agnostic systems combined with unsupervised learning to discover effective representations correlated with target properties.
5. Improved sensing materials and principles for enhanced durability, sensitivity and limit of detection.
6. Miniaturised and decentralised systems for portable and remote chemical analysis in the IoT era.
7. Plug-and-play sensor fusion combining heterogeneous sensing principles for broadband chemical sensing.

The need for interoperability stems from the fact that there are countless varieties of e-tongue and e-nose hardware configurations reported in literature and available commercially. These configurations differ in terms of sensing materials, sensing principles, measurement methodologies, data analysis and results reporting. By virtue of this fragmentation, today it is not easily possible to leverage combined data originating from different instruments. It is likely that a variety of e-tongues and e-noses will continue to exist and be developed in future, which means that interoperability will best be achieved at the level of model output and results reporting with respect to benchmarks or similarities between reference samples. Reporting results from trained models in a common format for qualitative and quantitative analysis along with uncertainty measures is considered to be key to allow exchange and continuity of data across different e-tongue and e-nose platforms over time.

Similarly, many e-tongue and e-nose developers have carefully devised schemes to optimally operate their tools, in particular with respect to reliability, reproducibility and robustness which require particular attention when leveraging nonlinear multivariate models for evaluation of measurement data. On the other hand, these different measurement schemes often result in a range of methodologies being applied to conduct measurements, train and validate models, and benchmark the e-tongue and e-nose performance. While results for a specific tool and use case may appear promising, comparing results from different tools is very difficult due to this heterogeneity in operation. Another consequence of this heterogeneity is that there is no standard procedure for implementing and benchmarking e-tongues and e-noses, which results in tool-specific recommendations and may lead to overengineering for particular use cases and lack of generality.

Finally, it may be argued that significant advances in technologies leveraging machine learning have always been spurred by the availability of common data sets that enable researchers and technologists to test their hypotheses and enhance training of their systems. Due to the heterogeneity of e-tongues and e-noses as mentioned above, such universal data sets are not yet widespread in the field. In particular for food applications, the complexity of matrices and formulations is vast, and data sets available for training are considered essential to improve the quality of validation and usefulness of the technology.

In terms of future perspectives for e-nose and e-tongue technology in digital sensory science, it is expected that future tools will increasingly leverage more and more data and unsupervised techniques to find correlations of surrogate representations directly with target properties such as sensory attributes, rather than rely on individual correlations of sensor responses with chemical composition and sensory descriptors. The integration of novel durable sensing materials could boost sensitivity and lower the detection limit for compounds that strongly affect sensory perception. The aforementioned trends in AI deployment, miniaturisation and IoT may fuel the development of portable prototypes for fast and decentralised analysis to support industrial food processes. Lastly, simultaneous sample analysis using e-tongues, e-noses and other sensor fusion would support the assessment of more complete sensory descriptions through sensor interaction with both volatile and nonvolatile compounds. Integration of multiple systems has already been explored and yielded promising results and improved sensory profiling compared to independent aroma and taste assessment (Borràs et al., 2015; Calvini & Pigani, 2022). In this context, future developments to combine heterogeneous sensing principles on the same platform may be expected. Such efforts include hardware design for integration of different materials as well as testing protocols for analysis of gas and liquid phases. Moreover, progress on data fusion techniques to combine multiple sensing modalities could generate more robust data pipelines compared to current analytical approaches. For widespread usage, however, all technological advances should always be accompanied by the considerations outlined above with respect to interoperability and benchmarking.

10.5 Conclusion

Industrial adoption of e-tongues and e-noses has the potential to greatly accelerate sensory analysis tasks, for example in food safety and food innovation. While numerous demonstrations of digital sensory analysis leveraging e-tongues and e-noses have been reported for many specific food use cases, a common framework and best practices for this technology in terms of interoperability, training, validation and benchmarking are considered essential to further promote widespread adoption. Two paradigm shifts may be imminent upon the introduction of data-driven e-tongues and e-noses for sensory science. First, the insight that a range of sensory tasks outsourced to systems that are largely automated and have a higher throughput compared to human sensory panels can transform the way in which processes are implemented for food safety and food innovation. Second, the shift from chemical analysis targeting recognition of specific molecules to a data-driven analysis of complex matrices with training by examples can radically change requirements with respect to fundamental processes such as instrument calibration, data exchange and interpretation. Advances in AI deployment, connected devices and device miniaturisation, as well as increasing community efforts to generate public datasets may all be considered enablers that will help drive the definition and adoption of data-driven e-tongues and e-noses in future.

References

Adley, C. C. (2014). Past, present and future of sensors in food production. *Foods*, 3(3), 491−510. Available from https://doi.org/10.3390/foods3030491.

Ali, M. M., Hashim, N., Abd Aziz, S., & Lasekan, O. (2020). Principles and recent advances in electronic nose for quality inspection of agricultural and food products. *Trends in Food Science and Technology*, 99, 1−10. Available from https://doi.org/10.1016/j.tifs.2020.02.028.

Aouadi, B., Zaukuu, J. L. Z., Vitális, F., Bodor, Z., Fehér, O., Gillay, Z., Bazar, G., & Kovacs, Z. (2020). Historical evolution and food control achievements of near infrared spectroscopy, electronic nose, and electronic tongue—Critical overview. *Sensors*, 20(19), 1−42. Available from https://doi.org/10.3390/s20195479.

Astill, J., Dara, R. A., Campbell, M., Farber, J. M., Fraser, E. D. G., Sharif, S., & Yada, R. Y. (2019). Transparency in food supply chains: A review of enabling technology solutions. *Trends in Food Science and Technology*, 91, 240−247. Available from https://doi.org/10.1016/j.tifs.2019.07.024.

Baldwin, E. A., Bai, J., Plotto, A., & Dea, S. (2011). Electronic noses and tongues: Applications for the food and pharmaceutical industries. *Sensors*, 11(5), 4744−4766. Available from https://doi.org/10.3390/s110504744.

Bengio, Y., Courville, A., & Vincent, P. (2013). Representation learning: A review and new perspectives. *IEEE Transactions on Pattern Analysis and Machine Intelligence*, 35(8), 1798−1828. Available from https://doi.org/10.1109/TPAMI.2013.50.

Bleibaum, R. N., Stone, H., Tan, T., Labreche, S., Saint-Martin, E., & Isz, S. (2002). Comparison of sensory and consumer results with electronic nose and tongue sensors for apple juices. *Food Quality and Preference*, 13(6), 409−422. Available from https://doi.org/10.1016/S0950-3293(02)00017-4.

Borràs, E., Ferré, J., Boqué, R., Mestres, M., Aceña, L., & Busto, O. (2015). Data fusion methodologies for food and beverage authentication and quality assessment—A review. *Analytica Chimica Acta*, 891, 1−14. Available from https://doi.org/10.1016/j.aca.2015.04.042.

Calvini, R., & Pigani, L. (2022). Toward the development of combined artificial sensing systems for food quality evaluation: A review on the application of data fusion of electronic noses, electronic tongues and electronic eyes. *Sensors*, 22(2), 577. Available from https://doi.org/10.3390/s22020577.

Ciosek, P., & Wróblewski, W. (2007). Sensor arrays for liquid sensing—Electronic tongue systems. *Analyst*, 132(10), 963−978. Available from https://doi.org/10.1039/b705107g.

Ciosek, P., Augustyniak, E., & Wróblewski, W. (2004). Polymeric membrane ion-selective and cross-sensitive electrode-based electronic tongue for qualitative analysis of beverages. *Analyst*, 129, 639−644. Available from https://doi.org/10.1039/b401390e.

Cortina, M., Duran, A., Alegret, S., & Del Valle, M. (2006). A sequential injection electronic tongue employing the transient response from potentiometric sensors for anion multidetermination. *Analytical and Bioanalytical Chemistry*, 385(7), 1186−1194. Available from https://doi.org/10.1007/s00216-006-0530-2.

Cui, S., Wang, J., Yang, L., Wu, J., & Wang, X. (2015). Qualitative and quantitative analysis on aroma characteristics of ginseng at different ages using E-nose and GC−MS combined with chemometrics. *Journal of Pharmaceutical and Biomedical Analysis*, 102, 64−77. Available from https://doi.org/10.1016/j.jpba.2014.08.030.

Dai, Y., Zhi, R., Zhao, L., Gao, H., Shi, B., & Wang, H. (2015). Longjing tea quality classification by fusion of features collected from E-nose. *Chemometrics and Intelligent Laboratory Systems*, 144, 63−70. Available from https://doi.org/10.1016/j.chemolab.2015.03.010.

Danezis, G. P., Tsagkaris, A. S., Brusic, V., & Georgiou, C. A. (2016). Food authentication: State of the art and prospects. *Current Opinion in Food Science*, 10, 22−31. Available from https://doi.org/10.1016/j.cofs.2016.07.003.10.1016/j.cofs.2016.07.003.

Delarue, J., Lawlor, J. B., & Goeaux, M. (2015). *Rapid sensory profiling techniques: Applications in new product development and consumer research*. Elsevier. Available from https://doi.org/10.1016/C2013-0-16502-6.

Di Rosa, A. R., Leone, F., Cheli, F., & Chiofalo, V. (2017). Fusion of electronic nose, electronic tongue and computer vision for animal source food authentication and quality assessment—A review. *Journal of Food Engineering, 210*, 62–75. Available from https://doi.org/10.1016/j.jfoodeng.2017.04.024.

Dong, W., Zhao, J., Hu, R., Dong, Y., & Tan, L. (2017). Differentiation of Chinese robusta coffees according to species, using a combined electronic nose and tongue, with the aid of chemometrics. *Food Chemistry, 229*, 743–751. Available from https://doi.org/10.1016/j.foodchem.2017.02.149.

Du, D., Wang, J., Wang, B., Zhu, L., & Hong, X. (2019). Ripeness prediction of postharvest kiwifruit using a MOS e-nose combined with chemometrics. *Sensors, 19*(2), 419. Available from https://doi.org/10.3390/s19020419.

Gabrieli, G., Muszynski, M., Thomas, E., Labbe, D., & Ruch, P. W. (2022). Accelerated estimation of coffee sensory profiles using an AI assisted electronic tongue. *Innovative Food Science and Emerging Technologies, 82*, 103205. Available from https://doi.org/10.1016/j.ifset.2022.103205.

Gallardo, J., Alegret, S., & Del Valle, M. (2005). Application of a potentiometric electronic tongue as a classification tool in food analysis. *Talanta, 66*(5), 1303–1309. Available from https://doi.org/10.1016/j.talanta.2005.01.049.

Gardner, J. W., & Bartlett, P. N. (1994). A brief history of electronic noses. *Sensors and Actuators B, 19*, 18–19. Available from https://doi.org/10.1016/0925-4005(94)87085-3.

Guo, J., Cheng, Y., Luo, D., Wong, K.-Y., Hung, K., & Li, X. (2021). ODRP: A deep learning framework for odor descriptor rating prediction using electronic nose. *IEEE Sensors Journal, 21*(13), 15012–15021. Available from https://doi.org/10.1109/JSEN.2021.3074173.

Han, F., Huang, X., Teye, E., Gu, F., & Gu, H. (2014). Nondestructive detection of fish freshness during its preservation by combining electronic nose and electronic tongue techniques in conjunction with chemometric analysis. *Analytical Methods, 6*(2), 529–536. Available from https://doi.org/10.1039/C3AY41579A.

Hong, X., & Wang, J. (2014). Detection of adulteration in cherry tomato juices based on electronic nose and tongue: Comparison of different data fusion approaches. *Journal of Food Engineering, 126*, 89–97. Available from https://doi.org/10.1016/j.jfoodeng.2013.11.008.

Ishihara, S., Ikeda, A., Citterio, D., Maruyama, K., Hagiwara, M., & Suzuki, K. (2005). Smart chemical taste sensor for determination and prediction of taste qualities based on a two-phase optimized radial basis function network. *Analytical Chemistry, 77*(24), 7908–7915. Available from https://doi.org/10.1021/ac0510686.

ISO 8586-2. (1994). Sensory Analysis: Methodology-General Guidance for Establishing a Sensory Profile ISO 8586-2: Sensory analysis. General guidance for the selection, training and monitoring of assessors. Part 2: Experts. International Organization for Standardization, Geneva.

ISO 13299. (2016). Sensory analysis—General guidance for the selection, training and monitoring of assessors—Part 2: Experts ISO 13299: Sensory Analysis – Methodology – General guidance for establishing a sensory profile. International Organization for Standards. Available from https://www.iso.org/standard/58042.html (last accessed 2023-04-13).

Jordan, M. I., & Mitchell, T. M. (2015). Machine learning: Trends, perspectives, and prospects. *Science, 349*(6245), 255–260. Available from https://doi.org/10.1126/science.aaa8415.

Judal, A., & Bhadania, A. (2008). Advances in artificial electronic sensing techniques for quality measurements in dairy and food industry. National seminar on "Indian Dairy Industry-Opportunities and Challenges," *Technical Articles*, 233–240. Available from https://www.dairyknowledge.in/sites/default/files/ch27_0.pdf (last accessed 2023-04-13).

Karakaya, D., Ulucan, O., & Turkan, M. (2020). Electronic nose and its applications: A survey. *International Journal of Automation and Computing, 17*(2), 179–209. Available from https://doi.org/10.1007/s11633-019-1212-9.

Khaydukova, M., Kirsanov, D., Pein-Hackelbusch, M., Immohr, L. I., Gilemkhanova, V., & Legin, A. (2017). Critical view on drug dissolution in artificial saliva: A possible use of in-line e-tongue measurements. *European Journal of Pharmaceutical Sciences, 99*, 266–271. Available from https://doi.org/10.1016/j.ejps.2016.12.028.

Längkvist, M., & Loutfi, A. (2011). Unsupervised feature learning for electronic nose data applied to bacteria identification in blood. NIPS 2011 Workshop on Deep Learning and Unsupervised Feature Learning. Available from http://urn.kb.se/resolve?urn=urn:nbn:se:oru:diva-24197 (last accessed 2023-04-13).

Legin, A., Rudnitskaya, A., Lvova, L., Vlasov, Y., Di Natale, C., & D'Amico, A. (2003). Evaluation of Italian wine by the electronic tongue: Recognition, quantitative analysis and correlation with human sensory perception. *Analytica Chimica Acta, 484*(1), 33–44. Available from https://doi.org/10.1016/S0003-2670(03)00301-5.

Liu, H., Chu, R., & Tang, Z. (2015). Metal oxide gas sensor drift compensation using a two-dimensional classifier ensemble. *Sensors, 15*(5), 10180–10193. Available from https://doi.org/10.3390/s150510180.

Lorenz, J. K., Reo, J. P., Hendl, O., Worthington, J. H., & Petrossian, V. D. (2009). Evaluation of a taste sensor instrument (electronic tongue) for use in formulation development. *International Journal of Pharmaceutics, 367*(1–2), 65–72. Available from https://doi.org/10.1016/j.ijpharm.2008.09.042.

Luo, Y., Wei, S., Chai, Y., & Sun, X. (2016). Electronic nose sensor drift compensation based on deep belief network. 2016 35th Chinese Control Conference (CCC). 3951–3955. https://doi.org/10.1109/ChiCC.2016.7553969.

Ma, Z., Luo, G., Qin, K., Wang, N., & Niu, W. (2018). Weighted domain transfer extreme learning machine and its online version for gas sensor drift compensation in E-nose systems. *Wireless Communications and Mobile Computing*, e2308237. Available from https://doi.org/10.1155/2018/2308237.

Marques, C., Correia, E., Dinis, L.-T., & Vilela, A. (2022). An overview of sensory characterization techniques: From classical descriptive analysis to the emergence of novel profiling methods. *Foods, 11*(3), 255. Available from https://doi.org/10.3390/foods11030255.

Meister, M. (2015). On the dimensionality of odor space. *Elife, 4*, e07865. Available from https://doi.org/10.7554/eLife.07865.

Men, H., Shi, Y., Jiao, Y., Gong, F., & Liu, J. (2018). Electronic nose sensors data feature mining: A synergetic strategy for the classification of beer. *Analytical Methods, 10*(17), 2016–2025. Available from https://doi.org/10.1039/C8AY00280K.

Müller-Maatsch, J., & van Ruth, S. M. (2021). Handheld devices for food authentication and their applications: A review. *Foods, 10*(12), 2901. Available from https://doi.org/10.3390/foods10122901.

Nallon, E. C., Schnee, V. P., Bright, C., Polcha, M. P., & Li, Q. (2016a). Chemical discrimination with an unmodified graphene chemical sensor. *ACS Sensors, 1*(1), 26–31. Available from https://doi.org/10.1021/acssensors.5b00029.

Nallon, E. C., Schnee, V. P., Bright, C. J., Polcha, M. P., & Li, Q. (2016b). Discrimination enhancement with transient feature analysis of a graphene chemical sensor. *Analytical Chemistry, 88*(2), 1401–1406. Available from https://doi.org/10.1021/acs.analchem.5b04050.

Newman, J., Harbourne, N., O'Riordan, D., Jacquier, J. C., & O'Sullivan, M. (2014). Comparison of a trained sensory panel and an electronic tongue in the assessment of bitter dairy protein hydrolysates. *Journal of Food Engineering, 128*, 127–131. Available from https://doi.org/10.1016/j.jfoodeng.2013.12.019.

Patel, H. K., Patel, P. H., & Patel, H. (2019). Innovative application electronic nose and electronic tongue techniques for food quality estimation. *International Journal of Recent Technology and Engineering, 8*, 318–323. Available from http://www.doi.org/10.35940/ijrte.B1506.078219.

Pérez-Ràfols, C., Serrano, N., Ariño, C., Esteban, M., & Díaz-Cruz, J. M. (2019). Voltammetric electronic tongues in food analysis. *Sensors, 19*(19), 4261. Available from https://doi.org/10.3390/s19194261.

Podrażka, M., Baczyńska, E., Kundys, M., Jeleń, P. S., & Witkowska Nery, E. (2017). Electronic tongue—a tool for all tastes? *Biosensors, 8*(1), 3.

Refaeilzadeh, P., Tang, L., & Liu, H. (2009). Cross-validation. *Encyclopedia of Database Systems, 5*, 532–538.

Rehman, A. U., & Bermak, A. (2018). Drift-insensitive features for learning artificial olfaction in e-nose system. *IEEE Sensors Journal, 18*(17), 7173–7182.

Rodríguez Méndez, M. L. (2016). *Electronic noses and tongues in food science* (pp. 1–316). Elsevier. Available from https://doi.org/10.1016/C2013-0-14449-2.

Rodríguez-Méndez, M. L., De Saja, J. A., González-Antón, R., García-Hernández, C., Medina-Plaza, C., García-Cabezón, C., & Martín-Pedrosa, F. (2016). Electronic noses and tongues in wine industry. *Frontiers in Bioengineering and Biotechnology, 4*, 1–12. Available from https://doi.org/10.3389/fbioe.2016.00081.

Ross, C. F. (2021). Considerations of the use of the electronic tongue in sensory science. *Current Opinion in Food Science, 40*, 87–93. Available from https://doi.org/10.1016/j.cofs.2021.01.011.

Rudnitskaya, A., Legin, A., Makarychev-Mikhailov, S., Goryacheva, O., & Vlasov, Y. (2001). Quality monitoring of fruit juices using an electronic tongue. *Analytical Sciences, 17*, i309–i312. Available from https://doi.org/10.14891/analscisp.17icas.0.i309.0.

Rudnitskaya, A. (2018). Calibration update and drift correction for electronic noses and tongues. *Frontiers in Chemistry*, 433. Available from https://doi.org/10.3389/fchem.2018.00433.

Schaffer, C. (1993). Selecting a classification method by cross-validation. *Machine Learning, 13*(1), 135–143. Available from https://doi.org/10.1007/BF00993106.

3. Digitalization in instrumental, neurological, psychological and behavioural methods: Current applications and opportunities

Sharma, H. J., Sonwane, N. D., & Kondawar, S. B. (2015). Electrospun SnO_2/polyaniline composite nanofibers based low temperature hydrogen gas sensor. *Fibers and Polymers, 16*(7), 1527−1532. Available from https://doi.org/10.1007/s12221-015-5222-0.

Shi, Y., Gong, F., Wang, M., Liu, J., Wu, Y., & Men, H. (2019). A deep feature mining method of electronic nose sensor data for identifying beer olfactory information. *Journal of Food Engineering, 263*, 437−445. Available from https://doi.org/10.1016/j.jfoodeng.2019.07.023.

Shi, Y., Yuan, H., Xiong, C., Zhang, Q., Jia, S., Liu, J., & Men, H. (2021). Improving performance: A collaborative strategy for the multi-data fusion of electronic nose and hyperspectral to track the quality difference of rice. *Sensors and Actuators B, 333*, 129546. Available from https://doi.org/10.1016/j.snb.2021.129546.

Sierra-Padilla, A., García-Guzmán, J. J., López-Iglesias, D., Palacios-Santander, J. M., & Cubillana-Aguilera, L. (2021). E-Tongues/noses based on conducting polymers and composite materials: Expanding the possibilities in complex analytical sensing. *Sensors, 21*(15), 4976. Available from https://doi.org/10.3390/s21154976.

Sobrino-Gregorio, L., Tanleque-Alberto, F., Bataller, R., Soto, J., & Escriche, I. (2019). Using an automatic pulse voltammetric electronic tongue to verify the origin of honey from Spain, Honduras, and Mozambique. *Journal of the Science of Food and Agriculture, 100*, 212−217. Available from https://doi.org/10.1002/jsfa.10022.

Sobrino-Gregorio, L., Bataller, R., Soto, J., & Escriche, I. (2018). Monitoring honey adulteration with sugar syrups using an automatic pulse voltammetric electronic tongue. *Food Control, 91*, 254−260. Available from https://doi.org/10.1016/j.foodcont.2018.04.003.

Sousa, M. E. B. C., Dias, L. G., Veloso, A. C. A., Estevinho, L., Peres, A. M., & Machado, A. A. S. C. (2014). Practical procedure for discriminating monofloral honey with a broad pollen profile variability using an electronic tongue. *Talanta, 128*, 284−292. Available from https://doi.org/10.1016/j.talanta.2014.05.004.

Su, P.-G., & Peng, Y.-T. (2014). Fabrication of a room-temperature H_2S gas sensor based on PPy/WO_3 nanocomposite films by in-situ photopolymerization. *Sensors and Actuators B, 193*, 637−643. Available from https://doi.org/10.1016/j.snb.2013.12.027.

Tahara, Y., & Toko, K. (2013). Electronic tongues-a review. *IEEE Sensors Journal, 13*(8), 3001−3011. Available from https://doi.org/10.1109/JSEN.2013.2263125.

Tao, Y., Xu, J., Liang, Z., Xiong, L., & Yang, H. (2018). Domain correction based on kernel transformation for drift compensation in the E-nose system. *Sensors, 18*(10), 3209. Available from https://doi.org/10.3390/s18103209.

Toko, K. (1996). Taste sensor with global selectivity. *Materials Science and Engineering C, 4*(2), 69−82. Available from https://doi.org/10.1016/0928-4931(96)00134-8.

Tudor Kalit, M., Marković, K., Kalit, S., Vahčić, N., & Havranek, J. (2014). Application of electronic nose and electronic tongue in the dairy industry. *Mljekarstvo, 64*, 228−244. Available from https://doi.org/10.15567/mljekarstvo.2014.0402.

Varsani, P., Moseley, R., Jones, S., James-Reynolds, C., Chinellato, E., & Augusto, J. C. (2018). *Sensorial computing*. In *New directions in third wave human-computer interaction: Volume 1-technologies*, (pp. 265−284). Springer. Available from https://doi.org/10.1007/978-3-319-73356-2_15.

Vergara, A., Vembu, S., Ayhan, T., Ryan, M. A., Homer, M. L., & Huerta, R. (2012). Chemical gas sensor drift compensation using classifier ensembles. *Sensors and Actuators B, 166*, 320−329. Available from https://doi.org/10.1016/j.snb.2012.01.074.

Verma, M., Asmita, S., & Shukla, K. K. (2015). A regularized ensemble of classifiers for sensor drift compensation. *IEEE Sensors Journal, 16*(5), 1310−1318. Available from https://doi.org/10.1109/JSEN.2015.2497277.

Viejo, C. G., Fuentes, S., Godbole, A., Widdicombe, B., & Unnithan, R. R. (2020). Development of a low-cost e-nose to assess aroma profiles: An artificial intelligence application to assess beer quality. *Sensors and Actuators B, 308*, 127688. Available from https://doi.org/10.1016/j.snb.2020.127688.

Vlasov, Y., & Legin, A. (1998). Non-selective chemical sensors in analytical chemistry: From "electronic nose" to "electronic tongue.". *Fresenius' Journal of Analytical Chemistry, 361*(3), 255−260. Available from https://doi.org/10.1007/s002160050875.

Vlasov, Y., Legin, A., Rudnitskaya, A., Di Natale, C., & D'Amico, A. (2005). Nonspecific sensor arrays ("electronic tongue") for chemical analysis of liquids (IUPAC Technical Report). *Pure and Applied Chemistry, 77*(11), 1965−1983. Available from https://doi.org/10.1351/pac200577111965.

Waldrop, M. E., & Ross, C. F. (2014). Sweetener blend optimization by using mixture design methodology and the electronic tongue. *Journal of Food Science, 79*(9), S1782−S1794. Available from https://doi.org/10.1111/1750-3841.12575.

3. Digitalization in instrumental, neurological, psychological and behavioural methods:
Current applications and opportunities

Wei, G., Li, G., Zhao, J., & He, A. (2019). Development of a LeNet-5 gas identification CNN structure for electronic noses. *Sensors, 19*(1), 217. Available from https://doi.org/10.3390/s19010217.

Wijaya, D. R., Sarno, R., & Zulaika, E. (2021). DWTLSTM for electronic nose signal processing in beef quality monitoring. *Sensors and Actuators B, 326*, 128931. Available from https://doi.org/10.1016/j.snb.2020.128931.

Yan, J., Guo, X., Duan, S., Jia, P., Wang, L., Peng, C., & Zhang, S. (2015). Electronic nose feature extraction methods: A review. *Sensors, 15*(11), 27804−27831. Available from https://doi.org/10.3390/s151127804.

Yan, K., & Zhang, D. (2016). Correcting instrumental variation and time-varying drift: A transfer learning approach with autoencoders. *IEEE Transactions on Instrumentation and Measurement, 65*(9), 2012−2022. Available from https://doi.org/10.1109/TIM.2016.2573078.

Ye, Z., Liu, Y., & Li, Q. (2021). Recent progress in smart electronic nose technologies enabled with machine learning methods. *Sensors, 21*(22), 7620. Available from https://doi.org/10.3390/s21227620.

Zhang, H., Ye, W., Zhao, X., Teng, R.K. F., & Pan, X. (2018). A novel convolutional recurrent neural network based algorithm for fast gas recognition in electronic nose system. 2018 IEEE International Conference on Electron Devices and Solid State Circuits (EDSSC). 1−2. https://doi.org/10.1109/EDSSC.2018.8487105.

Zhang, L., & Tian, F. (2014). Performance study of multilayer perceptrons in a low-cost electronic nose. *IEEE Transactions on Instrumentation and Measurement, 63*(7), 1670−1679. Available from https://doi.org/10.1109/TIM.2014.2298691.

Zhao, X., Li, P., Xiao, K., Meng, X., Han, L., & Yu, C. (2019). Sensor drift compensation based on the improved LSTM and SVM multi-class ensemble learning models. *Sensors, 19*(18), 3844. Available from https://doi.org/10.3390/s19183844.

Zhi, R., Zhao, L., & Zhang, D. (2017). A framework for the multi-level fusion of electronic nose and electronic tongue for tea quality assessment. *Sensors, 17*(5), 1007. Available from https://doi.org/10.3390/s17051007.

3. Digitalization in instrumental, neurological, psychological and behavioural methods: Current applications and opportunities

11

Leveraging neuro-behavioural tools to enhance sensory research

Kathryn Ambroze and Michelle M. Niedziela

HCD Research, Flemington, NJ, USA

11.1 Introduction

The way consumers perceive a product impacts every action they take towards (or away) from it. From packaging to flavour, hedonics to purchase intent, having a meaningful experience with a product is a major component in driving consumer satisfaction. Paradoxically, the focus of much sensory and consumer research tends to be more product-focused instead of consumer-centric. From sterile testing environments to treating consumers like machines, drivers of consumer perceptions and behaviours can be easily missed.

Traditionally, marketers and sensory consumer researchers have relied on consumers' self-reported reactions to products and services. Often these surveys take place in central location testing (CLT) facilities with high levels of experimental control (sterile and quiet testing rooms, isolated and measured samples, etc.). This level of control has helped to optimise product development by focusing on isolating and extenuating minute differences among products and prototypes, but has also lost focus of the naturalistic consumer experience. Additionally, the traditional approach of measuring self-reported product liking has failed to predict product performance in the market (Chandon et al., 2005; Graves, 2010; King et al., 2010; Thomson & Coates, 2021; Thomson & Crocker, 2015; Thomson, 2006). This product-focused approach falls flat in capturing the holistic consumer experience that negatively impacts researchers' ability to make predictions in the market.

Market and consumer researchers are motivated to apply newer and more in-depth emotional and perceptual measures to better understand the consumer experience. With more accessible neuro- and neuro-digital tools (cheaper headsets, greater popularity, etc.), the once very academic field of consumer neuroscience has become more mainstream and accessible in consumer and market research, especially in the areas of advertising,

Digital Sensory Science
DOI: https://doi.org/10.1016/B978-0-323-95225-5.00011-0

marketing, product research and measuring consumer emotions. Finding the best tool to measure emotions is highly sought-after, since it presents an opportunity to better connect consumers with products, concepts or packaging.

The advancements in digitisation in neuroscientific and psychological (neuro-behavioural) tools, through programs, websites or online resources, have excited market and consumer researchers as a potential solution to fill the gap between the product and consumer experience. In this chapter, we will address how digital neuro-psychological technologies and methodologies allow researchers to access a more agile consumer-centric approach to product testing. We will explore how a perceptual assessment can help product developers and consumer researchers better understand the consumer experience by introducing diverse types of technologies that can be used in sensory research. Finally, we will provide case studies demonstrating the utility of a digital-perceptual-emotional consumer-centric approach to product testing.

11.2 Digital neuro-behavioural methods

Neuro-behavioural approaches in market and sensory consumer research are becoming increasingly important not only in designing consumer products, but also in understanding how our bodies respond emotionally and physically to product experiences (from product use to marketing communications). Our brains dictate what we like and how we behave. As we interact with the world around us, we take in information nonconsciously (through our senses — taste, smell, sight, sound and touch). As we find meaning and importance in these inputs, we become consciously aware of them, deciding how we will react both behaviourally and emotionally (Fig. 11.1). One of the reasons for the success of added neuro-behavioural

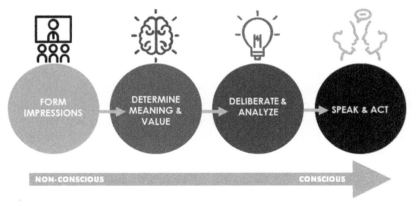

FIGURE 11.1   We interpret and interact with the word around us by accounting for stimuli in our environment via its representation in the brain. These internal impressions translate into physiological and behavioural responses. External stimuli trigger neural responses, forming impressions in the brain without conscious awareness. These neural impressions are then given meaning and value as they become further analysed and our conscious self becomes aware, forming decisions and behaviours. *Source: From Niedziela, M. M., Ambroze, K., Bajec, M., & Gutkowski, A. (2022). What the health? Sensory cues in wellness products–perceptions to reality. Science Talks, 2.*

insight is that it is widely recognised that measuring liking alone fails to predict product performance in the market (King et al., 2010; Thomson & Crocker, 2015; Thomson, 2007). Adding an applied neuroscience approach and behavioural science research to better predict consumer behaviour has added a new and valuable dimension to consumer and market research where, traditionally, researchers have relied on self-reported reactions.

Researchers interested in using neuro-tools are often looking to uncover implicit or nonconscious emotional reactions from consumers. Given the complex nature of emotion, finding a comprehensive methodology to measure this phenomenon is challenging. Although the literature lacks a consensus on a singular theory or definition of emotion, multiple components of these theories have widespread acceptance. For example, mood or emotional changes in physiological arousal, motivation, expressive motor behaviours, action tendencies and subjective feelings are well documented (Scherer, 2005). Yet, the information collected from these tools, especially when used alone as a singular measurement, is limited and can only emphasise specific aspects of the overall experience which result in an emotion or mood change.

11.2.1 Physiological measures

The more popular and recognisable neuro-behavioural approaches, frequently referred to as neuroscience or consumer neuroscience tools, measure direct physiological changes from the central and peripheral nervous systems. The consumer neuroscience toolbox has many great options for exploring the consumer experience from measuring skin temperature (SKT) and heart rate (HR) to more advanced technologies such as EEG (electroencephalogram) and fMRI (functional magnetic resonance imaging) (Agarwal, 2015; Genco et al., 2013). However, it is important to recognise that nonconscious activity is not easy to measure. The human brain is far too complicated to reduce to a simple algorithm read by one device to deliver a straightforward answer or metric. This is not to say that physiological measures cannot determine something about consumers' reactions and decision processes, but that it is not mind reading. Physiological measurements provide nuanced and complicated results that can be heavily influenced by research design and other external factors.

Each physiological tool within consumer neuroscience has its strengths and weaknesses, along with situations where certain measures are best applied and times where the same measures are not appropriate. Further, not all these tools can or should be used digitally. Table 11.1 is an overview of the most common tools using physiological measurement currently in practice for consumer research. While this is not an exhaustive list by any means, it gives an idea of the technologies currently (as of this publication) available and most popular. Notably, neurotechnology does not imply digitalisation, though digitalisation is often heavily involved in the data analysis of neurotech output with, for example algorithms for identifying key psychological states.

11.2.1.1 Peripheral measures

The peripheral nervous system (PNS) measurement tools (GSR (galvanic skin response), HR/HRV (heart rate/heart rate variability), facial electromyography (fEMG)) are often called the 'gold standard' of psychophysiological measurements (Fig. 11.2).

TABLE 11.1 An overview of neurological tools utilised to measure emotions discussed in this chapter with methodological citations.

	Measure	Psychophysiological states	References
Peripheral nervous system	Galvanic skin response (GSR)	Physiological arousal Engagement Intensity Excitement	Boucsein (2012), Dawson et al. (2007), Gouizi et al. (2011)
	Heart rate/heart rate variability (HR/HRV)	Motivation Emotional arousal Attention Physical activity Mental activity	de Geus et al. (2019), Poels and Dewitte (2006), Quintana and Heathers (2014), Basilio Vescio et al. (2018), Williams et al. (2015)
	Facial electromyography (fEMG)	Positive/negative emotional valence	Larsen et al. (2003), Magnée et al. (2007)
Central nervous system	Electroencephalogram (EEG)	Motivation Emotional valence Affective states Cognitive processes Long-term recall Attention	Beres (2017), Biasiucci et al. (2019), Britton et al. (2016), Liu et al. (2011), Macy (2015), Nunez and Srinivasan (2006)
	Functional near-infrared spectroscopy (fNIRS)	Affective states Cognitive processes	Balconi et al. (2015), Gruber et al. (2020), Jackson and Kennedy (2013), Villringer et al. (1993)
	Functional magnetic resonance imaging (fMRI)	Affective states Cognitive processes	Gross (2008), Singleton (2009)

From HCD Research. (2022). HCD Research. Retrieved April 28, 2023, from https://www.hcdi.net/.

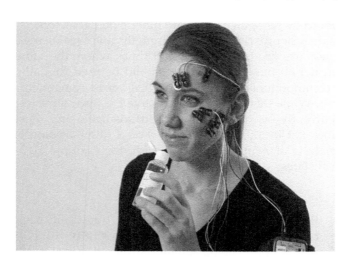

FIGURE 11.2 Facial electromyography (fEMG), a physiological measure detecting muscle activity as emotional valence. Electrodes are placed on two major muscle groups in the face, the corrugator supercilii group which is associated with frowning and the zygomaticus major muscle group which is associated with smiling (Larsen et al., 2003). *Source: From: Niedziela, M. M., Ambroze, K., Bajec, M., & Gutkowski, A. (2022). What the health? Sensory cues in wellness products—perceptions to reality. Science Talks, 2.*

These tools are considered gold standard due to their simplicity and direct correlation to what they measure (Genco et al., 2013; Kreibig, 2010). For example, increases in GSR are directly and positively correlated to increases in arousal (Dzedzickis et al., 2020; see Table 11.1). The technology for measuring these autonomic responses is also relatively simple and inexpensive, ranging from individual electrodes placed on the skin to wearable smart watches and even webcam-based measurements. While these tools come from clinical measurement, advances in metrics and wearable technologies continue to develop for these tools, making their use and data interpretation more efficient and applicable to consumer research.

11.2.1.2 *Central measures*

Central nervous system (CNS) measurement tools (EEG, fNIRS (functional near-infrared spectroscopy), and fMRI) are more consistent with what people think stereotypical neuroscience research looks like, directly measuring brain activity (De Araujo et al., 2003; Rolls & Baylis, 1994; Small & Prescott, 2005). These methods all monitor brain activity, though through different approaches. Functional MRI detects changes in blood flow by measuring magnetisation, while EEG measures changes in electrical activity across the scalp (Dzedzickis et al., 2020; Singleton, 2009). However, use of these tools in consumer neuroscience has not been without problems. Its use in interpreting consumer behaviour is often plagued with improper research design (Bell, 2015; Ferro, 2013; Harrell, 2019; Varan et al., 2015). Further, fMRI studies are notoriously expensive and difficult to perform in the confines of consumer research (limited mobility, accessibility, etc.). For context, a single MRI scan ranges from $400 to $10,500, according to the publication Imaging Technology News (2021). The nature of this equipment is very academic and clinical, requiring expertise to execute properly. However, advances in EEG technology, like development of user-friendly software and affordable electrodes, have made entrance into using these tools for consumer research much easier, as with ready-to-use platforms such as iMotions (https://imotions.com/) and consumer-grade, minimal electrode EEG sets such as Muse (https://choosemuse.com/), OpenBCI (https://openbci.com/), NeuroSky (http://neurosky.com/) or Emotiv (https://www.emotiv.com/) headsets (Kulke et al., 2020). Even with these user-friendly platforms, the lack of literature providing guidance regarding proper research design, instrument usage and data interpretation, and analysis about these tools for practitioners and researchers can lead to disappointing results.

11.2.1.3 *Wearables*

The methods to create and quantify physiological data have rapidly expanded into wearable technology (consider Apple Watch (https://www.apple.com/shop/buy-watch/apple-watch), FitBit (https://www.fitbit.com/global/us/home), Empatica (https://www.empatica.com/)). Wearable technology is one of the major advancements within consumer science to take measurements typically conducted in CLT centres, and brings it into intimate settings, like the home. This technology, if used correctly, gives researchers remote access to consumer lives, promoting an analysis of more real-life thoughts, feelings and experiences. Wearables integrate intelligent computers into accessories, like clothing, jewellery or other accessories. The wearable designs often include computing devices or sensors to gauge several types of health metrics (Cooper, 2022). If implemented and executed appropriately, wearable devices can collect and store in-the-moment data of consumer reactions, revealing its utility by promoting a contextual understanding of an experience ((Ambroze, 2022; Ferreira et al., 2021); see Case Study — Using Wearables to Measure Physiological Response).

Case Study

Using Wearables to Measure Physiological Response

Evaluating a naturalistic experience makes it easier to gain a comprehensive understanding of the consumer lifestyle because the behaviours and attitudes generated towards the product give authentic results. Wearable physiological measurement technology is one of the major advancements within consumer science to take measurements typically conducted in CLT centres, and brings it into intimate settings, like the home. This technology, if used correctly, may provide remote access to consumer naturalistic, physiological responses, promoting an analysis of more real-life thoughts, feelings and experiences.

The Challenge:

The challenge of this case study was to find the best way to measure in-the-moment, physiological changes of smokers to help identify potential cues or precursors of nicotine craving in a naturalistic, at-home environment.

The Solution:

Fully understanding a sensory experience is extremely complicated and can be heavily influenced by environmental factors. Sensory and market research, often explored in isolated, controlled research settings, can lack real world context. To explore psychophysiological nicotine craving in real-world settings, research-grade wearable devices (Empatica E4 wristbands, https://www.empatica.com/) were used to detect HR, GSR and SKT changes over time.

The Results:

Data revealed SKT and HR increase before and peak during smoking followed by an immediate decrease in SKT post smoking (Fig. 11.3). GSR showed increases in arousal followed immediately with sharp decreases before cigarette use, increasing again for 30 seconds after the smoking session started. Exploring these variations in experiences provides a deeper understanding of how something like context influences behaviour, both physically and mentally. Studying emotional spaces like stress and cravings is complex, but to fully encapsulate an experience studying both the physiology and the emotional component in context reveals how tightly these concepts are linked and influence each other.

Physiological changes due to nicotine presence and cravings were monitored at home with output showing that HR increased prior to smoking, followed by gradual decline with smoking. GSR arousal also increased and then immediately dipped before

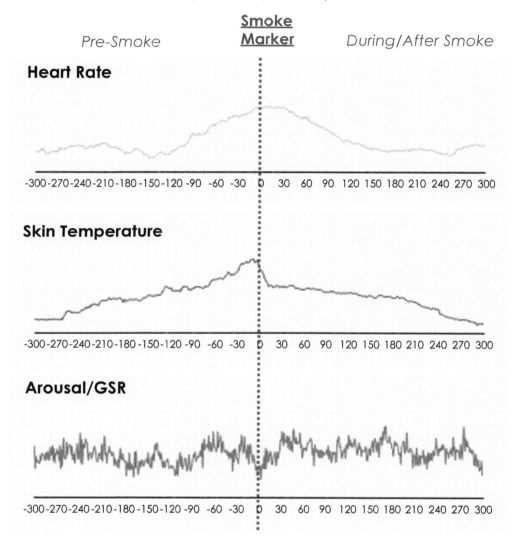

Cigarette Experience
(second-by-second)

Smoke Marker

Pre-Smoke **Smoke Marker** *During/After Smoke*

Heart Rate

-300 -270 -240 -210 -180 -150 -120 -90 -60 -30 0 30 60 90 120 150 180 210 240 270 300

Skin Temperature

-300 -270 -240 -210 -180 -150 -120 -90 -60 -30 0 30 60 90 120 150 180 210 240 270 300

Arousal/GSR

-300 -270 -240 -210 -180 -150 -120 -90 -60 -30 0 30 60 90 120 150 180 210 240 270 300

FIGURE 11.3 Wearable second-by-second data of heart rate (top), skin temperature (middle) and galvanic skin response (bottom) tracking physiological changes pre- and postsmoking collected via the Empatica E4 wristband from one individual participant. The initiation of smoking is annotated by the dotted red line. *Source: Unpublished from HCD Research. (2022). HCD Research. Retrieved April 28, 2023, from https://www.hcdi.net/.*

cigarette use and then increased again for 30 seconds after smoking. Evaluating the changes among the tested experiences provided a deeper understanding of the physiological experience of craving in a naturalistic setting.

11.2.2 Psychological tools

Tools from psychological sciences have added new dimensions to traditional consumer and sensory research, and are also considered a large part of the consumer neuroscience approach, particularly with the use of more digitised psychological tools. Though few studies have investigated the relationship between sensory and emotions deeply, instead relying on combining Quantitative Descriptive Analysis with basic emotional survey questions. In a seminal study of nine sensorially differentiated, unbranded, dark chocolates, Thomson et al. (2010) found that specific sensory characteristics could be associated with different emotional effects (e.g. cocoa was associated with powerful and energetic, bitter with confident, adventurous and masculine, and creamy and sweet with fun, comforting and easygoing). Many studies have found that psychological emotional measures may discriminate beyond traditional liking assessments (King et al., 2010; Spinelli et al., 2014) and recently also using implicit measurement (Mojet et al., 2015). For further discussion on implicit measurements, please see Chapter 13 on Added value of implicit measures in sensory and consumer science by René A. de Wijk and Lucas P.J.J. Noldus. Academic psychological literature can provide a plethora of potential psychological measurements that can be easily incorporated into consumer surveys to further explore and discriminate product experiences (Table 11.2).

These psychological methodologies can easily be digitised for mobile or smart technology applications to make them more accessible for both more naturalistic settings, whether for in-home studies, virtual reality environments, or other contexts outside of the typically CLT setting. Further they can be easily integrated into existing surveys,

TABLE 11.2 An overview of common psychological tools utilised to measure emotions discussed in this chapter with methodological citations.

Psychological methodology	Popular psychological assessments in consumer research
Implicit association	IAT (Mojet et al., 2015)
	EAST (Goodall, 2011)
	AMP (Payne et al., 2005)
Emotional profiling	PANAS (Watson et al., 1988)
	POMS (McNair et al., 1971)
	EmoSemio (Spinelli et al., 2014)
	EsSense Profile (King et al., 2010)
	GEOS (Chrea et al., 2009)
Visual self-report	SAM (Bradley & Lang, 1994)
	PrEmo (Desmet, 2003)
	VAS/LMS (Hayes et al., 2013; Noel & Dando, 2015)

AMP, Affect Misattribution Procedure; *EAST*, Extrinsic Affective Simon Task; *GEOS*, Geneva Emotion and Odor Scale; *IAT*, Implicit Association Testing; *PANAS*, Positive and Negative Affect Schedule; *POMS*, Profile of Mood States; *SAM*, Self Assessment Manikin; *VAS*, Visual Analogue Scale; *LMS*, Labeled Magnitude Scale.
From Niedziela, M. M., Ambroze, K., Bajec, M., & Gutkowski, A. (2022). What the health? Sensory cues in wellness products–perceptions to reality. Science Talks, 2.

seamlessly, helping researchers to investigate consumer perceptions and experiences beyond traditional self-report.

Case Study

Consumer Perceptions via Implicit Go-No Go Association Task (GNAT)

Leveraging the consumer expectations can be challenging. Promises made in marketing campaigns need to align with product experiences to ensure consumer satisfaction and brand harmony (the alignment among sensory, marketing and development teams). Evaluating target words and concepts associated with products or product experience exposes consumer perception and reveals conflict between marketing mix and product specifications. Understanding the product experience provides insights to help develop and modify products or choose prototypes to better meet consumer expectations.

The Challenge:

Assessing consumer perceptions of fragrance experiences to better discriminate among prototypes where traditional self-report methodologies had failed. Emotional and descriptive attributes from marketing directives were often challenging for

consumers to articulate and identify. A company was interested in adjusting the fragrance of one of their vacation-inspired candles, specifically how well prototype fragrances performed against a benchmark on key marketing directives.

The Solution:

Implicit reaction time (IRT) measurements, specifically utilising the GNAT approach, were used to assess consumer perceptions and associations of key marketing concepts and words. Response latencies were compared among fragrance protypes. Winning prototypes needed to perform for better than or parity with a benchmark fragrance at achieving strong associations to key marketing descriptives.

The Results:

The results revealed mental associations across fragrances, suggesting #734 shared the same highly associated descriptors as the benchmark (*fruity, exciting* and *fun*) as well as an additional key marketing association of *fresh* (Fig. 11.4).

The information from this output provided an understanding of consumer perceptions of the tested fragrances in addition to a traditional consumer survey. Having this insight helped the client further differentiate among the three fragrances and choose #734 as the winner. The ability to easily assess consumer perceptions of difficult concepts is key to achieving brand-product harmony.

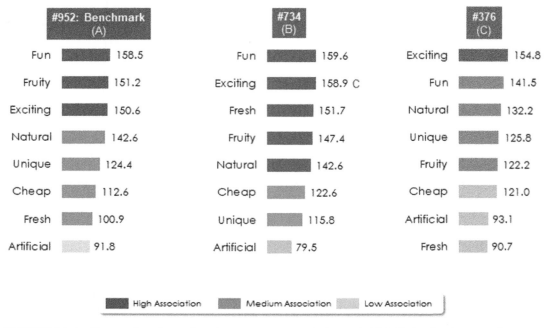

FIGURE 11.4 The reaction time values from go-no go association task (GNAT) for the three test fragrances are broken down into high, medium and low associations in descending order. The uppercase letters indicate significant differences at 95%. Statistical differences (indicated with red letters) are interpreted from the timed reaction scores, exposing the perceptions driven by the different fragrances. *Source: Unpublished from HCD Research. (2022). HCD Research. Retrieved April 28, 2023, from https://www.hcdi.net/.*

Case Study

Timed Reaction as a More Efficient Screening Tool

Descriptive analysis is a sensory methodology that provides quantitative word descriptions of products based on perceptions verbalised by a group of qualified subjects, but can be a long and costly process (Stone & Sidel, 2003). It is the most sophisticated source of product information available to researchers for providing a complete and quantitative word description of a product's sensory characteristics to help in making business and product market decisions.

The Challenge:

The need for more agile descriptive analysis methodologies has encouraged companies to explore, expand and compare approaches to build a more efficient way to uncover key differences among high volume samples. HCD Research Inc. (https://www.hcdi.net/) teamed with Sensory Spectrum, Inc. (https://www.sensoryspectrum.com/) to determine if methodologies from consumer neuroscience may help to create a new, agile approach.

The Solution:

Implicit measurement, such as reaction-time testing of trained panellists, was used

---

(cont'd)

---

to address the need for more cost-effective and less time-consuming descriptive panel evaluations for rapid, quick-check results. Orange juice samples were evaluated with both a descriptive analysis (DA) and the Go/No-Go Implicit Test, and compared for similarities in conclusions and recommendations.

The Results:

When assessing DA compared to the reaction time (RT) generated maps (Fig. 11.5), the results show the methods were aligned for larger product differences. For example, the DA and RT perceptual maps showed Juicy Juice as an outlier among orange juice samples with a profile of distilled orange notes and sweet aromatics.

This case study demonstrates how implicit RT testing can be utilised as a screening tool to capture automatic responses in cognitive tasks. Although the RT methodology is not as detailed or robust as full descriptive profiling, RT may be a valuable alternative screening tool providing greater throughput and less time-consuming research compared to tradi-tional screening methods. Implicit RT creates greater throughput and is less time-

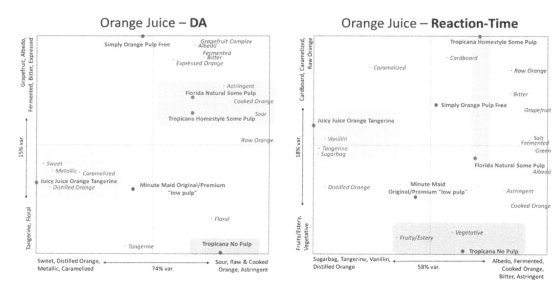

FIGURE 11.5 The descriptive analysis (DA) perceptual map (left) and the reaction-time (RT) perceptual map (right) reveal the brands sensory associations among the samples. The variability in the DA data can be explained by two dimensions (89% of the variability). Variability in the RT data can be explained by two dimensions (90% of the variability). Only two out of the three dimensions for the RT data are shown here. *Source: From Gibbons, S. (2021). A more efficient panel method: Rapid descriptive analysis with reaction-timed response. In Pangborn Sensory Science Symposium.*

consuming compared to traditional screening methods. Digitising panel work by using implicit RT testing as a screening methodology has advantages that warrant consideration when making decisions about the product experience.

11.2.3 Behavioural tools

Behavioural measurement, or observational research, refers to the study of nonexperimental situations in which behaviour is observed and recorded. People are studied as they perform a task, allowing researchers to see what consumers really do when confronted with various choices or situations. Several types of observational research are described below, and each have their strengths and weaknesses, but in general, these approaches can be less hypothetical than other methods as they capture what people are doing as opposed to what they say they will do or have done (Table 11.3). Behavioural coding, facial coding and eye tracking, all standard behavioural methodologies in consumer neuroscience, continue to be improved upon with digital and machine learning advances in their technology helping researchers make better predictions with data.

11.2.4 Digital technologies

There is currently a digital revolution where digital technologies, from smart devices to the internet, are quickly becoming the backbone of daily life (Poslad, 2009). This growth is also beginning to expand into research applications with investigators becoming more interested in finding ways to access consumers in their natural environments (homes) to get more realistic, observational measures of their interactions with products. Neuro-behaviourally, digital technologies provide an opportunity to explore drivers of behaviours and emotions in a more naturalistic environment, such as the consumers' home or in-store.

11.2.4.1 Smart devices

Smart-home and other internet-connected devices have become quite convenient and accessible, collecting vast amounts of information on consumers and their homes. While designed to work predominantly while interacting with digital assistants, smart devices

TABLE 11.3 An overview of behavioural tools, though not exhaustive, utilised to measure emotions discussed in this chapter with popular methodological resources.

Behavioural methodologies	Behavioural tool resources
Behavioural coding	Noldus' Observer XT (https://www.noldus.com/)
	Interact (https://www.mangold-international.com/)
	BORIS (http://www.boris.unito.it)
Facial coding	FaceReader (https://www.noldus.com/)
	Realeyes (https://www.realeyesit.com/)
	Affectiva (https://www.affectiva.com/)
Eye tracking	Tobii (https://www.tobiipro.com/)
	ViewPoint Systems (https://viewpointsystem.com/)
	Pupil Labs (https://pupil-labs.com/)

From HCD Research. (2022). HCD Research. Retrieved April 28, 2023, from https://www.hcdi.net/.

can collect data both while in and out of use, introducing both privacy and ethical concerns (Whittaker, 2019). Smart speakers are smart devices with loudspeaker and voice command capabilities, which include an integrated virtual assistant for interactive actions and hands-free activation (Van Gils, 2020). Aigora (https://www.aigora.ai/) has introduced the use of smart speaker surveys for consumers to respond to survey questions hands-free, in the moment, and in their own homes (Ennis, 2020; Ennis et al., 2020). For further discussion on smart speaker surveys, see Chapter 17 entitled, "Voice-activated technology in sensory and consumer research: a new frontier" by Tian Yu, Janavi Kumar, Hamza Diaz, Natalie Stoer and John Ennis. For example, a consumer can answer questions regarding the fragrance or the packaging of a laundry detergent, while simultaneously using the product as they normally would. Smart labels, or smart tags, on the other hand have also been integrated into in-home research to provide detailed information consumer behaviour and interaction with products (Banterle et al., 2012). Accessing consumers in their personal environment through these smart devices allows measurement and access to different emotions and behaviours leading to understanding of products within a more naturalistic context.

11.2.4.2 *Immersive technologies*

Virtual reality (VR) and augmented reality (AR) are two digital technologies that are changing screen-use, creating new and exciting interactive experiences. VR uses a headset to provide a 360-degree computer-generated simulated experience to interact with, while AR is a form of mixed reality that adds new components into a space of preexisting objects in the real world (Bonetti et al., 2017; Krevelen & Poelman, 2010). Further, digital or electronic activation of sensory modalities such as virtual tongues or noses has been introduced (Kerruish, 2019; Ranasinghe et al., 2017). Digital realities have been more recently of interest for creating or evoking context within sensory and consumer research as a better setting to measure consumer acceptability than lab or CLT-based environments (Wang et al., 2021). Since the AR survey can be completed anywhere with a mobile device, there is an opportunity to help consumers evaluate products or experiences either at home or in a store (a prospect recently being explored by HCD Research, https://www.hcdi.net/; see Case Study: Utilising AR to Understand Consumer Experiences). AR surveys push the boundaries between naturalistic and controlled research, revealing top-of-mind answers from consumers through the modern twist on traditional survey methods.

Case Study

Utilising AR to Understand Consumer Experiences

The shopping environment has a major influence on consumer behaviour. AR surveys provide in-the-moment responses to give a more authentic understanding of drivers of decision-making for context-specific environments (HCD Research, 2021). AR surveys merge physical products with digital landscapes, so consumers can evaluate survey questions about products and experiences easily through a smart phone or tablet while still being able to

---

(cont'd)

interact with the contextual environment. To gain a realistic understanding and better predict consumer perception and behaviour, it is crucial to investigate what consumers in context without disrupting that experience.

The Challenge:

Evaluating responses in the naturalistic space where emotions are evoked, or decisions are made, has been difficult for researchers to do without interrupting consumer behaviour. Virtual and laboratory environments do not evoke the same perceptions, emotions or responses to those made in more naturalistic settings (Sinesio et al., 2019).

The Solution:

To compare how perceptions vary between store environments, HCD Research, Inc. created an AR-based survey to capture the in-the-moment responses to the shopping experiences in two retail stores: Walmart and Target. The digital survey was superimposed into participants viewing through their cell phone camera so that they could view the environment and products while rating their experiences. Participants rated feelings towards the stores pre- and postshopping as well as towards three consumer products within those stores (Bounty Paper Towels, Noosa Yogurt and Burt's Bees Lip Balm) using the Self Assessment Manikin (SAM), a nonverbal pictorial assessment of affective states, and survey questions about the price, purchase intent, and overall shopping experience.

The Results:

The SAM scores gathered from the AR survey suggest that the shoppers had more negative emotions postshopping compared to preshopping, more positive experiences with Target than Walmart (Fig. 11.6). Data suggest that feelings products can be influenced by emotional reactions to the in-store shopping experience.

The shopping environment has a major influence on consumer behaviour and perception. AR surveys push the boundaries between naturalistic and controlled research, revealing top-of-mind answers from consumers through the modern twist on traditional survey methods. The simplicity of this approach serves as a nondisruptive shop-along, but can also be incorporated into at-home use tests. As more and more consumers experience the world through screens, having tools like the AR surveys embedded in these mediums can efficiently capture more authentic, real-time responses in a format that has already become an essential part of daily living.

11.3 Future directions and challenges in digitalising neuro-behavioural research

Technology advancements are fast-paced and impossible to keep up with. Buying the latest tech gadget today, within 6 months, will be outdated with a newer model. While it is important to be willing and able to change with the times by staying on top of the game with the latest tools and methodologies, these changes come with new nuisances and

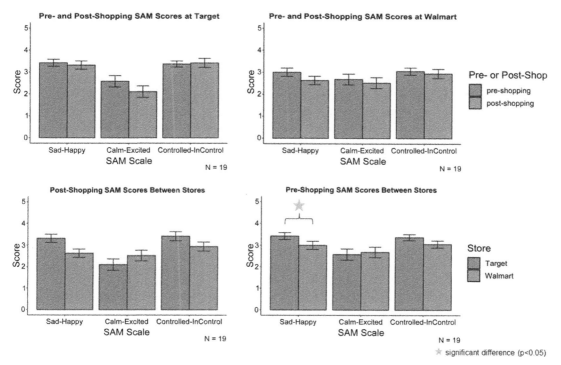

FIGURE 11.6 A two-sample t-test was performed to compare the means of the SAM scores of pre- and post-shopping experiences (top right, top left), as well as the scores from between stores (bottom right, bottom left) for Walmart and Target. The star represents a significant difference between scores ($P < .05$). *Source: From Niedziela, M. M., Ambroze, K., Bajec, M., & Gutkowski, A. (2022). What the health? Sensory cues in wellness products–perceptions to reality. Science Talks, 2.*

complexities requiring continuous learning and proficiency to truly make actionable progress. With more accessible tools (cheaper headsets, greater popularity, etc.), tech that was once very academic has become more mainstream in consumer research. These advancements, digitalisation and mobilisation of methodologies, while exciting, have also introduced new challenges.

While the interest in adding technologies, such as neuro-behavioural tools, has grown significantly in consumer research and specifically in sensory research, a challenge remains on how to make the academic approach more consumer research friendly. Even though there are many tools proposed for neuro-behavioural research and introduced in this chapter, the number of tools that can be practically used for consumer research is much lower. The technologies themselves can be quite challenging and require a certain level of expertise to run, analyse and interpret correctly. Efficiencies through digitalisation can be attractive, but may introduce experimental noise, bias in data and other confounding factors. Concerns regarding data privacy, security and safety will also be a major hurdle in neuro-behavioural research, rooted in distrust that companies will sell user data to other businesses for a profit. Addressing these shortcomings and anxieties surrounding brain-based technologies is necessary to successfully promote these technological advances in the future.

Most noninvasive neuro-behavioural methodologies require in person measurement, with some exceptions that can be executed online or via wearable technologies. There are advantages, disadvantages and challenges involved in integrating technology into research. Electrodes and monitors often must be placed precisely and directly onto a person or near enough a person to accurately capture changes in physiological or behavioural activity. While self-administration, relying on research participants to correctly apply technologies to themselves, is possible, it presents the potential for incorrect application (i.e. incorrect placement of equipment such as electrodes, straps or headcaps) and difficulty in troubleshooting problems with equipment. Stimulation devices do have safety concerns about the amount of current that can be put in a headband or overuse of a device, making it preferable to have a knowledgeable technician available and on-site for proper setup and any potential safety or technological issues (Mackenzie, 2021). Yet, even with perfect applications, getting recordings of high-resolution brain signals through several biological intervening layers is a challenge. Noise and crosstalk are inevitable when interpreting electrical signals in the brain, which is why other neuro-based approaches, like light, magnetism and ultrasounds, are being explored (Anonymous, 2018). It should also be noted that all technologies exist on a spectrum of quality with some struggling to meet the rigour needed for research studies because of intentional design decision catering to a consumer audience.

Remote or mobile enabled technologies present an additional hurdle: selection bias. Selection bias stems from an absence of comparability between groups being studied, where proper randomisation cannot be achieved and therefore fails to ensure that a representative sample of the population is obtained (Kempf-Leonard, 2004). This bias can be introduced in several ways, including excluding participants without home internet, mobile or computer access (Greenacre, 2016). Further, a sharp increase in participant dropout occurs when asked for webcam access, or other sorts of invasions to personal space, particularly when trying to obtain eye-tracking, facial coding or behavioural user experience data from a participant's personal computer, where participants may feel their privacy is at risk.

Advances using CNS measurement technology have been sensationalised, but with that uptick in interest, with several companies working to create technologies for tracking cognitive performance, monitor emotions and control both virtual and physical objects via machine learning of trained mental commands with brain—computer interface (BCI), brain—machine interface (BMI) and direct neural interface (DNI) technologies (Teunisse et al., 2019; Wexler & Reiner, 2019). EMOTIV (https://www.emotiv.com/) is a bioinformatics company advancing the understanding of the human brain using affordable, consumer-grade EEG technology. Muse offers wearable brain-sensing headbands to moderate and provide feedback as a wellness and relaxation tool (https://choosemuse.com/). One of these most popularised companies founded by Elon Musk is Neuralink (https://neuralink.com/), which is developing micron-scale threads inserted into areas of the brain to control movement or feedback sensory information. Each of these technologies envisions its application to be the pioneer of future innovations in this space but are riddled with foundational scepticism. Direct-to-consumer and adapted-for-general-public neurotechnologies may come at an ethical cost; between unsubstantiated claims, questionable technology, and core ethical issues of transparency, rights, and responsibility, consumer neurotechnologies, while exciting, should be approached with caution (McCall et al., 2019). However, it is important to note that interest and investment in consumer-grade neuro-behavioural technology, overall, may be a positive step towards creating more accurate algorithms and more affordable technologies.

With the continued growth of neuro-behavioural technology, it is important to build foundational regulation surrounding brain and cognition privacy surrounding the data these tools collect. Incorporating brains, behaviours and machinery into consumer use or consumer research invites the misuse or exploration of data if data privacy is presumed. And while consumers may find it exciting to watch brainwaves, for-profit enterprises may overextend the technologies' ability for a flashy headline or inappropriate use of information. Neural and cognitive data of users should be understood at a foundational level to empower those interested in partaking in these types of devices. Consent must be fully informed to be effective because limited awareness can result in the sharing data unethically. Goering et al. (2021) suggest developing a framework to establish 'Neurorights' and an international commission for the responsible development of neuro-behavioural tools to address these types of concerns. These concerns for data privacy in recent years has increased in both the United States and internationally, which can disincentivise consumers from being interested in participating in using any neuro-behavioural technology.

Many neuro-behavioural tools utilise some form or combination of algorithms, metrics and other digital efficiencies, such as with artificial intelligence (AI) or machine learning (ML), used for data cleaning, normalisations, analytics and modelling, to increase the utility of neuro-applications in consumer research. For example, recent reviews of ML algorithms used with EEG and other neuro-physiological data report numerous easy analytic techniques for preprocessing, extracting and classifying various brain activities (Azlan & Low, 2014; Guozhen et al., 2016; Kim et al., 2013; Xie & Oniga, 2020; Yu et al., 2019). Human perceptual artificial intelligence, however, is incredibly complex and can present a multitude of challenges resulting in confounding errors (Chaney et al., 2018; Grimes & Schulz, 2002; Macpherson et al., 2021; Thompson, 2021).

Market, consumer and sensory research are experiencing a paradigm change in the way they study consumers. Technology has already had a significant impact on sensory and consumer research, creating a new generation of faster and easier-to-use tools that help brands discover what consumers think and experience. The need for faster and more robust insights has been growing due to continued pressure on budgets and timelines; and has only been accelerated by the coronavirus pandemic. Used alone, each of these tools offers a powerful, tech-driven solution to create better answers to familiar questions but integrated together with traditional sensory and consumer research methodologies, they combine to create an advantaged capability. Proper utilisation of the tools and methods discussed in this chapter helps brands gain deeper understanding of natural, in-the-moment behaviours and responses to identify how they can better serve their consumers' needs.

These technologies have become an essential part of the global consumer insights machine, enabling brands to keep up with accelerating change in consumer behaviour. Tech-driven sensory and consumer research may already seem like an extreme paradigm shift, but this is just the beginning. The application of Big Data and AI powered tools will continue to mature and integrate all sources of data to accelerate and spin-off into multiple use-cases and applications. The opportunities are many, and the future is full of promise.

References

Agarwal, S. (2015). Introduction to neuromarketing and consumer neuroscience [review of introduction to neuro-marketing and consumer neuroscience]. *Journal of Consumer Marketing*, 32(4), 302–303. Available from https://doi.org/10.1108/jcm-08-2014-1118.

Ambroze, K. (2022). Wearable Tech - Fit for MRX? Daily Research News Online No. 32684. Retrieved January 31, 2022, from https://www.mrweb.com/drno/news32684.htm?fbclid=IwAR2xu8f6x5-NvvIt0xwZ1Ay5nPlEyyO8h1VImid0Q8sHQd68sxi5NWfFVM0.

Anonymous. (2018). The next frontier: When thoughts control machines. *The Economist*, 426(9073).

Azlan, W. A., & Low, Y. F. (2014). Feature extraction of electroencephalogram (EEG) signal—A review. *IECBES 2014, Conference Proceedings—2014 IEEE Conference on Biomedical Engineering and Sciences: Miri, Where Engineering in Medicine and Biology and Humanity Meet. Feature extraction of electroencephalogram (EEG) signal—A review*. Institute of Electrical and Electronics Engineers Inc. Malaysia. (pp. 801–806), 9781479940844. Available from https://doi.org/10.1109/IECBES.2014.7047620.

Balconi, M., Grippa, E., & Vanutelli, M. E. (2015). Resting lateralized activity predicts the cortical response and appraisal of emotions: An fNIRS study. *Social Cognitive and Affective Neuroscience*, 10(12), 1607–1614. Available from https://doi.org/10.1093/scan/nsv041.

Banterle, A., Cavaliere, A., & Ricci, E. C. (2012). Food labelled information: An empirical analysis of consumer preferences. *International Journal on Food System Dynamics*, 3(2), 156–170. Available from http://www.centma-press.org.10.18461/ijfsd.v3i2.325.

Bell, V. (2015, June 28). Marketing has discovered neuroscience, but the results are more glitter than gold. Retrieved September 09, 2020, from https://www.theguardian.com/science/2015/jun/28/vaughan[1]bell-neuroscience-marketing-advertising.

Beres, A. M. (2017). Time is of the essence: A review of electroencephalography (EEG) and event-related brain potentials (ERPs) in language research. *Applied Psychophysiology Biofeedback*, 42(4), 247–255. Available from http://www.wkap.nl/journalhome.htm/1090-0586.10.1007/s10484-017-9371-3.

Biasiucci, A., Franceschiello, B., & Murray, M. M. (2019). Electroencephalography. *Current Biology*, 29(3), R80–R85.

Bonetti, F., Warnaby, G., & Quinn, L. (2017). Augmented reality and virtual reality in physical and online retailing: A review, synthesis and research agenda. *Augmented Reality and Virtual Reality*, 119–132. Available from https://doi.org/10.1007/978-3-319-64027-3_9.

Boucsein, W. (2012). *Electrodermal activity*. Springer Science & Business Media.

Bradley, M. M., & Lang, P. J. (1994). Measuring emotion: The self-assessment manikin and the semantic differential. *Journal of Behavior Therapy and Experimental Psychiatry*, 25(1), 49–59. Available from https://doi.org/10.1016/0005-7916(94)90063-9.

Britton, J. W., Frey, L. C., Hopp, J. L., Korb, P., Koubeissi, M. Z., Lievens, W. E., & St, E. L. (2016). Electroencephalography (EEG): An introductory text and atlas of normal and abnormal findings in adults, children, and infants. *American Epilepsy Society*.

Chandon, P., Morwitz, V. G., & Reinartz, W. J. (2005). Do intentions really predict behavior? Self-generated validity effects in survey research. *Journal of Marketing*, 69(2), 1–14. Available from https://doi.org/10.1509/jmkg.69.2.1.60755.

Chaney, A.J., Stewart, B.M., & Engelhardt, B.E. (2018). How algorithmic confounding in recommendation systems increases homogeneity and decreases utility. Unpublished content. Proceedings of the 12th ACM Conference on Recommender Systems, pp. 224–232.

Chrea, C., Grandjean, D., Delplanque, S., Cayeux, I., Le Calvé, B., Aymard, L., Velazco, M. I., Sander, D., & Scherer, K. R. (2009). Mapping the semantic space for the subjective experience of emotional responses to odors. *Chemical Senses*, 34(1), 49–62. Available from https://doi.org/10.1093/chemse/bjn052.

Cooper, L. (2022). How health and fitness trackers are about to get a lot more granular; many people have become accustomed to devices quantifying their steps or heart rate. That's just the beginning. *Wall Street Journal*.

Dawson, M. E., Schell, A. M., & Filion, D. L. (2007). *The electrodermal system*. In J. T. Cacioppo, L. G. Tassinary, & G. G. Berntson (Eds.), *Handbook of psychophysiology* (pp. 159–181). Cambridge University Press. Available from https://doi.org/10.1017/CBO9780511546396.007.

de Geus, E. J. C., Gianaros, P. J., Brindle, R. C., Jennings, J. R., & Berntson, G. G. (2019). Should heart rate variability be "corrected" for heart rate? Biological, quantitative, and interpretive considerations. *Psychophysiology, 56*(2). Available from http://onlinelibrary.wiley.com/journal/10.1111/(ISSN)1469-8986.10.1111/psyp.13287.

De Araujo, I. E. T., Rolls, E. T., Kringelbach, M. L., McGlone, F., & Phillips, N. (2003). Taste-olfactory convergence, and the representation of the pleasantness of flavour, in the human brain. *European Journal of Neuroscience, 18*(7), 2059–2068. Available from https://doi.org/10.1046/j.1460-9568.2003.02915.x.

Desmet, P. (2003). *Measuring emotion: Development and application of an instrument to measure emotional responses to products* (pp. 111–123). Springer Nature. Available from https://doi.org/10.1007/1-4020-2967-5_12.

Dzedzickis, A., Kaklauskas, A., & Bucinskas, V. (2020). Human emotion recognition: Review of sensors and methods. *Sensors, 20*(3), 592. Available from https://doi.org/10.3390/s20030592.

Ennis, J. (Host). (2020, October 12). Janavi Kumar - A Learning Mindset (No. 56) [Audio Podcast episode]. In *AigoraCast*. Aigora. Available from https://www.aigora.ai/post/janavi-kumar-a-learning-mindset.

Ennis, J., Kumar, J., Niedziela, M., & Pierce-Feldmeye, A. (2020). SSP Making the Most of Data Diversity in Sensory Science.

Ferreira, J. J., Fernandes, C. I., Rammal, H. G., & Veiga, P. M. (2021). Wearable technology and consumer interaction: A systematic review and research agenda. *Computers in Human Behavior, 118*. Available from https://www.journals.elsevier.com/computers-in-human-behavior.10.1016/j.chb.2021.106710.

Ferro, S. (2013). Why neuromarketing is a neuroscam. Retrieved September 09, 2020, from https://www.popsci.com/science/article/2013-07/why-neuromarketing-neuroscam/.

Genco, S. J., Pohlmann, A. P., & Steidl, P. (2013). *Neuromarketing for dummies*. John Wiley & Sons.

Goering, S., Klein, E., Specker Sullivan, L., Wexler, A., Agüeray Arcas, B., Bi, G., Carmena, J. M., Fins, J. J., Friesen, P., Gallant, J., Huggins, J. E., Kellmeyer, P., Marblestone, A., Mitchell, C., Parens, E., Pham, M., Rubel, A., Sadato, N., Teicher, M., ... Yuste, R. (2021). Recommendations for responsible development and application of neurotechnologies. *Neuroethics, 14*(3), 365–386. Available from http://www.springer.com/philosophy/ethics/journal/12152.10.1007/s12152-021-09468-6.

Goodall, C. E. (2011). An overview of implicit measures of attitudes: Methods, mechanisms, strengths, and limitations. *Communication Methods and Measures, 5*(3), 203–222. Available from https://doi.org/10.1080/19312458.2011.596992.

Gouizi, K., Bereksi Reguig, F., & Maaoui, C. (2011). Emotion recognition from physiological signals. *Journal of Medical Engineering and Technology, 35*(6-7), 300–307. Available from https://doi.org/10.3109/03091902.2011.601784.

Graves, P. (2010). *Consumerology: The market research myth, the truth about consumers and the psychology of shopping*. Boston: Nicholas Brealey.

Greenacre, Z. A. (2016). The importance of selection bias in internet surveys. *Open Journal of Statistics, 06*(03), 397–404. Available from https://doi.org/10.4236/ojs.2016.63035.

Grimes, D. A., & Schulz, K. F. (2002). Bias and causal associations in observational research. *Lancet, 359*(9302), 248–252. Available from http://www.journals.elsevier.com/the-lancet/0.10.1016/S0140-6736(02)07451-2.

Gross, J. J. (2008). Emotion regulation. *Handbook of emotions* (3).

Gruber, T., Debracque, C., Ceravolo, L., Igloi, K., Marin Bosch, B., Frühholz, S., & Grandjean, D. (2020). Human discrimination and categorization of emotions in voices: A functional near-infrared spectroscopy (fNIRS) study. *Frontiers in Neuroscience, 14*. Available from https://www.frontiersin.org/journals/neuroscience#0.10.3389/fnins.2020.00570.

Guozhen, Z., Jinjing, S., Yan, G., Yongjin, L., Lin, Y., & Tao, W. (2016). Advances in emotion recognition based on physiological big data. *Journal of Computer Research and Development, 53*(1), 80.

HCD Research. (2021, June 24). *Augmented reality market research: Connect with consumers in the durable goods and CPG markets*. HCD Research Inc. Retrieved January 20, 2022, from https://www.hcdi.net/post/augmented-reality-market-research-connect-with-consumers-in-the-durable-goods-and-cpg-markets.

Harrell, E. (2019). Neuromarketing: What you need to know. *Harvard Business Review, 97*(4), 64–70.

Hayes, J. E., Allen, A. L., & Bennett, S. M. (2013). Direct comparison of the generalized visual analog scale (gVAS) and general labeled magnitude scale (gLMS). *Food Quality and Preference, 28*(1), 36–44. Available from https://doi.org/10.1016/j.foodqual.2012.07.012.

Jackson, P. A., & Kennedy, D. O. (2013). The application of near infrared spectroscopy in nutritional intervention studies. *Frontiers in Human Neuroscience, 7*, 473. Available from https://doi.org/10.3389/fnhum.2013.00473.

Kempf-Leonard, K. (2004). *Encyclopedia of social measurement* (pp. 1–3000). United States: Elsevier Inc. Available from http://www.sciencedirect.com/science/book/9780123693983.

Kerruish, E. (2019). Arranging sensations: Smell and taste in augmented and virtual reality. *The Senses and Society, 14*(1), 31–45. Available from https://doi.org/10.1080/17458927.2018.1556952.

Kim, M. K., Kim, M., Oh, E., & Kim, S. P. (2013). A review on the computational methods for emotional state estimation from the human EEG. *Computational and Mathematical Methods in Medicine, 2013*. Available from http://www.hindawi.com/journals/cmmm/0.10.1155/2013/573734.

King, S. C., Meiselman, H. L., & Carr, B. T. (2010). Measuring emotions associated with foods in consumer testing. *Food Quality and Preference, 21*(8), 1114–1116. Available from https://doi.org/10.1016/j.foodqual.2010.08.004.

Kreibig, S. D. (2010). Autonomic nervous system activity in emotion: A review. *Biological Psychology, 84*(3), 394–421. Available from https://doi.org/10.1016/j.biopsycho.2010.03.010.

Krevelen, D. W. F., & Poelman, R. (2010). A survey of augmented reality technologies, applications and limitations. *International Journal of Virtual Reality, 9*(2), 1–20.

Kulke, L., Feyerabend, D., & Schacht, A. (2020). A comparison of the affectiva iMotions facial expression analysis software with EMG for identifying facial expressions of emotion. *Frontiers in Psychology, 11*, 329. Available from http://www.frontiersin.org/Psychology.10.3389/fpsyg.2020.00329.

Larsen, J. T., Norris, C. J., & Cacioppo, J. T. (2003). Effects of positive and negative affect on electromyographic activity over zygomaticus major and corrugator supercilii United States. *Psychophysiology, 40*(5), 776–785. Available from https://doi.org/10.1111/1469-8986.00078.

Liu, Y., Sourina, O., & Nguyen, M.K. (2011). 2011/08 Lecture Notes in Computer Science (including subseries Lecture Notes in Artificial Intelligence and Lecture Notes in Bioinformatics). Real-time EEG-based emotion recognition and its applications. Singapore. Unpublished content. 6670. 10.1007/978-3-642-22336-5_13. 16113349. 256-277.

Mackenzie, R. (2021, August 31). *Privacy in the brain: The ethics of neurotechnology.* Neuroscience from Technology Networks. Retrieved February 9, 2022, from https://www.technologynetworks.com/neuroscience/articles/privacy-in-the-brain-the-ethics-of-neurotechnology-353075.

Macpherson, T., Churchland, A., Sejnowski, T., DiCarlo, J., Kamitani, Y., Takahashi, H., & Hikida, T. (2021). Natural and artificial intelligence: A brief introduction to the interplay between AI and neuroscience research. *Neural Networks, 144*, 603–613. Available from http://www.elsevier.com/locate/neunet.10.1016/j.neunet.2021.09.018.

Macy, A. (2015, July 15). Electroencephalography (EEG) Pt. 1. Retrieved February 22, 2019, from https://blog.biopac.com/electroencephalography-eeg/.

Magnée, M. J. C. M., De Gelder, B., Van Engeland, H., & Kemner, C. (2007). Facial electromyographic responses to emotional information from faces and voices in individuals with pervasive developmental disorder. *Journal of Child Psychology and Psychiatry, 48*(11), 1122–1130. Available from https://doi.org/10.1111/j.1469-7610.2007.01779.x.

McCall, I. C., Lau, C., Minielly, N., & Illes, J. (2019). Owning ethical innovation: Claims about commercial wearable brain technologies. *Neuron, 102*(4), 728–731. Available from http://www.cell.com/neuron/home.10.1016/j.neuron.2019.03.026.

McNair, D., Lorr, M., & Doppleman, L. (1971). *POMS manual for the profile of mood states.* San Diego, CA: Educational and Industrial Testing Service.

Mojet, J., Dürrschmid, K., Danner, L., Jöchl, M., Heiniö, R. L., Holthuysen, N., & Köster, E. (2015). Are implicit emotion measurements evoked by food unrelated to liking? *Food Research International, 76*(2), 224–232. Available from http://www.elsevier.com/inca/publications/store/4/2/2/9/7/0.10.1016/j.foodres.2015.06.031.

Noel, C., & Dando, R. (2015). The effect of emotional state on taste perception. *Appetite, 95*, 89–95. Available from http://www.elsevier.com/inca/publications/store/6/2/2/7/8/5/index.htt.10.1016/j.appet.2015.06.003.

Nunez, P. L., & Srinivasan, R. (2006). *Electric fields of the brain: The neurophysics of EEG.* Oxford University Press.

Payne, B.K., Cheng, C.M., Govorun, O., & Stewart, B.D. (2005). Affect Misattribution Procedure (AMP) [Database record]. APA PsycTests.

Poels, K., & Dewitte, S. (2006). How to capture the heart? Reviewing 20 years of emotion measurement in advertising. *Journal of Advertising Research, 46*(1), 18–37. Available from https://doi.org/10.2501/S0021849906060041.

Poslad, S. (2009). *Ubiquitous computing smart devices, smart environments and smart interaction. Smart environments and smart interaction.* Wiley.

Quintana, D. S., & Heathers, J. A. J. (2014). Considerations in the assessment of heart rate variability in biobehavioral research. *Frontiers in Psychology, 5,* 805. Available from http://journal.frontiersin.org/Journal/10.3389/fpsyg.2014.00805/full.10.3389/fpsyg.2014.00805.

Ranasinghe, N., Nguyen, T.N. T., Liangkun, Y., Lin, L.Y., Tolley, D., & Do, E.Y. L. (2017). Vocktail: A virtual cocktail for pairing digital taste, smell, and color sensations. Association for Computing Machinery, Inc Singapore. Unpublished content. 2017/10/23 MM 2017 - Proceedings of the 2017 ACM Multimedia Conference. 1139–1147. 10.1145/3123266.3123440. 9781450349062..

Rolls, E. T., & Baylis, L. L. (1994). Gustatory, olfactory, and visual convergence within the primate orbitofrontal cortex. *Journal of Neuroscience, 14*(9), 5437–5452. Available from http://www.jneurosci.org.10.1523/jneurosci.14-09-05437.1994.

Scherer, K. R. (2005). What are emotions and how can they be measured? *Social Science Information, 44*(4), 695–729. Available from https://doi.org/10.1177/0539018405058216.

Sinesio, F., Moneta, E., Porcherot, C., Abbá, S., Dreyfuss, L., Guillamet, K., . . . McEwan, J. A. (2019). Do immersive techniques help to capture consumer reality? *Food Quality and Preference, 77,* 123–134. Available from https://doi.org/10.1016/j.foodqual.2019.05.004.

Singleton, M. J. (2009). Functional magnetic resonance imaging. *The Yale Journal of Biology and Medicine, 82*(4), 233.

Small, D. M., & Prescott, J. (2005). Odor/taste integration and the perception of flavor United States. *Experimental Brain Research, 166*(3-4), 345–357. Available from https://doi.org/10.1007/s00221-005-2376-9, 00144819.

Spinelli, S., Masi, C., Dinnella, C., Zoboli, G. P., & Monteleone, E. (2014). How does it make you feel? A new approach to measuring emotions in food product experience. *Food Quality and Preference, 37,* 109–122. Available from https://doi.org/10.1016/j.foodqual.2013.11.009.

Stone, H., & Sidel, J. L. (2003). *Sensory evaluation | descriptive analysis* (pp. 5152–5161). Elsevier BV. Available from https://doi.org/10.1016/b0-12-227055-x/01065-8.

Teunisse, W., Youssef, S., & Schmidt, M. (2019). Human enhancement through the lens of experimental and speculative neurotechnologies. *Human Behavior and Emerging Technologies, 1*(4), 361–372. Available from http://onlinelibrary.wiley.com/journal/25781863.10.1002/hbe2.179.

Thompson, J. A. F. (2021). Forms of explanation and understanding for neuroscience and artificial intelligence. *Journal of Neurophysiology, 126*(6), 1860–1874. Available from https://journals.physiology.org/doi/full/10.1152/jn.00195.2021.10.1152/jn.00195.2021.

Thomson, D. (2007). SensoEmotional optimisation of food products and brands. *Consumer-Led Food Product Development* (pp. 281–303). United Kingdom: Elsevier Ltd. Available from http://www.sciencedirect.com/science/book/9781845690724.10.1533/9781845693381.2.281.

Thomson, D. M. H., & Coates, T. (2021). *Concept profiling – navigating beyond liking* (pp. 381–438). Elsevier BV. Available from https://doi.org/10.1016/b978-0-12-821124-3.00012-0.

Thomson, D. M. H., & Crocker, C. (2015). Application of conceptual profiling in brand, packaging and product development. *Food Quality and Preference, 40,* 343–353. Available from https://doi.org/10.1016/j.foodqual.2014.04.013.

Thomson, D. M. H., Crocker, C., & Marketo, C. G. (2010). Linking sensory characteristics to emotions: An example using dark chocolate. *Food Quality and Preference, 21*(8), 1117–1125. Available from https://doi.org/10.1016/j.foodqual.2010.04.011.

Thomson, M. (2006). Human brands: Investigating antecedents to consumers' strong attachments to celebrities. *Journal of Marketing, 70*(3), 104–119. Available from https://doi.org/10.1509/jmkg.70.3.104.

van Gils, J. (2020, January 11). *How smart speakers will transform the way we live!* iGadgetsworld. Retrieved February 12, 2022, from https://www.igadgetsworld.com/smart-speakers-history-future/.

Varan, D., Lang, A., Barwise, P., Weber, R., & Bellman, S. (2015). How reliable are neuromarketers' measures of advertising effectiveness: Data from ongoing research holds no common truth among vendors. *Journal of Advertising Research, 55*(2), 176–191. Available from http://www.journalofadvertisingresearch.com/ArticleCenter/default.asp?ID = 104851&Type = Article&Med = PDF.10.2501/JAR-55-2-176-191.

Vescio, B., Salsone, M., Gambardella, A., & Quattrone, A. (2018). Comparison between electrocardiographic and earlobe pulse photoplethysmographic detection for evaluating heart rate variability in healthy subjects in short-and long-term recordings. *Sensors, 18*(3), 844. Available from https://doi.org/10.3390/s18030844.

3. Digitalization in instrumental, neurological, psychological and behavioural methods:
Current applications and opportunities

Villringer, A., Planck, J., Hock, C., Schleinkofer, L., & Dirnagl, U. (1993). Near infrared spectroscopy (NIRS): A new tool to study hemodynamic changes during activation of brain function in human adults. *Neuroscience Letters, 154*(1-2), 101−104. Available from https://doi.org/10.1016/0304-3940(93)90181-J.

Wang, Q. J., Barbosa Escobar, F., Alves Da Mota, P., & Velasco, C. (2021). Getting started with virtual reality for sensory and consumer science: Current practices and future perspectives. *Food Research International, 145*. Available from http://www.elsevier.com/inca/publications/store/4/2/2/9/7/0.10.1016/j.foodres.2021.110410.

Watson, D., Clark, L. A., & Tellegen, A. (1988). Development and validation of brief measures of positive and negative affect: The PANAS scales. *Journal of Personality and Social Psychology, 54*(6), 1063−1070. Available from https://doi.org/10.1037/0022-3514.54.6.1063.

Wexler, A., & Reiner, P. B. (2019). Oversight of direct-to-consumer neurotechnologies. *Science, 363*(6424), 234−235. Available from http://science.sciencemag.org/content/363/6424/234/tab-pdf.10.1126/science.aav0223.

Whittaker, Z. (2019, December 11). Smart device makers won't say if they give governments user data. TechCrunch. Retrieved February 12, 2022, from https://techcrunch.com/2019/12/11/smart-home-tech-user-data-government/.

Williams, D. W. P., Cash, C., Rankin, C., Bernardi, A., Koenig, J., & Thayer, J. F. (2015). Resting heart rate variability predicts self-reported difficulties in emotion regulation: A focus on different facets of emotion regulation. *Frontiers in Psychology, 6*. Available from http://journal.frontiersin.org/article/10.3389/fpsyg.2015.00261/full.10.3389/fpsyg.2015.00261.

Xie, Y., & Oniga, S. (2020). A review of processing methods and classification algorithm for EEG signal. *Carpathian Journal of Electronic and Computer Engineering, 13*(1), 23−29. Available from https://doi.org/10.2478/cjece-2020-0004.

Yu, M., Zhang, D., Zhang, G., Zhao, G., Liu, Y. J., Han, Y., & Chen, G. (2019). A review of EEG features for emotion recognition. *Scientia Sinica Informationis, 49*(9), 1097−1118. Available from https://doi.org/10.1360/n112018-00337.

CHAPTER

12

Emerging biometric methodologies for human behaviour measurement in applied sensory and consumer science

Danni Peng-Li[1,2,3], Qian Janice Wang[1,2,4] and Derek Victor Byrne[1,2]

[1]Food Quality Perception and Society Science Team, iSENSE Lab, Department of Food Science, Faculty of Technical Sciences, Aarhus University, Aarhus, Denmark [2]Sino-Danish College (SDC), University of Chinese Academy of Sciences, Beijing, P.R. China [3]Neuropsychology and Applied Cognitive Neuroscience Laboratory, CAS Key Laboratory of Mental Health, Institute of Psychology, Chinese Academy of Sciences, Beijing, P.R. China [4]Department of Food Science, University of Copenhagen, Frederiksberg, Denmark

12.1 Introduction

This chapter will cover some of the most popular cutting-edge technologies used for measuring human perception and behaviour in applied sensory and consumer science. After reading this chapter, you will have a general understanding of the methods traditionally used in sensory and consumer science and why they in isolation are incomplete. You will gain deeper insights into the basic principles of the different biometric technologies, including what they can measure and how they can measure it. You will be introduced to state-of-the-art research cases employing these technologies. Finally, you will get an idea of the future potential applications that these emerging technologies enable.

12.1.1 Explicit versus implicit measures

A plethora of methods are available for measuring and quantifying human food perception and behaviour. In the field of sensory and consumer science, these have historically been dominated by subjective self-report tools, such as questionnaires and rating scales

Digital Sensory Science
DOI: https://doi.org/10.1016/B978-0-323-95225-5.00012-2

(Lawless & Heymann, 2010; Meilgaard et al., 2006). They are often referred to as explicit procedures, since they require consumers' explicit verbalisation (Fitzsimons et al., 2002). Two of the most widely applied measures are the visual analogue scale (VAS) and the Likert scale (Kuhlmann et al., 2017). Such measures are flexible, inexpensive and relatively easy to administer and comprehend. They may be insightful for capturing simple attitudes, yet they can be naturally prone to biases and highly influenced by individual differences if not sensorily trained. As the responses are usually collected poststimulus exposure, they rely on conscious introspection, deliberate information processing and memory (Fitzsimons et al., 2002). However, our perception, behaviours and choices are not necessarily based on information processing alone nor are they always possible to explicitly express or even remember (Dijksterhuis & Byrne, 2005). In fact, much of human perception and behaviour are contextually and environmentally cue-induced and outside our interoceptive awareness (Bechara & Damasio, 2005).

In recent years, a tendency of integrating more objective methods has emerged to circumvent the limitations of conventional subjective methods. Such methods utilise techniques that can implicitly monitor nonconscious and typically automated and habitual consumer behaviours. Therefore they are often labelled as implicit measures with higher experimental reliability for capturing consumers' more "true" responses (de Wijk & Noldus, 2021).

Examples of implicit, typically biometric, measurements include measures that reflect central nervous system (or brain) activity (e.g. electroencephalography; EEG), measures that reflect autonomic (peripheral) nervous system (ANS) activity (e.g. electrodermal activity; EDA), behavioural measures (e.g. eye-tracking; ET) and expressive measures (e.g. facial expression analysis; FEA). As these implicit responses are captured in real-time, they allow researchers to explore the temporal dynamics of behaviour. This also implies that they can yield extensive time series data, especially if the sampling rate is high (hundreds of data points per second). The data generated from these measures are therefore more challenging to analyse, not immediately interpretable and require significantly more theoretical knowledge and practical expertise. Moreover, the equipment itself can, although this can vary, be very expensive, albeit pricing is becoming more competitive with increasing demand (de Wijk & Noldus, 2021).

Yet, the application of biosensors enables the exploration of completely new avenues of cutting-edge research. To gain deeper insights into the underlying, sometimes unconscious, psychophysiological drivers of human perception and behaviour, the employment of such measures is indispensable and unparalleled. Of course, no tool in isolation — neither explicit nor implicit — can capture the full continuum of perceptual and behavioural processes. Instead, each measure should serve as complementary pieces of information that in combination can provide a more nuanced and holistic understanding of human perception and behaviour.

The implementation of biometric and neuroscientific methods, which by nature are objective and implicit, is popularly referred to as consumer neuroscience or neuromarketing (Ariely & Berns, 2010). It is a subdiscipline of various fields — from neuroscience and psychology to behavioural economy and market research — with the common aim to better understand why consumers make the choices they make and how can one optimally be able to forecast future purchase decisions. This trend has also started emerging into the sensory and consumer science community with an increasing number of studies utilising these methods to better understand the attentional, emotional and cognitive behaviours of the consumers in real time (Gunaratne et al., 2019; Kaneko et al., 2021; Kytö et al., 2019; Pedersen et al., 2021). Hence, consumer neuroscience brings tremendous value to both the product designers as well as marketers and other practitioners in the product development process. A discussion of the practical applications will be elucidated later in the chapter. However, first, we will go through each emerging method, including the underlying psychological, physiological and technological mechanisms. Here, we will exclusively focus on ET, EDA, FEA and EEG research, as these are the most popular and widespread methods to study food-related sensory and consumer science (see Table 12.1 for an overview of selected studies). Of course, other biometric

TABLE 12.1 Overview of selected studies employing state-of-the-art implicit tools, including eye-tracking (ET), electrodermal activity (EDA), facial expression analysis (FEA) and electroencephalography (EEG), to study human behaviour (HR) in sensory and consumer science.

Modality	Outcome measure	Manipulation/ condition	References
ET	Visual attention	Background decoration	Zhang and Seo (2015)
ET	Visual attention	Weight group, food type	Potthoff and Schienle (2020)
ET	Visual attention	Visual priming	Manippa et al. (2019)
ET	Visual attention, food choice	Food labelling	Peschel et al. (2019)
ET	Visual attention, food choice	Background music	Peng-Li et al. (2021)
ET	Visual attention, food choice	Background music	Peng-Li, Byrne, et al. (2020)
ET	Visual attention, food choice	Food brand, food type, food labelling	Bialkova et al. (2020)
ET	Visual attention, food choice	Food labelling	Bialkova et al. (2014)
ET	Visual attention, food choice	Cognitive style, food labelling	Mawad et al. (2015)
ET	Visual attention, food choice	Food placement, signage	Clement et al. (2015)
ET	Visual attention, reaction time	Food type	Motoki et al. (2018)
ET, EDA, FEA	Visual attention, emotional arousal, food choice, food reward, food intake	Food type	Pedersen et al. (2021)
EDA, HR	Emotional arousal	Food expectation	Verastegui-Tena et al. (2019)
EDA, HR, respiration rate	Emotional arousal, food perception	Background music	Kantono et al. (2019)

(Continued)

3. Digitalization in instrumental, neurological, psychological and behavioural methods:
Current applications and opportunities

TABLE 12.1   (Continued)

Modality	Outcome measure	Manipulation/condition	References
EDA, FEAs, HR	Emotional arousal, emotional valence, food liking	Food brand	de Wijk et al. (2021)
EDA, FEA, HR, skin temperature	Emotional arousal, emotional valence	Food type	de Wijk et al. (2012)
EDA, FEA, HR, skin temperature	Emotional arousal, emotional valence, food liking, purchase intent	Food type	Samant and Seo (2020)
EDA, FEA, HR, skin temperature	Emotional arousal, emotional valence, sensory-specific satiety	Hunger status	He et al. (2017)
EDA, FEA, HR, skin temperature, pulse volume amplitude	Emotional arousal, emotional valence, food liking	Food type	Danner et al. (2014)
FEA	Emotional valence	Food type	Zeinstra et al. (2009)
FEA	Emotional valence, food liking, food desire	Food type	Leitch et al. (2015)
FEA	Emotional valence, food liking, food acceptance	Food type	Edwards et al. (2022)
FEA	Emotional valence, food liking	Food type	Galler et al. (2022)
EEG, EDA	Cognitive load, emotional motivation emotional arousal, food cravings	Background noise, cognitive strategy	Peng-Li, Alves Da Mota, et al. (2022)
EEG	Food valuation, reward sensitivity	Food attitude	van Bochove et al. (2016)
EEG	Cortical arousal, body shape dissatisfaction	Weight group	Hume et al. (2015)
EEG	Food perception, cross-modal processing	Food type	Domracheva and Kulikova (2020)
EEG	Attention bias, emotional motivation	Food type	McGeown and Davis (2018)
EEG	Cognitive focus, emotional arousal, attention, food craving	Cognitive strategy, food type	Meule et al. (2013)
EEG	Food wanting, anticipation	Food type	Tashiro et al. (2019)
EEG	Motivational salience, attentional allocation, food desire	Food type	Biehl et al. (2020)
EEG	Attention allocation, food craving, willingness to consume	Food labelling	Schubert et al. (2021)
EEG	Food reward	Food type	Qiu et al. (2020)
EEG	Food reward, attention allocation, food choice	Food type	Kirsten et al. (2022)

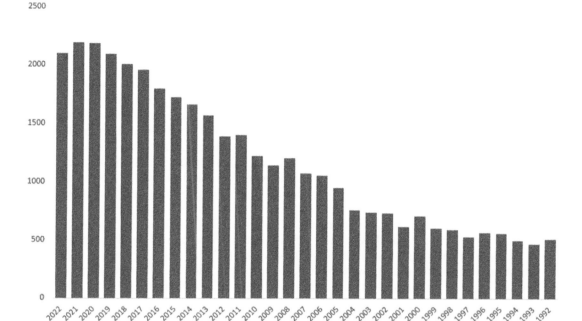

FIGURE 12.1 The total number of publications per year from 1992 to 2022 using the search term "eye tracking" OR "eyetracking" OR "eye-tracking" OR "eye movements" OR "eye gaze" based on data from PubMed (https://pubmed. ncbi.nlm.nih.gov/?term=%22eye+tracking%22+OR+%22eyetracking%22+OR+%22eye-tracking%22+OR+%22eye+movements%22+OR+%22eye+gaze%22&filter=years.1992-2022&timeline=expanded).

measures are also available on the market and already widely used in clinical contexts, such as heart rate (HR) (Billman et al., 2015) and respiration rate (AL-Khalidi et al., 2011), but these are outside the scope of this chapter.

12.1.2 Eye-tracking

ET is one of the most popular emerging experimental techniques across multiple scientific fields and has undergone a dramatic surge in the past years (Fig. 12.1). This popularity is a result of increasing accessibility, affordability and technological improvements (Carter & Luke, 2020). In simple terms, ET technology enables measuring the activity of the eye, such as eye movements, and is often referred to as the most direct method of measuring visual attention (Clement, 2007). In order to better understand the technology behind ET and its applications, we will first introduce the fundamentals of visual attention.

12.1.2.1 Visual attention

Attention can be defined as selectivity in perception (Orquin & Loose, 2013). Traditionally, attention has been viewed as the brain's filtering mechanism by which only relevant sensory inputs are selected for further perceptual processing to avoid information overload during decision-making (Hauser & Salinas, 2014). However, contemporary models in vision research have established that attention also affects perception (Carrasco, 2011). That is, by directing the fovea of the eye and overt visual attention towards a specific stimulus, the visual processing of that stimulus increases in the visual cortex of the brain (Vossel et al., 2014). Although covertly attending the stimulus outside the fovea (i.e. without moving the eyes) is possible, the empirical consensus is that eye movements are strongly correlated with visual attention (Just & Carpenter, 1976). Accordingly, if the stimulus is outside the perceptual span of the fovea, it is in most cases not processed. Therefore fixating on a stimulus ought to augment the visual features and/or spatial location of the stimulus, thereby increasing the information during a decision. These are, however, only some of the different dichotomies encompassing visual attention. In addition to overt versus covert attention and feature-based versus spatial attention, visual attention can also be divided between *bottom-up versus top-down* processes of visual attention (Corbetta & Shulman, 2002).

Bottom-up and top-down attention are also commonly defined as exogenous and endogenous attention or goal-driven and stimulus-driven attention. They have been extensively studied through visual search tasks (Belkaid et al., 2017) — a perceptual task involving attentional processes that allow the participant to scan the visual environment in order to identify specific target objects amongst distractors. Like other behavioural processes, bottom-up and top-down attention are manifested in distinct neural systems (Macaluso, 2010). The ventral frontoparietal-driven bottom-up processes of attention are mainly guided by visual saliency through which attention is involuntarily and promptly allocated to salient features (e.g. contrast, colour and shape) in the perceptual span (Moore & Zirnsak, 2017). In contrast, dorsal frontoparietal top-down control of attention is deliberately deployed and deals with voluntary and goal-directed processes. The nature of such processes entails that top-down processes increase the signal-to-noise ratio of relevant information by discarding possibly salient and distracting but irrelevant sensory stimuli (Noudoost et al., 2010). Correspondingly, top-down attention occurs when the consumer is focused on stimuli relevant to the particular goal.

12.1.2.2 *The technology*

Modern eye-trackers enable recording the motion and gaze location of the eye (Clement, 2007). In general, there are two types of eye-trackers that essentially run on the same technology — screen-based and mobile eye-trackers. For simplicity, we will in this chapter cover the former. The majority of these eye-trackers track gaze behaviour using infrared light in combination with optical sensors (usually a high-resolution camera), and due to corneal reflection in the pupil centre, the eye-tracker is able to track respondents' eye movements and presumably visual attention (Fig. 12.2). As mentioned in Section 12.1.2.1, the eye movements reflect overt (but not covert) visual attention, that is when the fovea is directed towards the stimuli of interest (Orquin & Holmqvist, 2019).

To quantify eye movements and visual attention, specific gaze behaviour is decomposed into certain metrics. Most of these metrics are based on gaze points and fixations (Fig. 12.3). Gaze points represent the basic unit of measure, that is one gaze point is equivalent to one raw sample. For example, if the eye-tracker has a sampling rate of 60 Hz, every gaze point equals a sixtieth of a second. A fixation is when a sequence of gaze points

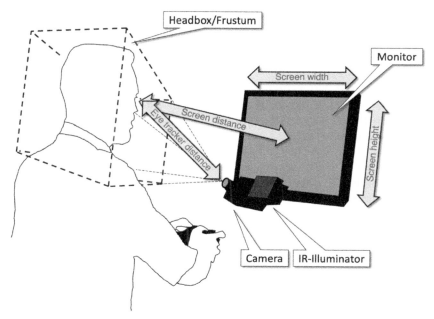

FIGURE 12.2 Illustration of screen-based ET setup (Holmqvist et al., 2023). *Source: Holmqvist, K., Örbom, S.L., Hooge, I.T.C. et al. (2023). Eye tracking: empirical foundations for a minimal reporting guideline. Behavior Research Methods, 55, 364–416. https://doi.org/10.3758/s13428-021-01762-8.*

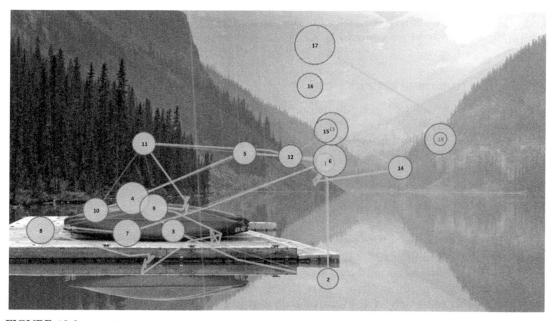

FIGURE 12.3 Simple visualisation of fixations and saccades (iMotions, 2017a). Fixations are the periods in which the eye is fixated/locked onto a specific entity (represented by the circles). In between these fixations are the saccades which denote the spatial 'jumps' the eye makes from one location to the next (represented by the connecting lines). *Source: From iMotions. (2017a). Eye tracking the complete pocket guide. https://imotions.com/blog/eye-tracking/.*

occurs in close temporal and spatial proximity and denotes when our eyes are locked/fixated onto a specific entity. The duration of single fixation varies depending on the visual stimuli, the purpose and complexity of the task and individual differences, but in general, the last between 100 and 300 milliseconds (iMotions, 2017a). The spatial 'jumps' between one fixation to another are called saccades. They are ballistic eye movements that suppress the visual input of the viewer, making us virtually blind during saccade (Rayner, 2009). The duration and velocity of a saccade are directly related to the distance travelled. Similar to fixations, the properties of a saccade are influenced by the contexts. In reading, the size of a standard saccade is equal to 2 degrees of rotation with an approximate duration of 30 milliseconds. However, in scene perception, they are often around 5 degrees of rotation and can last up to 50 milliseconds (Abrams et al., 1989).

Some of the most prevalent ET metrics include fixation duration, fixation count, fixation revisit count and time to first fixation. Fixation duration refers to the average duration of fixations in a given area of interest (AOI). Fixation count denotes the number of fixations in an AOI. Fixation revisit count is the number of fixation-based returns to an AOI. Time to first fixation simply indicates the time whereby the first fixation in an AOI occurs. Obviously, the choice of metric should depend on the specific research question, hypothesis and experimental design (Orquin & Holmqvist, 2018). While time to first fixation might be an appropriate metric when exploring bottom-up processes of visual attention in, for example visual search tasks, fixation revisit count might be better suited for examining tasks related to top-down visual attention.

Nonetheless, regardless of the ET metric chosen, any of these implicit measures are far more superior than self-reported questionnaires or basic visual search tasks when exploring visual attention in relation to human behaviour and perception. An accurate, precise and quantifiable subjective estimation and description of the temporal and spatial dynamics in one's visual field during a food valuation or decision is practically impossible.

12.1.3 Electrodermal activity

How can we measure our emotions before, during and after the consumption of food? Electrodermal activity, or EDA for short, is a noninvasive, emerging, biometric technique for implicitly measuring human emotions. It is often referred to as galvanic skin response or skin conductance response and is a response to sympathetic nervous system activity. With increased sympathetic activity, sweat production is elevated, leading to heightened skin conductance (Verastegui-Tena et al., 2019; Wang et al., 2018). EDA is directly related to arousal level, such that higher EDA corresponds to elevated emotional arousal, and lower EDA is associated with decreased emotional arousal.

12.1.3.1 Emotional arousal

Emotional arousal is essentially an ANS response and the product of downstream hormonal activity and neurotransmission in deeper subcortical brain structures (Arnsten & Li, 2004). It has been involved in several contemporary theories and models of emotions (LeDoux, 2012) with dimensional models being especially popular in the consumer science community (Kantono et al., 2019). According to the circumplex model of affect (Russell, 1980), independent neural systems do not subserve every emotion as portrayed in theories

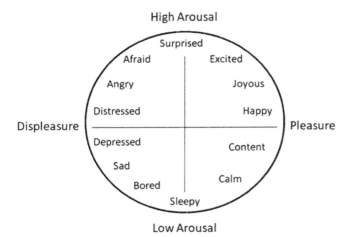

FIGURE 12.4 Graphical representation of the circumplex model of affect. The horizontal axis represents the valence dimension, and the vertical axis represents the arousal or activation dimension (Tseng et al., 2014). *Source: From Tseng, A., Bansal, R., Liu, J., Gerber, A. J., Goh, S., Posner, J., Colibazzi, T., Algermissen, M., Chiang, I.-C., Russell, J. A., & Peterson, B. S. (2014). Using the circumplex model of affect to study valence and arousal ratings of emotional faces by children and adults with autism spectrum disorders.* Journal of Autism and Developmental Disorders, 44(6), 1332–1346. https://doi.org/ 10.1007/s10803-013-1993-6.

of basic and discrete emotions. Instead, affective states arise from at least two orthogonally related dimensions, one attributed to a valence/pleasantness continuum (low valence—high valence) and one to an arousal/activation continuum (low arousal—high arousal; Fig. 12.4). In this view, every emotion can therefore be understood as a linear combination of valence and arousal (Posner et al., 2005). Excitement for food, for example, is the product of high arousal and moderately high valence, whereas food of low saliency could generate the dimensionally inverse, that is boredom towards the food (low arousal and moderately low valence). Consequently, the intensity of emotional arousal is critical in understanding and determining consumer behaviour.

12.1.3.2 The technology

Different devices for measuring EDA exist on the market — some more ambulatory than others. Nevertheless, the basic principle of EDA is the temporal dynamics of sweat production in the sweat glands causing electrical changes in the skin that are detected by attached electrodes and amplified through a signal amplifier (Tyler et al., 2015). The raw EDA consists of two main signals — tonic and phasic activity (Boucsein, 2012). Tonic activity corresponds to the electrodermal level and reflects slower alterations (tens of seconds to minutes) in the skin that is heavily dependent on individual differences, including skin hydration and autonomic response. This signal is less informative in isolation and usually filtered away during signal processing. The phasic signal on the other hand is measured in terms of electrodermal response 'on top of' the phasic signal and reflects faster fluctuations (1—5 seconds) that are sensitive to arousing stimuli (iMotions, 2017b). The phasic electrodermal response is typically the signal of interest in psychological tests. After preprocessing of the raw data (i.e. downsampling and filtering), this signal can be decomposed into specific EDA metrics. The most widespread metrics of interest exploit the EDA peaks, either peak amplitude or the number of peaks (Fig. 12.5). The former is usually measured in microvolts, while the latter depends on a predetermined voltage threshold (Boucsein, Fowles, et al., 2012). The relatively convenient use of an EDA system can offer unique insights into the autonomic emotional drivers of food perception and behaviour.

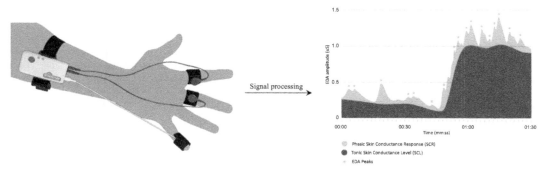

FIGURE 12.5   Example of electrodermal activity (EDA) setup and data representation including tonic and phasic signal as well as EDA peaks (Hernando-Gallego & Artés-Rodríguez, 2015). *Source: Adapted from Hernando-Gallego F., & Artés-Rodríguez, A. (2015).* Individual performance calibration using physiological stress signals. *http://arxiv.org/abs/1507.03482. iMotions. (2017b). Galvanic skin response. The complete pocket guide.*

12.1.4 Facial expression analysis

Our faces are the most obvious social indicator of emotions. Instant, spontaneous and to some degree involuntary facial movements and expressions are the result of psychological emotional responses. These emotional reactions can be captured through simple facial recordings. With the increasingly more sophisticated computer-based algorithms, researchers are now able to quantify these emotions with high validity and reliability. In comparison to EDA, facial coding algorithms and artificial intelligence (AI) can measure the valence of one's emotions in addition to the arousal level, making FEA in combination with EDA extremely useful for decoding and classifying our emotional responses.

12.1.4.1 Emotional valence

Emotional valence, also referred to as hedonic tone, denotes the (un)pleasantness of the emotional reaction (Barrett, 1998). In contrast to arousal, indicating the intensity of the emotions, valence thus refers to the directionality of that emotion, that is being positive or negative (Fig. 12.4). While positive emotions, such as joy, are induced by pleasant and attractive events, situations and foods, negative emotions, including disgust, are evoked by unpleasant and aversive events, situations, etc. (Jaeger et al., 2021). These emotional responses are hence regulated by the brain areas associated with affect and reward and is naturally linked to the rewarding and hedonic eating experiences as denoted by the liking of and preference for the food (Rogers & Hardman, 2015) (see also Section 12.1.5.2, Reward valuation). Valence in combination with arousal comprise the fundamental constructs of the full circumplex model of affect (Russell, 1980), which sets the foundation for a more holistic understanding of consumers emotional responses to food.

12.1.4.2 The technology

In recent years, a number of automatic facial coding procedures based on advanced AI technology have been developed to identify and accurately measure human emotions and expressions. Overall, these procedures are based on state-of-the-art computer vision and

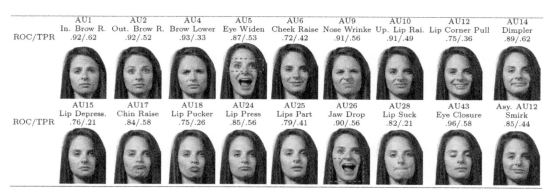

FIGURE 12.6 Definitions and illustrative, posed examples of the facial actions we detected and analyzed. Example images were collected by Affectiva, Inc. *Source: From McDuff, D. (2016). Discovering facial expressions for states of amused, persuaded, informed, sentimental and inspired.* Proceedings of the 18th ACM International Conference on Multimodal Interaction, *71−75. https://doi.org/10.1145/2993148.2993192.*

machine learning algorithms fed by thousands of facial expressions recordings in the database from people around the world. To access this facial FEA, simple cameras from laptops, tablets and smartphones or individual webcams mounted to computer screens can be used to collect videos recordings of consumers' faces during the exposure of food stimuli. Such simple practical setups overcome the challenges that other more expensive biosensors often implicate, as these can be applied inside and outside the laboratory, including naturalistic settings of the consumers, such as homes, workplaces, etc.

Despite some differences between the FEA algorithms, they are basically running on the same computational steps to quantify the emotional response. These include face detection, feature detection and feature classification (see McDuff, 2016 for a detailed description of the algorithm). First, the algorithm is simply capturing the face and its position in the video frame. Next, facial landmarks, including eyes, lips and nose tips are detected followed by an internal face model adjustment and scaling to match the person's face. Ultimately, spatial position and orientation as well as feature information go into a deep learning classification algorithm which translate all the information into specific action units, facial expressions and emotional states (Fig. 12.6).

12.1.5 Electroencephalography

Our behaviours are essentially governed by the brain and its plethora of neural processes. Electroencephalography, more known as EEG, can capture the brain activity underlying some of these processes. EEG is a neurophysiological technique that can measure electrical activity in the brain (Allen et al., 2004). It is popularly conceptualised as a brain wave decoder and a more mobile alternative to more expensive neuroimaging modalities, such as functional magnetic resonance imaging (fMRI). It is a very versatile instrument that can measure various perceptual, sensory, emotional and cognitive processes in the brain. In consumer sensory and consumer neuroscience, two of the most widespread neuropsychological constructs comprise cognitive load and reward valuation.

12.1.5.1 Cognitive load

Neural computations in higher cortical brain areas and mental processes of high cognitive demand are often referred to as cognitive load. That is, the degree to which a decision or task is processed and regulated depends on its complexity as well as the level of expertise of the decision or task-maker (Antonenko et al., 2010). The notion of cognitive load is one way to conceptualise the degree of mental engagement in a decision and task. It refers to the used amount of working memory recourses, that is the capacity to process and hold novel information temporarily (Sweller, 1988). In contrast to long-term memory, which practically has no processing capacity limit, working memory (and cognitive load) is often regarded as the bottleneck of learning and executive tasks performance. Not surprisingly, it has been demonstrated to correlate with emotional arousal (Granholm et al., 1996).

According to cognitive load theory, three domains exist – intrinsic, extraneous and germane cognitive load – each with distinct functionalities and demands on the brain. Intrinsic cognitive load refers to the inherent level of difficulty of the specific task, problem or decision. For example, solving 2 + 2 is cognitively less demanding than advanced calculus. Extraneous cognitive load essentially denotes any distractors that decelerate the speed by which working memory consolidates into long-term memory, or in other words, prevent information from becoming knowledge. This could be the way the inherent problem is described or simply sensory distractors such as background noise. Finally, germane cognitive load is essentially the process of memory consolidation of new intrinsic information through the connection of existing learned material. When learning a new language, germane cognitive load is likely occurring, since we often compare and contrast words in the new language with words from our native language (Krahnke & Krashen, 1983). These are, however, psychological constructs which in isolation do not necessarily comply with biology or real-life scenarios. Instead, an integration of the three is most likely the case.

12.1.5.2 Reward processing

Reward of food comprises at least three neural and behavioural constructs – motivation, pleasure and learning – where most computational models and behavioural instruments have concentrated on the latter, for example through Pavlovian instrumental transfer paradigms (da Costa et al., 2020). However, both constructs of incentive salience/motivation and hedonic pleasure are just as important for a well-functioning reward system (Fig. 12.7). In fact, compared to learning, motivational 'wanting' and hedonic 'liking' might even be stronger predictors of appetite control and food choice (Berridge & Kringelbach, 2015).

Although 'wanting' and 'liking' are closely related and often occur in juxtaposition, they are neurologically and psychologically discriminable (Peng-Li, Andersen, et al., 2022). According to the incentive saliency theory (Robinson et al., 2013), the anticipatory and motivational behaviours of 'wanting' are governed by the mesolimbic dopaminergic system, which, through Pavlovian learning presentations in the brain, is attributed to a rewarding stimulus. This stimulus then becomes a highly valued motivational target and the root of cue-induced cravings for the consumer (Zijlstra et al., 2009). The fulfilment of these cravings, that is by choosing to consume the desired food, generates the actual hedonic experience, pleasure and 'liking' (Egecioglu et al., 2011). In other words, while

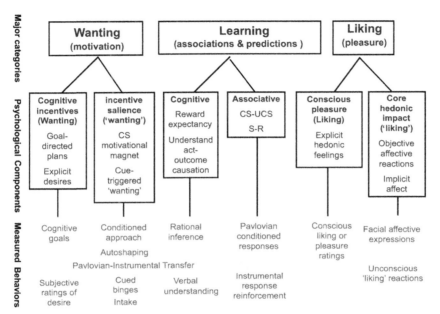

FIGURE 12.7 Heatmap from American versus Chinese participants (Zhang & Seo, 2015). Typical examples of culture-induced variation in visual attention towards the food item (Zhang & Seo, 2015). An American participant (A) focused more on the food item (i.e. donut), while a Chinese participant (B) explored the background context (i.e. tableware items and table decoration) more than the food item. *Source: From Berridge, K. C. (2009). "Liking" and "wanting" food rewards: Brain substrates and roles in eating disorders.* Physiology and Behavior, 97(5), 537–550. *https://doi.org/10.1016/j.physbeh.2009.02.044.*

'wanting' denotes the anticipatory reward, 'liking' is the constitution of consummatory reward valuation and is associated with 'hedonic hotspots' in the endogenous opioid system and cortical sites of the brain (Nguyen et al., 2021).

12.1.5.3 The technology

The fundamental principle behind EEG can be attributed to the electrodes which are placed on the scalp typically via a cap and connected to an amplifier. These electrodes record the macroscopic activity of cortical neurons in the brain (Bazzani et al., 2020). The number of electrode channels varies from a few to hundreds depending on the system, usually according to the international 10–20 system of electrode position (Picton, 1991). Each electrode measures the changes in electrical currents based on a population of cortical pyramidal neurons underneath the surface area perpendicular, oriented to the given electrode (Luck, 2005). In many cases, these electrical fluctuations can be examined in response to a time-locked sensory, cognitive or motor event. Such measurement is referred to as event-related potentials and is essentially the direct measurable electrophysiological response to a given stimulus, resulting in an excellent temporal resolution of milliseconds (Delorme & Makeig, 2004).

However, short time-locked stimuli are not always the optimal choice of measure when studying human behaviour, since not all behavioural responses are triggered by a single event. Sometimes, longer-lasting continuous and spontaneous neural representations are experimentally superior. Fortunately, the EEG signal can also be decomposed into various frequency spectra representing the oscillatory dynamics in the brain (Aoh et al., 2019). This requires mathematical transformations of the preprocessed data (i.e. after downsampling, re-referencing, filtering, artefact rejection, etc.) from the original time-domain (milliseconds) into the frequency domain (Hz) using the Fourier-transformation (Cohen, 2014). In studies investigating human behaviour and decision-making, the frequency bands of interest are most commonly confined to theta (4−8 Hz), alpha (8−12 Hz) and beta (12−25 Hz), yet delta (1−3 Hz) and gamma (∼40 Hz) frequencies have also been examined for specific purposes (Fig. 12.8). The energy of the signal within each of these frequency bands is often referred to as the power spectral density (PSD) and measured in watts per hertz or decibel. Elevated PSD resembles spectral amplification/synchronisation and lowered PSD signifies spectral attenuation/desynchronisation (Aldayel et al., 2020; Nagel, 2019; Tashiro et al., 2019).

Changes in the PSD have been correlated with mental effort and cognitive load. While theta and alpha activity in frontal and parietal regions correlates with cognitive load and task

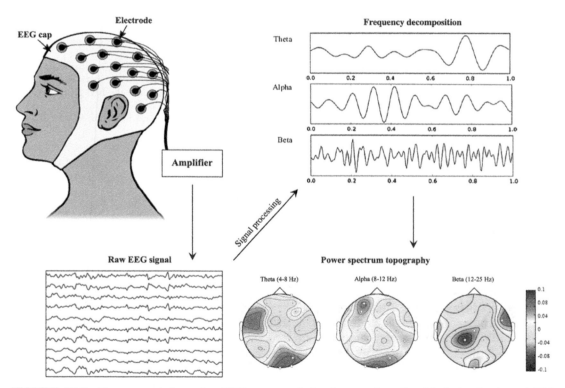

FIGURE 12.8 Electroencephalography (EEG) setup and signal processing. Simplified visualisation of EEG setup and basic processing steps. Please see the online version to view the color image of the figure. *Source: From Aldayel, M., Ykhlef, M., & Al-Nafjan, A. (2020). Deep learning for EEG-based preference classification in neuromarketing. Applied Sciences, 10(4), 1525−1525. https://doi.org/10.3390/app10041525. Nagel, S. (2019). Towards a home-use BCI: Fast asynchronous control and robust non-control state detection. https://doi.org/10.15496/publikation-37739.*

difficulty (Antonenko et al., 2010; Brouwer et al., 2012), including focused attention and sensory processing (Cabañero et al., 2019), beta activity has been linked to Zhang et al. (2008).

Furthermore, lateralised investigation of the PSD has recently been of great interest in consumer neuroscience and neuromarketing. Here, experimental convenience is highly valued and thus only a few electrodes are usually applied. They are often applied on the frontal F_3 and F_4 positions according to the 10−20 system in order to investigate frontal asymmetries (Ramsøy et al., 2018), especially in the alpha frequency range (Smith et al., 2017). Frontal alpha asymmetry (FAA) is an index of so-called approach behaviour, including 'wanting' and 'liking', and shown to converge with blood-oxygen-level-dependent (BOLD) activity in brain areas associated with reward valuation from fMRI studies (Gorka et al., 2015). In particular, greater right (vs left) frontal hemispheric alpha power is indexed by a positive FAA score, indicating reward anticipation, hedonic pleasure and, in general, positive valence, while a negative FAA score is related to avoidance and withdrawal behaviour, that is negative valence (Fischer et al., 2018).

In short, EEG with its functional and continuous neural representations is an excellent tool and measure for studying the underlying brain dynamics of human perception and behaviour Through PSD analysis, it offers an implicit, objective and nuanced quantification of cognitive load and reward processes which is not restrained by verbalisation, introspection or any other subjective limitations as in conventional consumer behaviour methods.

12.2 Research cases

12.2.1 Eye-tracking to investigate bottom-up effects of visual attention

In a supermarket environment, the consumers are exposed to a myriad of food products and often faced with several similar alternatives within the same food category. This implies that the products are competing to attract the most attention to the consumers. Placement of the product may drive the attention (Clement et al., 2015; Romero & Biswas, 2016), but extrinsic product-specific factors such as package design and labelling also plays a central role − not only for attention but also for the ultimate decision-making (Mai et al., 2016; Symmank, 2019).

How can we study these processes? How can we optimise attention capture to the product through packaging design? And how can we increase the choice likelihood of the product? These are questions that are all questions that we could try to answer through standard self-report questionnaires, but with ET technology, we would obtain much deeper quantifiable insights into the visual attention of the consumers. This is exactly what Peschel et al. (2019) did.

In particular, they investigated if manipulation of product labelling, that is an organic logo, could affect especially bottom-up effects of visual attention as well as choice of that product. To do so, they carried out a combined ET and choice experiment. Here, participants were exposed to three product categories, with each product's logo being manipulated on the basis of size and visual saliency (Fig. 12.9).

They found that the manipulation of the product indeed affected visual attention as determined by fixation likelihood (Fig. 12.10). However, unexpectedly, they did not find that increasing the saliency of the logo alone did not increase attention capture. Instead,

FIGURE 12.9 Examples of chocolate choice sets from Peschel et al. (2019). At the top is a close-up of the Danish organic label. In the middle is an example of a chocolate choice set with alternative A carrying the manipulate Danish organic logo. Below are examples of the four conditions investigated in our experiment: Condition 1: small, low saliency; Condition 2: small, high saliency; Condition 3: large, low saliency; Condition 4: large, high saliency. Option C is the 'no choice' option. *Source: From Peschel, A. O., Orquin, J. L., & Mueller Loose, S. (2019). Increasing consumers' attention capture and food choice through bottom-up effects. Appetite, 132, 1–7. https://doi.org/10.1016/j.appet.2018.09.015.*

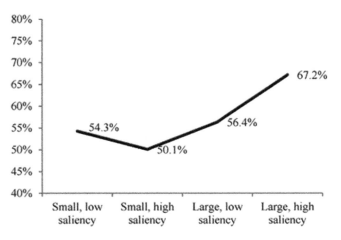

FIGURE 12.10 Results of fixation likelihood of the organic logo on a descriptive level from Peschel et al. (2019). *Source: From Peschel, A. O., Orquin, J. L., & Mueller Loose, S. (2019). Increasing consumers' attention capture and food choice through bottom-up effects. Appetite, 132, 1–7. https://doi.org/10.1016/j.appet.2018.09.015.*

the combined effect of saliency and size significantly increased fixation likelihood. That is, increasing both the size and visual saliency of the organic logo correspondingly increased the visual attention to same. Furthermore, this bigger and more salient logo also increased the actual choice likelihood of the product compared to that of the small and low salient logo. This effect was predominantly driven by the bottom-up attention capture to that product, indicating that the products you fixate at more are also the products that you are more likely to choose.

These ET results could be extremely useful to food marketers and policy makers to better promote healthier and sustainable products in the supermarket. Simply by adjusting certain features of the product labelling, and not necessarily the whole packaging format, can thus improve the attention to it and ultimately the performance of it.

Other ET studies have even employed the technology to investigate contextual influences of consumer behaviours not related to the product itself (Seo, 2020). These include everything from social contexts and table decorations to store ambience (e.g. lightning, background music, etc.). As ET also overcomes the methodological limitations involved in cross-cultural studies, such as interpretation and self-report differences (Masuda et al., 2012; Peng-Li, Chan, et al., 2020), some studies have explored cultural differences in food/product valuation.

For instance, Zhang and Seo (2015) investigated the influence of background saliency of food plating and ethnicity on visual attention (Zhang & Seo, 2015). Specifically, they recruited a group of North American and a group of Chinese participants for an ET study in which the participants were instructed to look at a series of food images with varying backgrounds (Fig. 12.11). No additional tasks were given, thus solely bottom-up effects were explored.

Across both ethnicities, visual attention towards the food itself decreased with increased background saliency (i.e. more decorations). That is, with increased background saliency, participants' fixation on the food items became significantly slower (i.e. higher entry time) and shorter (i.e. lower fixation time). Moreover, with a more salient surrounding background, participants fixated back and forth between the food itself and background setting more frequently as indicated by the increased revisit count. Essentially, these findings imply that more complicated and salient table setting, and decorations can lessen the focus of the food itself and make consumers pay more attention to the surroundings. Such strategy might especially be relevant and applicable in the context of unhealthy foods, as increased background saliency could steer away the focus to the food.

Despite these similar behaviours between Chinese and American participants, the study also highlighted a number of cultural differences (Fig. 12.12). First, the American group fixated at the food itself more immediately compared to Chinese participants. Secondly,

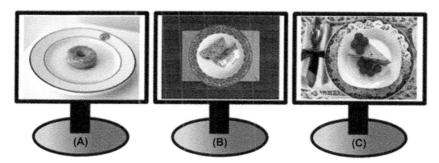

FIGURE 12.11 Examples of pictures of food items with low (A), medium (B) and high (C) levels of background saliency (Zhang & Seo, 2015). *Source: From Zhang, B., & Seo, H.-S. (2015). Visual attention toward food-item images can vary as a function of background saliency and culture: An eye-tracking study.* Food Quality and Preference, *41, 172−179. https://doi.org/10.1016/j.foodqual.2014.12.004.*

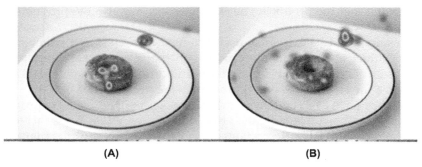

(A)   **(B)**

FIGURE 12.12   Heatmap from American versus Chinese participants (Zhang & Seo, 2015). Typical examples of culture-induced variation in visual attention towards the food item (Zhang & Seo, 2015). An American participant (A) focused more on the food item (i.e. donut), while a Chinese participant (B) explored the background context (i.e. tableware items and table decoration) more than the food item. *Source: From Zhang, B., & Seo, H.-S. (2015). Visual attention toward food-item images can vary as a function of background saliency and culture: An eye-tracking study.* Food Quality and Preference, *41, 172−179. https://doi.org/10.1016/j.foodqual.2014.12.004.*

the American participants also tended to fixate at food with a low-saliency background significantly longer than their Chinese counterparts. In the same vein, the background setting (i.e. tableware items and table decoration) attracted more attention than the food itself for the Chinese consumers, whereas the opposite pattern happened for the American group. These findings suggest that background saliency influenced Chinese participants more than Americans and emphasise the importance of considering cultural differences in food research.

Altogether, these studies exemplify that ET experiments contribute with an additional layer of consumer insights that would be practically impossible to obtain with conventional sensory and consumer science methods. Hence, practitioners ought to exploit these methodological advances in order to optimise their advertisements, store architecture and packaging designs of the products.

12.2.2 Electrodermal activity and facial expression analysis to study emotional responses of food expectation and acceptance

More than half of newly launched food products fail on market (de Wijk et al., 2012). This unfortunate reality occurs even when many of these products are evaluated by trained sensory panels and tested by the consumers before they hit the marketplace. These product tests are often based on explicit sensory analytical profiling and liking assessments (Meilgaard et al., 2006), indicating that such tests have relatively low predictive validity of product performance. Again, actual consumer behaviour and food purchases may to a large extent be manifested in unconscious and implicit motives, acceptances and expectations, rather than rational reasoning and deliberate information processing. The actual underlying reasons for food preference and avoidance can be somewhat difficult to explicitly articulate, even when they may be the strongest drivers of consumer behaviour and food choice. So why do we choose one food over the other? Because we expect that particular food to give us pleasure — they generate some kind of emotional response.

Using EDA measurements, Verastegui-Tena et al. (2019) investigated consumers' underlying emotional reactions to degrees of food expectations (Verastegui-Tena et al., 2019). In particular, they assessed how these ANS responses changed when tasting juice samples that were as expected, juice samples that differed slightly or juice samples that differed significantly from the manipulated expectations. They also explored whether these responses diverged from those obtained when there was no manipulation of expectations.

The participants, therefore, completed two separate sessions of the experiment. In one session, the expectations were manipulated by presenting an image of an ingredient followed by a juice with a flavour that was as expected, differed slightly or differed significantly from that of the image. In another session, no explicit manipulation was provided, and the participants simply tasted the juices followed by the image presentation. The images were the same in both sessions. Participants also rated the juice samples in terms of sensory properties before and after tasting them in order to ensure that they actually perceived the confirmations as well as small and large disconfirmations. These ratings showed that the participants perceived the designed confirmation and disconfirmation of expectations as intended, except for sourness and taste intensity.

The actual EDA results also revealed that the overall skin conductance level between the two sessions differed, such that the emotional arousal increased when explicit manipulation of expectations was given but decreased when it was absent. This suggests that the expectations play an important role in the way we emotionally react to food and therefore also the way we value the food. The authors also argued that this arousal difference could be reflected in changes in attention such that when participants are more attentive, they also respond to more stimuli. That is, EDA responses capturing elevated emotional arousal states might be due to an increase in attention necessary for processing the information related to the (dis)confirmation of expectations. In conclusion, their findings demonstrate that biometric EDA data can serve as a complementary measure to self-report information when investigating emotional responses of food expectations.

Akin to food expectations, acceptance of food might be difficult to explicitly verbalise in an experimental setting – especially for children – and thus implicit physiological and expressive measurements can be advantageous. By means of EDA and FEA, de Wijk et al. (2012) explored 30 children and young adults' responses to liked and disliked foods. The food items were selected based on individual self-reported questionnaire prior to the food evaluation. The experiment consisted of three sensory evaluations in which the participants were instructed to visually inspect the food, smell the food and taste the food.

The results showed that the visual inspection of disliked food items induced higher EDA, indicating increased arousal states, as well as greater facial micro-expressions of negative emotions. Emotional states of higher arousal and negative valence, such as anger, anxiety and distress might therefore reflect disliked foods, but positive emotions will not necessarily manifest in liked foods (Zeinstra et al., 2009) at least during first sight of the food. Furthermore, EDA also varied between the type of sensory encounter of the food, with relatively low EDA during first visual encounter as compared to both olfactory and gustatory stimulations of the food. Moreover, during the food tasting, EDA decreased for young adults, whereas it increased for children (Fig. 12.13).

This study highlights the usability and practical value of exploring ANS responses and implicit expressive (and behavioural) measures in relation to food acceptance. It

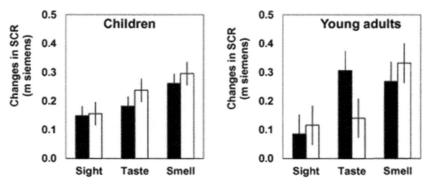

FIGURE 12.13   Changes in electrodermal activity (EDA)/skin conductance response (SCR) in children and young adults (de Wijk et al., 2012). Results from de Wijk et al. (2012) show the effect of instruction (sight, smell or taste) on changes in EDA/SCR in children and young adults. Filled bars reflect liked foods and open bars disliked foods. Error bars indicate standard errors. *Source: From de Wijk, R. A., Kooijman, V., Verhoeven, R. H. G., Holthuysen, N. T. E., & de Graaf, C. (2012). Autonomic nervous system responses on and facial expressions to the sight, smell, and taste of liked and disliked foods. Food Quality and Preference, 26(2), 196–203. https://doi.org/10.1016/j.foodqual.2012.04.015.*

demonstrates that physiological emotional responses vary with food preference as well as age group during different sensory stimulations. These FEA and EDA measurements enable the possibility to uncover the underlying unconscious reactions to specific food products across different sensory exposures, that is sight, smell and taste for better consumer–product interactions. With such insights, marketers might be in better position for predicting food acceptance of the consumers.

12.2.3 Electroencephalography and electrodermal activity to explore the neurophysiology of food cravings

Food pleasure and cravings are strong drivers of our eating behaviours (Petit et al., 2016). These are affected by our mood, feelings and the ability to internally regulate our valuation of the food. However, at the same time, atmospheric sensory distractions, such as ambient noise, are also influencing our cognitive resources and emotional states necessary for controlling these behaviours. This implies that the underlying mechanisms of food-related decision-making are based on an integration of exteroceptive sensory inputs and interoceptive bodily states (Papies et al., 2020) that translate our somatic signals into feelings of anticipation, desires or cravings (Peng-Li, Sørensen, et al., 2020). Such complex neuropsychological downstream factors can be challenging, if not impossible, to measure through traditional sensory and consumer science methods. Again, in such context, implicit psychophysiological examinations are superior.

One example of this is a recent multimodal EEG and EDA study by Peng-Li, Alves Da Mota, et al. (2022). Here, the authors explored the underlying neurophysiology of food cravings and how ambient noise and cognitive self-regulation influenced this (Peng-Li, Alves Da Mota, et al., 2022). In particular, they focused on cognitive load and emotional arousal and motivation as biobehavioural drivers of subjective food cravings. To test this, they employed the Regulation of Craving task — a task which can measure the specific causal effect of regulation strategies on craving for cigarettes, alcohol and/or foods (Sun & Kober, 2020). In this study, the authors exclusively focused on high-calorie food items as craving cues. The participants were exposed to one of these cues, preceded by the instruction to follow one of two decision perspectives: 'now' — focus on the immediate sensations and feelings associated with consuming the food (e.g. it will taste good and satisfy my cravings), or 'later' — focus on the long-term negative consequences associated with repeated consumption (e.g. it will increase my risk for weight gain and heart disease). Participants were then asked to rate their craving for the specific food they just saw on a VAS. All participants completed two sessions of the task — once with soft ambient restaurant noise (~50 dB) and once with loud ambient restaurant noise (~70 dB; Fig. 12.14).

The results demonstrated that thinking about future ('later') consequences versus immediate ('now') sensations associated with the food decreased cravings. Furthermore, loud (vs soft) noise increased theta activity in the frontal part of the brain — a strong indicator of cognitive overload — and interestingly, this only occurred during 'later' trials. Peng-Li

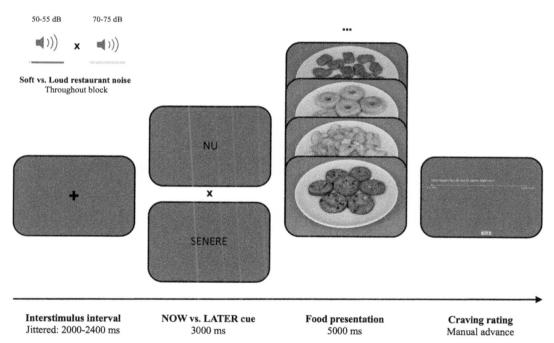

FIGURE 12.14   The adapted Regulation of Craving (ROC) task (Peng-Li, Alves Da Mota, et al., 2022). Before each trial, a jittered intertrial interval (fixation cross) is presented for 2–2.4 seconds. Then either a now or later cue (nu or senere in Danish) is shown for 3 seconds, followed by 5 seconds of exposure of a high-calorie food item. Finally, participants rate how much they want the presented food on a visual analogue scale (VAS) from 'not at all' to 'very much'. Either soft or loud noise is played in the background throughout the entire block. *Source: From Peng-Li, D., Alves Da Mota, P., Correa, C. M. C., Chan, R. C. K., Byrne, D. V., & Wang, Q. J. (2022). "Sound" decisions: The combined role of ambient noise and cognitive regulation on the neurophysiology of food cravings. Frontiers in Neuroscience, 16. https://doi.org/10.3389/fnins.2022.827021.*

et al. also found that the EDA signal peak probability was higher in the loud (vs soft) noise condition.

These findings provide direct support for the hypothesis that prospectively thinking about long-term consequences can effectively reduce food cravings, as demonstrated in Kober et al. (2010). At the same time, the results suggest that the underlying causal mechanisms of these self-regulated cravings may at least partially be explained through frontal brain oscillations. That is, analyses signified a partial mediation effect of decision perspective on self-reported cravings through frontal alpha power. This denotes that in particular augmented activity in the alpha frequency range is associated with increased cravings of high-calorie foods and potentially unhealthy eating behaviour. Additionally, irrespectively of behavioural ratings, the results showed that during delayed (vs immediate) gratification of food rewards, that is in later-trials, the PSD in both the theta and alpha frequency spectra as well as FAA were increased as an indication of higher emotional motivation and reward valuation (Fig. 12.15).

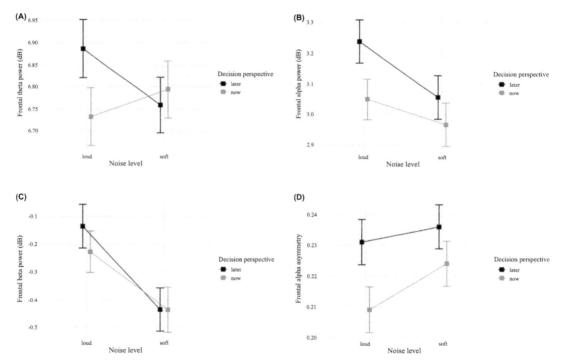

FIGURE 12.15 Electroencephalography (EEG) results (Peng-Li, Alves Da Mota, et al., 2022) visualised through interaction plots of (A) frontal theta power, (B) frontal alpha power, (C) frontal beta power frontal and (D) alpha asymmetry between noise level and decision perspective. Error bars represent standard error. *Source: From Peng-Li, D., Alves Da Mota, P., Correa, C. M. C., Chan, R. C. K., Byrne, D. V., & Wang, Q. J. (2022). "Sound" decisions: The combined role of ambient noise and cognitive regulation on the neurophysiology of food cravings. Frontiers in Neuroscience, 16. https://doi.org/10.3389/fnins.2022.827021.*

The finding that EDA peak detection was higher during the exposure to loud noise also indicated elevated arousal state (Kantono et al., 2019). Louder noise may lead to a more stressful mindset that in turn diminishes the cognitive resources requisite for processing and making more rational and healthy decisions (Caviola et al., 2021). In contrast, when consumers are not interrupted by loud restaurant noises, they are in a more relaxed psychological state, which places them in a better position of restraining and managing their irrational and unhealthy food choices (Peng-Li et al., 2021).

These findings indicate that the presence of loud ambient noise in conjunction with prospective thinking can lead to the highest emotional arousal and cognitive load both of which may be important drivers in regulating food cravings. The study exemplifies that these neurophysiological markers can generate unique information about people's sometimes unconscious behaviours which are induced by both bottom-up manipulations and top-down regulations. The study furthermore highlights the prospect of and need for considering both interoceptive states and exteroceptive cues, while employing different physiological measurements to more holistically, objectively and optimally study food-related decision-making that can result in an actual societal impact. This is not solely confined to the field of sensory and consumer neuroscience, but for any behavioural sciences, this seems applicable and highly pertinent.

12.3 Concluding remarks

In this chapter, we have introduced some of the most promising and emerging technologies in sensory and consumer science, encompassing ET, EDA and EEG. We have covered what they measure, how they measure it and finally, when and where they can be employed. ET serves as a useful tool when researchers and practitioners are interested in what consumers are visually attending to, that is before or during a food choice with high temporal and spatial precision. This is exceptionally valuable as consumers are often very bad at remembering past behaviour and therefore reporting retrospectively. Similarly, EDA and FEA provide psychophysiological insights into consumers' emotional and unconscious responses to food presentation, which would be otherwise impossible to acquire from traditional self-report questionnaires. Finally, EEG makes it possible to explore underlying neural correlates of various perceptual, emotional and cognitive processes happening in the brain. The technologies allow unbiased, implicit and dynamic responses, which in combination with conventional explicit measures facilitate a holistic understanding of consumer behaviour.

12.3.1 Theoretical and practical considerations

One should, however, also consider the challenges and limitations that these technologies involve prior to implementing them. Some of these comprise theoretical considerations, while others are of a more practical nature. Before conducting an experiment and concluding on the results, it is important to understand the theoretical foundation behind the technology — what it can do, but also what it cannot — in order to prevent meaningless inferences. All the above mentioned technologies are indeed able to measure implicit and unconscious responses, but none of them are magical mind readers. They can only measure what is captured, but one's empirical inferences should be based on the study design as a whole and not solely on arbitrary biometric measurements in isolation.

While ET is used for visual attention research, it does not necessarily always capture participants' attentional processes. In fact, it is capturing eye movements, which are computationally estimated based on the given eye-tracker's algorithms and the calibration process during the experiment. In a similar vein, EDA and FEA are widely used for capturing emotional responses. This is of course not incorrect, but one must keep in mind that EDA for example is measuring changes in the electrical conductivity between electrodes in response to an event. These are then decomposed into phasic (but not tonic) event-related EDA peaks (count, rate or amplitude), which are then inferred to be correlated with emotional arousal after signal processing, cleaning and decomposition. Likewise, FEA does not directly capture discrete emotions nor emotional valence per se. Instead, based on the particular type of FEA (e.g. Affectiva), the human-coded algorithm is classifying different facial expressions into what it 'believes' to be a smile, cheek raise, etc. This is then computed into a probability value of an emotion being present or not based on some weighting of the action units. Finally, EEG is a direct and sensitive measure of neurocortical activity. However, it might be difficult to interpret what the presence of augmented intensity of the specific frequency bands entails, even after complex signal amplification, processing and decomposition. In light of previous literature, they might be correlated with certain emotional and cognitive states and processes. For example, cognitive load tasks have induced increased frontal theta power. But does increased frontal theta power always entail higher cognitive workload? Or can frontal theta power occur even in the absence of cognitive load paradigms? Again, this is a problem of reverse inference (Hutzler, 2014) and important to keep in mind.

Challenges of a more practical nature includes the constraints related to the experimental setup that these technologies require. Collectively, they all require rather fixed and controlled laboratory setups, which naturally decrease the ecological validity. Some mitigation steps are more easily deployable, such as having consistent light conditions and nonobstructed eyes/faces for especially ET and FEA. Others are more comprehensive to implement and can cause detrimental effects on the outcome of the data. The most prominent might be attributed to movement artefacts, which are essentially relevant for all the technologies, but inherently crucial for especially EDA and EEG measurements, as these are very sensitive to noise and electrode connection. Although signal processing, including filtering and artefact rejection, can clean some of the raw data, highly noise-contaminated and poorly collected EDA and EEG data are difficult to extract meaningful and valid information out of. Therefore these measures should not always directly be implemented in traditional sensory and consumer tests. For instance, experiments involving tasting and especially chewing should be done with caution when collecting FEA or EEG data. The motion of chewing can affect the detection and classification of facial expression and generate electrical artefacts in the EEG signal. Having for example a reflection period after consumption of food, in which the relevant data are collected, can be beneficial. Thus the design of the experiment is critical as it can help circumvent some of these limitations.

12.3.2 Future perspectives

Despite these limitations, the current technical advances act as important steppingstones towards future research on promoting healthier and more sustainable eating and purchasing behaviours. By continuously addressing the psychophysiological aspects of these behaviours through reliable and replicable studies in combination with the increasing availability and usage of biosensors and neuroimaging, this could eventually become an integrative part of sensory and consumer science. With enough evidence and information gathered, new methodological standards could be set, which would push and improve the state-of-the-art research on multisensory influences on conscious and unconscious behaviours. This would generate a theoretical foundation, allowing more sound explorations of novel research directions.

Thus future studies should strive to implement relevant (neuro)physiological measures in combination with traditional methods to uncover undiscovered drivers of consumers' perception and behaviour. This would of course require and push more interdisciplinary research and interinstitutional collaboration for both idea generation as well as experimental execution. With the emerging popularity of these biometric technologies, it would be natural for sensory and consumer scientists to team up with, for example neuroscientists. This would not only support more robust EEG research but also potentially enable the exploitation of other more advanced techniques, such as fMRI.

fMRI measures offer the exploration of neurofunctional correlates of food-related behaviour on a much more spatially accurate scale (Smeets et al., 2019). This implies that neural activity in specific brain regions of interest can be precisely assessed in response to food cues or other sensory stimuli. With fMRI, researchers do not have to rely on mere electrophysical assumptions of source-specific neuronal behaviour based on electrode placement and mathematically transformed frequency data as in EEG. Instead, they can study and visualise the BOLD activity in task-relevant regions, such as in distinct visual cortices during visual presentation of food, in deeper limbic structures of the reward system when examining food valuation or in the various prefrontal cortices associated with certain cognitive functions. Some sensory and consumer studies are already using fMRI paradigms, but the evidence is at this point scarce. For an even stronger design, one could employ an EEG-compatible fMRI-system which would improve both spatial and temporal resolution of data and thereby improve the power of the research (Gorka et al., 2015).

In general, multimodal biometric measures, in combination with self-reported data, open up the possibility for new dimensions of human behaviour research. Researchers will be able to explore multiple aspects of the perceptual and behavioural manifestations of human–food interaction. Especially if the data streams from multiple sources are synchronised onto the same time series, causal inferences and statistical predictions become significantly more accessible and accurate. An increasing number of companies are realising this trend and thus facilitate such multimodal research. This allows researchers from academia and the commercial sector to tap into a specific point in time to explore a specific participant's time-locked perceptual, sensory and emotional response validated from several biosensors.

However, such comprehensive multimodal procedures in the laboratory will involve some challenges. That is, high experimental scrutiny and controllability often decrease the ecological validity. A naturalistic setting for the consumers is not quite a laboratory setting. At the same time, the restaurants and shopping environments are the natural habitats for multimodal biosensors. Fortunately, several technological developments are currently alleviating this limitation.

One way is to adopt virtual reality (VR). VR can generate immersive and realistic simulations of 3D environments, while at the same time offering the same degree of controllability as in conventional laboratory studies (for a review, see Wang et al. (2021)). With its increasing popularity follows technological enhancement, which makes contemporary VR systems more and more compatible with different biosensors, such as ET and EEG. In combination with ET gaze behaviour build into the VR-system and an EEG headset attached, the technology could monitor consumers' visual attention and brain correlates, not solely towards static food images but in various realistic eating or shopping contexts and without having the expense and necessity to physically move to the central location (Pennanen et al., 2020; Zhang et al., 2019).

In recent years, we have seen a push towards the development of the miniaturisation and commercialisation of mobile biosensors, including EDA, EEG, ET and HR devices, making them more suitable for real-life research scenarios (Folwarczny et al., 2019). In combination with live video recordings of, for example, FEA and tracking of GPS coordinates, it enables holistic monitoring of consumers' conscious and unconscious behaviours. Some of these nonintrusive wearable sensors could even be applied for 24/7 activity monitoring of the consumers (Smets et al., 2018). Smartwatches, in combination with smartphones, are already commercially available devices that allow tracking and feedback of multiple types of physiological responses and physical activities. One could imagine that these continuous measurements could also be applied for monitoring daily and weekly responses to food and beverages — from instantaneous emotional responses to longer-lasting habitual behaviours. This might set the future standards for tracking eating behaviour in sensory and consumer science, as smartwatches allow for uninterrupted, nonintrusive, unobtrusive, objective and ecologically valid human behaviour data.

Nevertheless, through this chapter, we have seen that sensory and consumer research is in fact inclining towards the employment of biometric technologies that enable real-time data collection of human behaviour. Researchers in the field have realised the insights such implicit and objective methods can yield — not in isolation but as complemental information to traditional self-report measures which can be prone to subjective biases. It is not surprising if this biometric path we are on will continue and follow the same trend as seen in the publication rate of ET studies for many years into the future.

Funding

The curation of this research collection was funded by the Sino Danish Centre.

Acknowledgements

The authors thank Food Quality Perception & Society Team and iSENSE Lab at the Department of Food Science at Aarhus University, Denmark.

Conflicts of interest

The authors declare no conflict of interest.

References

Abrams, R. A., Meyer, D. E., & Kornblum, S. (1989). Speed and accuracy of saccadic eye movements: Characteristics of impulse variability in the oculomotor system. *Journal of Experimental Psychology: Human Perception and Performance, 15*(3), 529–543. Available from https://doi.org/10.1037/0096-1523.15.3.529; http://doi.apa.org/getdoi.cfm?doi = 10.1037/0096-1523.15.3.529.

Aldayel, M., Ykhlef, M., & Al-Nafjan, A. (2020). Deep learning for EEG-based preference classification in neuromarketing. *Applied Sciences, 10*(4), 1525. Available from https://doi.org/10.3390/app10041525; https://www.mdpi.com/2076-3417/10/4/1525.

AL-Khalidi, F. Q., Saatchi, R., Burke, D., Elphick, H., & Tan, S. (2011). Respiration rate monitoring methods: A review. *Pediatric Pulmonology, 46*(6), 523–529. Available from https://doi.org/10.1002/ppul.21416.

Allen, J. J. B., Coan, J. A., & Nazarian, M. (2004). Issues and assumptions on the road from raw signals to metrics of frontal EEG asymmetry in emotion. *Biological Psychology, 67*(1-2), 183–218. Available from https://doi.org/10.1016/j.biopsycho.2004.03.007; https://linkinghub.elsevier.com/retrieve/pii/S0301051104000377.

Antonenko, P., Paas, F., Grabner, R., & van Gog, T. (2010). Using electroencephalography to measure cognitive load. *Educational Psychology Review, 22*(4), 425–438. Available from https://doi.org/10.1007/s10648-010-9130-y; http://link.springer.com/10.1007/s10648-010-9130-y.

Aoh, Y., Hsiao, H.-J., Lu, M.-K., Macerollo, A., Huang, H.-C., Hamada, M., Tsai, C.-H., & Chen, J.-C. (2019). Event-related desynchronization/synchronization in spinocerebellar ataxia type 3. *Frontiers in Neurology, 10*. Available from https://doi.org/10.3389/fneur.2019.00822; https://www.frontiersin.org/article/10.3389/fneur.2019.00822/full.

Ariely, D., & Berns, G. S. (2010). Neuromarketing: The hope and hype of neuroimaging in business. *Nature Reviews Neuroscience, 11*(4), 284–292. Available from https://doi.org/10.1038/nrn2795; http://www.nature.com/articles/nrn2795.

Arnsten, A. A., & Li, B. (2004). Neurobiology of executive functions: Catecholamine influences on prefrontal cortical functions. *Biological Psychiatry.* Available from https://doi.org/10.1016/j.bps.2004.08.019; https://linkinghub.elsevier.com/retrieve/pii/S0006322304009333.

Barrett, L. F. (1998). Discrete emotions or dimensions? The role of valence focus and arousal focus. *Cognition & Emotion, 12*(4), 579–599. Available from https://doi.org/10.1080/026999398379574; http://www.tandfonline.com/doi/abs/10.1080/026999398379574.

Bazzani, A., Ravaioli, S., Trieste, L., Faraguna, U., & Turchetti, G. (2020). Is EEG suitable for marketing research? A systematic review. *Frontiers in Neuroscience, 14*(December). Available from https://doi.org/10.3389/fnins.2020.594566; https://www.frontiersin.org/articles/10.3389/fnins.2020.594566/full.

Bechara, A., & Damasio, A. R. (2005). The somatic marker hypothesis: A neural theory of economic decision. *Games and Economic Behavior, 52*(2), 336–372. Available from https://doi.org/10.1016/j.geb.2004.06.010; https://linkinghub.elsevier.com/retrieve/pii/S0899825604001034.

Belkaid, M., Cuperlier, N., & Gaussier, P. (2017). Emotional metacontrol of attention: Top-down modulation of sensorimotor processes in a robotic visual search task. *PLoS One, 12*(9), e0184960. Available from https://doi.org/10.1371/journal.pone.0184960; https://dx.plos.org/10.1371/journal.pone.0184960.

Berridge, K. C., & Kringelbach, M. L. (2015). Pleasure systems in the brain. *Neuron, 86*(3), 646–664. Available from https://doi.org/10.1016/j.neuron.2015.02.018; https://linkinghub.elsevier.com/retrieve/pii/S0896627315001336.

Bialkova, S., Grunert, K. G., Juhl, H. J., Wasowicz-Kirylo, G., Stysko-Kunkowska, M., & van Trijp, H. C. M. (2014). Attention mediates the effect of nutrition label information on consumers' choice. Evidence from a choice experiment involving eye-tracking. *Appetite, 76*, 66–75. Available from https://doi.org/10.1016/j.appet.2013.11.021; https://linkinghub.elsevier.com/retrieve/pii/S0195666314000816.

Bialkova, S., Grunert, K. G., & van Trijp, H. (2020). From desktop to supermarket shelf: Eye-tracking exploration on consumer attention and choice. *Food Quality and Preference, 81*(May 2019), 103839. Available from https://doi.org/10.1016/j.foodqual.2019.103839; https://linkinghub.elsevier.com/retrieve/pii/S0950329319303696.

Biehl, S. C., Keil, J., Naumann, E., & Svaldi, J. (2020). ERP and oscillatory differences in overweight/obese and normal-weight adolescents in response to food stimuli. *Journal of Eating Disorders, 8*(1), 1–11. Available from https://doi.org/10.1186/s40337-020-00290-8.

Billman, G. E., Huikuri, H. V., Sacha, J., & Trimmel, K. (2015). An introduction to heart rate variability: Methodological considerations and clinical applications. *Frontiers in Physiology, 6*. Available from https://doi.org/10.3389/fphys.2015.00055.

Boucsein, W., Fowles, D. C., Grimnes, S., Ben-Shakhar, G., Roth, W. T., Dawson, M. E., & Filion, D. (2012). Publication recommendations for electrodermal measurements. *Psychophysiology*, *49*(8), 1017−1034. Available from https://doi.org/10.1111/j.1469-8986.2012.01384.x; https://onlinelibrary.wiley.com/doi/10.1111/j.1469-8986.2012.01384.x.

Boucsein, W. (2012). *Electrodermal activity*. Boston, MA: Springer US. Available from http://link.springer.com/10.1007/978-1-4614-1126-0; https://doi.org/10.1007/978-1-4614-1126-0.

Brouwer, A.-M., Hogervorst, M. A., van Erp, J. B. F., Heffelaar, T., Zimmerman, P. H., & Oostenveld, R. (2012). Estimating workload using EEG spectral power and ERPs in the n-back task. *Journal of Neural Engineering*, *9*(4), 045008. Available from https://doi.org/10.1088/1741-2560/9/4/045008; https://iopscience.iop.org/article/10.1088/1741-2560/9/4/045008.

Cabañero., Hervás., González., Fontecha., Mondéjar., & Bravo. (2019). Analysis of cognitive load using EEG when interacting with mobile devices. *Proceedings*, *31*(1), 70. Available from https://doi.org/10.3390/proceedings2019031070; https://www.mdpi.com/2504-3900/31/1/70.

Carrasco, M. (2011). Visual attention: The past 25 years. *Vision Research*, *51*(13), 1484−1525. Available from https://doi.org/10.1016/j.visres.2011.04.012; https://linkinghub.elsevier.com/retrieve/pii/S0042698911001544.

Carter, B. T., & Luke, S. G. (2020). Best practices in eye tracking research. *International Journal of Psychophysiology*, *155*, 49−62. Available from https://doi.org/10.1016/j.ijpsycho.2020.05.010; https://linkinghub.elsevier.com/retrieve/pii/S0167876020301458.

Caviola, S., Visentin, C., Borella, E., Mammarella, I., & Prodi, N. (2021). Out of the noise: Effects of sound environment on maths performance in middle-school students. *Journal of Environmental Psychology*, *73*, 101552. Available from https://doi.org/10.1016/j.jenvp.2021.101552; https://linkinghub.elsevier.com/retrieve/pii/S0272494421000050.

Clement, J., Aastrup, J., & Forsberg, S. C. (2015). Decisive visual saliency and consumers' in-store decisions. *Journal of Retailing and Consumer Services*, *22*, 187−194. Available from https://doi.org/10.1016/j.jretconser.2014.09.002; https://linkinghub.elsevier.com/retrieve/pii/S0969698914001295.

Clement, J. (2007). Visual influence on in-store buying decisions: An eye-track experiment on the visual influence of packaging design. *Journal of Marketing Management*, *23*(9-10), 917−928. Available from https://doi.org/10.1362/026725707X250395; http://www.tandfonline.com/doi/abs/10.1362/026725707X250395.

Cohen, M. (2014). *Analyzing neural time series data*. MIT Press.

Corbetta, M., & Shulman, G. L. (2002). Control of goal-directed and stimulus-driven attention in the brain. *Nature Reviews. Neuroscience*, *3*(3), 201−215. Available from https://doi.org/10.1038/nrn755; http://www.nature.com/articles/nrn755.

da Costa, R. Q. M., Furukawa, E., Hoefle, S., Moll, J., Tripp, G., & Mattos, P. (2020). An adaptation of pavlovian-to-instrumental transfer (PIT) methodology to examine the energizing effects of reward-predicting cues on behavior in young adults. *Frontiers in Psychology*, *11*(February), 1−9. Available from https://doi.org/10.3389/fpsyg.2020.00195.

Danner, L., Haindl, S., Joechl, M., & Duerrschmid, K. (2014). Facial expressions and autonomous nervous system responses elicited by tasting different juices. *Food Research International*, *64*, 81−90. Available from https://doi.org/10.1016/j.foodres.2014.06.003; https://linkinghub.elsevier.com/retrieve/pii/S0963996914003950.

de Wijk, R. A., Kooijman, V., Verhoeven, R. H. G., Holthuysen, N. T. E., & de Graaf, C. (2012). Autonomic nervous system responses on and facial expressions to the sight, smell, and taste of liked and disliked foods. *Food Quality and Preference*, *26*(2), 196−203. Available from https://doi.org/10.1016/j.foodqual.2012.04.015.

de Wijk, R. A., & Noldus, L. P. J. J. (2021). Using implicit rather than explicit measures of emotions. *Food Quality and Preference*, *92*(November 2020), 104125. Available from https://doi.org/10.1016/j.foodqual.2020.104125; https://linkinghub.elsevier.com/retrieve/pii/S0950329320303943.

de Wijk, R. A., Ushiama, S., Ummels, M. J., Zimmerman, P. H., Kaneko, D., & Vingerhoeds, M. H. (2021). Effect of branding and familiarity of soy sauces on valence and arousal as determined by facial expressions, physiological measures, emojis, and ratings. *Frontiers in Neuroergonomics*, *2*. Available from https://doi.org/10.3389/fnrgo.2021.651682; https://www.frontiersin.org/articles/10.3389/fnrgo.2021.651682/full.

Delorme, A., & Makeig, S. (2004). EEGLAB: An open source toolbox for analysis of single-trial EEG dynamics including independent component analysis. *Journal of Neuroscience Methods*, *134*(1), 9−21. Available from https://doi.org/10.1016/j.jneumeth.2003.10.009; https://linkinghub.elsevier.com/retrieve/pii/S0165027003003479.

Dijksterhuis, G. B., & Byrne, D. V. (2005). Does the mind reflect the mouth? Sensory profiling and the future. *Critical Reviews in Food Science and Nutrition*, *45*(7-8), 527−534. Available from https://doi.org/10.1080/10408690590907660.

Domracheva, M., & Kulikova, S. (2020). EEG correlates of perceived food product similarity in a cross-modal taste-visual task. *Food Quality and Preference, 85*, 103980. Available from https://doi.org/10.1016/j.foodqual.2020.103980; https://linkinghub.elsevier.com/retrieve/pii/S0950329320302494.

Edwards, K., Thomas, J., Higgs, S., & Blissett, J. (2022). The effect of models' facial expressions towards raw broccoli on food liking and desire to eat. *Appetite, 169*, 105499. Available from https://doi.org/10.1016/j.appet.2021.105499.

Egecioglu, E., Skibicka, K. P., Hansson, C., Alvarez-Crespo, M., Friberg, P. A., Jerlhag, E., Engel, J. A., Dickson, S. L., Anders Friberg, P., Jerlhag, E., Engel, J. A., & Dickson, S. L. (2011). Hedonic and incentive signals for body weight control. *Reviews in Endocrine and Metabolic Disorders, 12*(3), 141−151. Available from https://doi.org/10.1007/s11154-011-9166-4; http://link.springer.com/10.1007/s11154-011-9166-4.

Fischer, N. L., Peres, R., & Fiorani, M. (2018). Frontal alpha asymmetry and theta oscillations associated with information sharing intention. *Frontiers in Behavioral Neuroscience, 12*(August), 1−12. Available from https://doi.org/10.3389/fnbeh.2018.00166; https://www.frontiersin.org/article/10.3389/fnbeh.2018.00166/full.

Fitzsimons, G. J., Hutchinson, J. W., Williams, P., Alba, J. W., Chartrand, T. L., Huber, J., Kardes, F. R., Menon, G., Raghubir, P., Russo, J. E., Shiv, B., & Tavassoli, N. T. (2002). Non-conscious influences on consumer choice. *Marketing Letters, 13*(3), 269−279. Available from https://doi.org/10.1023/A:1020313710388; http://link.springer.com/10.1023/A:1020313710388.

Folwarczny, M., Pawar, S., Sigurdsson, V., & Fagerstrøm, A. (2019). Using neuro-IS/consumer neuroscience tools to study healthy food choices: A review. *Procedia Computer Science, 164*, 532−537. Available from https://doi.org/10.1016/j.procs.2019.12.216; https://linkinghub.elsevier.com/retrieve/pii/S187705091932263X.

Galler, M., Grendstad, Á. R., Ares, G., & Varela, P. (2022). Capturing food-elicited emotions: Facial decoding of children's implicit and explicit responses to tasted samples. *Food Quality and Preference, 99*, 104551. Available from https://doi.org/10.1016/j.foodqual.2022.104551.

Gorka, S. M., Phan, K. L., & Shankman, S. A. (2015). Convergence of EEG and fMRI measures of reward anticipation. *Biological Psychology, 112*, 12−19. Available from https://doi.org/10.1016/j.biopsycho.2015.09.007.

Granholm, E., Asarnow, R. F., Sarkin, A. J., & Dykes, K. L. (1996). Pupillary responses index cognitive resource limitations. *Psychophysiology, 33*(4), 457−461. Available from https://doi.org/10.1111/j.1469-8986.1996.tb01071.x; https://onlinelibrary.wiley.com/doi/10.1111/j.1469-8986.1996.tb01071.x.

Gunaratne, T. M., Fuentes, S., Gunaratne, N. M., Torrico, D. D., Gonzalez Viejo, C., & Dunshea, F. R. (2019). Physiological responses to basic tastes for sensory evaluation of chocolate using biometric techniques. *Foods, 8*(7), 243. Available from https://doi.org/10.3390/foods8070243; https://www.mdpi.com/2304-8158/8/7/243.

Hauser, C. K., & Salinas, E. (2014). *Perceptual decision making encyclopedia of computational neuroscience* (pp. 1−21). New York, NY: Springer New York. Available from http://link.springer.com/10.1007/978-1-4614-7320-6_317-1; https://doi.org/10.1007/978-1-4614-7320-6_317-1.

He, W., Boesveldt, S., Delplanque, S., de Graaf, C., & de Wijk, R. A. (2017). Sensory-specific satiety: Added insights from autonomic nervous system responses and facial expressions. *Physiology & Behavior, 170*, 12−18. Available from https://doi.org/10.1016/j.physbeh.2016.12.012; https://linkinghub.elsevier.com/retrieve/pii/S0031938416311519.

Hernando-Gallego F., & Artés-Rodríguez, A. (2015). *Individual performance calibration using physiological stress signals.* Available from http://arxiv.org/abs/1507.03482.

Holmqvist, K., Örbom, S. L., Hooge, I. T. C., Niehorster, D. C., Alexander, R. G., Andersson, R., Benjamins, J. S., Blignaut, P., Brouwer, A.-N., Chuang, L. L., Dalrymple, K. A., Drieghe, D., Dunn, M. J., Ettinger, U., Fiedler, S., Foulsham, T., van der Geest, J. N., Hansen, D. W., . . . Hessels, R. S. (2023). Eye tracking: empirical foundations for a minimal reporting guideline. *Behavior Research Methods, 55*, 364−416. Available from https://doi.org/10.3758/s13428-021-01762-8.

Hume, D. J., Howells, F. M., Karpul, D., Rauch, H. G. L., Kroff, J., & Lambert, E. V. (2015). Cognitive control over visual food cue saliency is greater in reduced-overweight/obese but not in weight relapsed women: An EEG study. *Eating Behaviors, 19*, 76−80. Available from https://doi.org/10.1016/j.eatbeh.2015.06.013.

Hutzler, F. (2014). Reverse inference is not a fallacy per se: Cognitive processes can be inferred from functional imaging data. *Neuroimage, 84*, 1061−1069. Available from https://doi.org/10.1016/j.neuroimage.2012.12.075.

iMotions. (2017a). *Eye tracking: The complete pocket guide.* https://imotions.com/blog/eye-tracking/.

iMotions. (2017b). *Galvanic skin response: The complete pocket guide.*

Jaeger, S. R., Roigard, C. M., & Chheang, S. L. (2021). The valence × arousal circumplex-inspired emotion questionnaire (CEQ): Effect of response format and question layout. *Food Quality and Preference, 90*, 104172.

Available from https://doi.org/10.1016/j.foodqual.2020.104172; https://linkinghub.elsevier.com/retrieve/pii/S0950329320304419.

Just, M. A., & Carpenter, P. A. (1976). Eye fixations and cognitive processes. *Cognitive Psychology, 8*(4), 441–480. Available from https://doi.org/10.1016/0010-0285(76)90015-3; https://linkinghub.elsevier.com/retrieve/pii/0010028576900153.

Kaneko, D., Stuldreher, I., Reuten, A. J. C., Toet, A., van Erp, J. B. F., & Brouwer, A.-M. (2021). Comparing explicit and implicit measures for assessing cross-cultural food experience. *Frontiers in Neuroergonomics, 2.* Available from https://doi.org/10.3389/fnrgo.2021.646280; https://www.frontiersin.org/articles/10.3389/fnrgo.2021.646280/full.

Kantono, K., Hamid, N., Shepherd, D., Lin, Y. H. T., Skiredj, S., & Carr, B. T. (2019). Emotional and electrophysiological measures correlate to flavour perception in the presence of music. *Physiology & Behavior, 199*(November 2018), 154–164. Available from https://doi.org/10.1016/j.physbeh.2018.11.012; https://linkinghub.elsevier.com/retrieve/pii/S0031938418310138.

Kirsten, H., Seib-Pfeifer, L.-E., & Gibbons, H. (2022). Effects of the calorie content of visual food stimuli and simulated situations on event-related frontal alpha asymmetry and event-related potentials in the context of food choices. *Appetite, 169,* 105805. Available from https://doi.org/10.1016/j.appet.2021.105805.

Kober, H., Mende-Siedlecki, P., Kross, E. F., Weber, J., Mischel, W., Hart, C. L., & Ochsner, K. N. (2010). Prefrontal-striatal pathway underlies cognitive regulation of craving. *Proceedings of the National Academy of Sciences of the United States of America, 107*(33), 14811–14816. Available from https://doi.org/10.1073/pnas.1007779107.

Krahnke, K. J., & Krashen, S. D. (1983). Principles and practice in second language acquisition. *TESOL Quarterly, 17*(2), 300. Available from https://doi.org/10.2307/3586656; https://www.jstor.org/stable/3586656?origin = crossref.

Kuhlmann, T., Dantlgraber, M., & Reips, U.-D. (2017). Investigating measurement equivalence of visual analogue scales and Likert-type scales in Internet-based personality questionnaires. *Behavior Research Methods, 49*(6), 2173–2181. Available from https://doi.org/10.3758/s13428-016-0850-x; http://link.springer.com/10.3758/s13428-016-0850-x.

Kytö, E., Bult, H., Aarts, E., Wegman, J., Ruijschop, R. M. A. J., & Mustonen, S. (2019). Comparison of explicit vs. implicit measurements in predicting food purchases. *Food Quality and Preference, 78*(June). Available from https://doi.org/10.1016/j.foodqual.2019.103733.

Lawless, H. T., & Heymann, H. (2010). *Sensory evaluation of food.* New York, NY: Springer New York. Available from https://doi.org/10.1007/978-1-4419-6488-5.

LeDoux, J. (2012). Rethinking the emotional brain. *Neuron, 73*(4), 653–676. Available from https://doi.org/10.1016/j.neuron.2012.02.004; https://linkinghub.elsevier.com/retrieve/pii/S0896627312001298.

Leitch, K. A., Duncan, S. E., O'Keefe, S., Rudd, R., & Gallagher, D. L. (2015). Characterizing consumer emotional response to sweeteners using an emotion terminology questionnaire and facial expression analysis. *Food Research International, 76,* 283–292. Available from https://doi.org/10.1016/j.foodres.2015.04.039.

Luck, S. (2005). *An introduction to the event-related potential technique.* MIT Press.

Macaluso, E. (2010). Orienting of spatial attention and the interplay between the senses. *Cortex; a Journal Devoted to the Study of the Nervous System and Behavior, 46*(3), 282–297. Available from https://doi.org/10.1016/j.cortex.2009.05.010; https://linkinghub.elsevier.com/retrieve/pii/S0010945209001622.

Mai, R., Symmank, C., & Seeberg-Elverfeldt, B. (2016). Light and pale colors in food packaging: When does this package cue signal superior healthiness or inferior tastiness? *Journal of Retailing, 92*(4), 426–444. Available from https://doi.org/10.1016/j.jretai.2016.08.002.

Manippa, V., van der Laan, L. N., Brancucci, A., & Smeets, P. A. M. (2019). Health body priming and food choice: An eye tracking study. *Food Quality and Preference, 72*(October 2018), 116–125. Available from https://doi.org/10.1016/j.foodqual.2018.10.006; https://linkinghub.elsevier.com/retrieve/pii/S0950329318301745.

Masuda, T., Wang, H., Ishii, K., & Ito, K. (2012). Do surrounding figures' emotions affect judgment of the target figure's emotion? Comparing the eye-movement patterns of European Canadians, Asian Canadians, Asian international students, and Japanese. *Frontiers in Integrative Neuroscience, 6.* Available from https://doi.org/10.3389/fnint.2012.00072; http://journal.frontiersin.org/article/10.3389/fnint.2012.00072/abstract.

Mawad, F., Maiche, A., Ares, G., Trías, M., & Giménez, A. (2015). Influence of cognitive style on information processing and selection of yogurt labels: Insights from an eye-tracking study. *Food Research International, 74,* 1–9. Available from https://doi.org/10.1016/j.foodres.2015.04.023.

McDuff, D. (2016). Discovering facial expressions for states of amused, persuaded, informed, sentimental and inspired. *Proceedings of the 18th ACM International Conference on Multimodal Interaction, 71—75*. Available from https://doi.org/10.1145/2993148.2993192.

McGeown, L., & Davis, R. (2018). Frontal EEG asymmetry moderates the association between attentional bias towards food and body mass index. *Biological Psychology, 136*(January), 151—160. Available from https://doi.org/10.1016/j.biopsycho.2018.06.001; https://linkinghub.elsevier.com/retrieve/pii/S0301051118300267.

Meilgaard, M. C., Carr, B. T., & Carr, B. T. (2006). *Sensory evaluation techniques*. CRC Press. Available from https://doi.org/10.1201/b16452.

Mento, M. A. (June, 2020): *Different Kinds of Eye Tracking Devices*. Bitbrain. https://www.bitbrain.com/blog/eye-tracking-devices.

Meule, A., Kübler, A., & Blechert, J. (2013). Time course of electrocortical food-cue responses during cognitive regulation of craving. *Frontiers in Psychology, 4*(September), 1—11. Available from https://doi.org/10.3389/fpsyg.2013.00669.

Moore, T., & Zirnsak, M. (2017). Neural mechanisms of selective visual attention. *Annual Review of Psychology, 68* (1), 47—72. Available from https://doi.org/10.1146/annurev-psych-122414-033400; http://www.annualreviews.org/doi/10.1146/annurev-psych-122414-033400.

Motoki, K., Saito, T., Nouchi, R., Kawashima, R., & Sugiura, M. (2018). Tastiness but not healthfulness captures automatic visual attention: Preliminary evidence from an eye-tracking study. *Food Quality and Preference, 64*, 148—153. Available from https://doi.org/10.1016/j.foodqual.2017.09.014; https://linkinghub.elsevier.com/retrieve/pii/S095032931730215X.

Nagel S. (2019). Towards a home-use BCI: Fast asynchronous control and robust non-control state detection. https://publikationen.uni-tuebingen.de/xmlui/bitstream/handle/10900/96356/Sebastian_Nagel_Dissertation.pdf?sequence = 1&isAllowed = y. https://doi.org/10.15496/publikation-37739.

Nguyen, D., Naffziger, E. E., & Berridge, K. C. (2021). Positive affect: Nature and brain bases of liking and wanting. *Current Opinion in Behavioral Sciences, 39*, 72—78. Available from https://doi.org/10.1016/j.cobeha.2021.02.013.

Noudoost, B., Chang, M. H., Steinmetz, N. A., & Moore, T. (2010). Top-down control of visual attention. *Current Opinion in Neurobiology, 20*(2), 183—190. Available from https://doi.org/10.1016/j.conb.2010.02.003; https://linkinghub.elsevier.com/retrieve/pii/S0959438810000255.

Orquin, J. L., & Holmqvist, K. (2018). Threats to the validity of eye-movement research in psychology. *Behavior Research Methods, 50*(4), 1645—1656. Available from https://doi.org/10.3758/s13428-017-0998-z; http://link.springer.com/10.3758/s13428-017-0998-z.

Orquin, J. L., & Holmqvist, K. (2019). *A primer on eye tracking methodology for behavioral science. s. A handbook of process tracing methods*. Routledge.

Orquin, J. L., & Loose, S. M. (2013). Attention and choice: A review on eye movements in decision making. *Acta Psychologica, 144*(1), 190—206. Available from https://doi.org/10.1016/j.actpsy.2013.06.003; https://linkinghub.elsevier.com/retrieve/pii/S0001691813001364.

Papies, E. K., Barsalou, L. W., & Rusz, D. (2020). Understanding desire for food and drink: A grounded-cognition approach. *Current Directions in Psychological Science, 29*(2), 193—198. Available from https://doi.org/10.1177/0963721420904958; http://journals.sagepub.com/doi/10.1177/0963721420904958.

Pedersen, H., Quist, J. S., Jensen, M. M., Clemmensen, K. K. B., Vistisen, D., Jørgensen, M. E., Færch, K., & Finlayson, G. (2021). Investigation of eye tracking, electrodermal activity and facial expressions as biometric signatures of food reward and intake in normal weight adults. *Food Quality and Preference*, 104248. Available from https://doi.org/10.1016/j.foodqual.2021.104248; https://linkinghub.elsevier.com/retrieve/pii/S0950329321001312.

Peng-Li, D., Alves Da Mota, P., Correa, C. M. C., Chan, R. C. K., Byrne, D. V., & Wang, Q. J. (2022). "Sound" decisions: The combined role of ambient noise and cognitive regulation on the neurophysiology of food cravings. *Frontiers in Neuroscience, 16*. Available from https://doi.org/10.3389/fnins.2022.827021; https://www.frontiersin.org/articles/10.3389/fnins.2022.827021/full.

Peng-Li, D., Andersen, T., Finlayson, G., Byrne, D. V., & Wang, Q. J. (2022). The impact of environmental sounds on food reward. *Physiology & Behavior, 245*, 113689. Available from https://doi.org/10.1016/j.physbeh.2021.113689; https://linkinghub.elsevier.com/retrieve/pii/S0031938421003760.

Peng-Li, D., Byrne, D. V., Chan, R. C. K., & Wang, Q. J. (2020). The influence of taste-congruent soundtracks on visual attention and food choice: A cross-cultural eye-tracking study in Chinese and Danish consumers.

3. Digitalization in instrumental, neurological, psychological and behavioural methods:
Current applications and opportunities

Food Quality and Preference, 85(May), 103962. Available from https://doi.org/10.1016/j.foodqual.2020.103962; https://linkinghub.elsevier.com/retrieve/pii/S0950329320302317.

Peng-Li, D., Chan, R. C. K., Byrne, D. V., & Wang, Q. J. (2020). The effects of ethnically congruent music on eye movements and food choice—A cross-cultural comparison between Danish and Chinese consumers. *Foods, 9*(8), 1109. Available from https://doi.org/10.3390/foods9081109; https://www.mdpi.com/2304-8158/9/8/1109.

Peng-Li, D., Mathiesen, S. L., Chan, R. C. K., Byrne, D. V., & Wang, Q. J. (2021). Sounds healthy: Modelling sound-evoked consumer food choice through visual attention. *Appetite, 164*(April), 105264. Available from https://doi.org/10.1016/j.appet.2021.105264; https://linkinghub.elsevier.com/retrieve/pii/S0195666321001719.

Peng-Li, D., Sørensen, T. A., Li, Y., & He, Q. (2020). Systematically lower structural brain connectivity in individuals with elevated food addiction symptoms. *Appetite, 155*, 104850. Available from https://doi.org/10.1016/j.appet.2020.104850; https://linkinghub.elsevier.com/retrieve/pii/S0195666320304104.

Pennanen, K., Närväinen, J., Vanhatalo, S., Raisamo, R., & Sozer, N. (2020). Effect of virtual eating environment on consumers' evaluations of healthy and unhealthy snacks. *Food Quality and Preference, 82*(October 2019), 103871. Available from https://doi.org/10.1016/j.foodqual.2020.103871; https://linkinghub.elsevier.com/retrieve/pii/S0950329319308328.

Peschel, A. O., Orquin, J. L., & Mueller Loose, S. (2019). Increasing consumers' attention capture and food choice through bottom-up effects. *Appetite, 132*(August 2018), 1–7. Available from https://doi.org/10.1016/j.appet.2018.09.015; https://linkinghub.elsevier.com/retrieve/pii/S0195666318308729.

Petit, O., Basso, F., Merunka, D., Spence, C., Cheok, A. D., & Oullier, O. (2016). Pleasure and the control of food intake: An embodied cognition approach to consumer self-regulation. *Psychology & Marketing, 33*(8), 608–619. Available from https://doi.org/10.1002/mar.20903; http://doi.wiley.com/10.1002/mar.20903.

Picton, T. (1991). American electroencephalographic society guidelines for standard electrode position nomenclature. *Journal of Clinical Neurophysiology, 8*(2), 200–202. Available from https://doi.org/10.1097/00004691-199104000-00007; http://journals.lww.com/00004691-199104000-00007.

Posner, J., Russell, J. A., & Peterson, B. S. (2005). The circumplex model of affect: An integrative approach to affective neuroscience, cognitive development, and psychopathology. *Development and Psychopathology, 17*(03). Available from https://doi.org/10.1017/S0954579405050340; http://www.journals.cambridge.org/abstract_S0954579405050340.

Potthoff, J., & Schienle, A. (2020). Time-course analysis of food cue processing: An eye-tracking investigation on context effects. *Food Quality and Preference, 84*. Available from https://doi.org/10.1016/j.foodqual.2020.103936; https://linkinghub.elsevier.com/retrieve/pii/S0950329320302056.

Qiu, R., Qi, Y., & Wan, X. (2020). An event-related potential study of consumers' responses to food bundles. *Appetite, 147*(May 2019). Available from https://doi.org/10.1016/j.appet.2019.104538.

Ramsøy, T. Z., Skov, M., Christensen, M. K., & Stahlhut, C. (2018). Frontal brain asymmetry and willingness to pay. *Frontiers in Neuroscience* (12 (March). Available from https://doi.org/10.3389/fnins.2018.00138; http://journal.frontiersin.org/article/10.3389/fnins.2018.00138/full.

Rayner, K. (2009). The 35th Sir Frederick Bartlett Lecture: Eye movements and attention in reading, scene perception, and visual search. *Quarterly Journal of Experimental Psychology, 62*(8), 1457–1506. Available from https://doi.org/10.1080/17470210902816461; http://journals.sagepub.com/doi/10.1080/17470210902816461.

Robinson, M. J. F., Robinson, T. E., & Berridge, K. C. (2013). Incentive salience and the transition to addiction. *Biological Research on Addiction. Elsevier*, 391–399. Available from https://doi.org/10.1016/B978-0-12-398335-0.00039-X; https://linkinghub.elsevier.com/retrieve/pii/B978012398335000039X.

Rogers, P. J., & Hardman, C. A. (2015). Food reward. What it is and how to measure it. *Appetite, 90*, 1–15. Available from https://doi.org/10.1016/j.appet.2015.02.032.

Romero, M., & Biswas, D. (2016). Healthy-left, unhealthy-right: Can displaying healthy items to the left (versus right) of unhealthy items nudge healthier choices? *Journal of Consumer Research, 43*(1), 103–112. Available from https://doi.org/10.1093/jcr/ucw008; https://academic.oup.com/jcr/article-lookup/doi/10.1093/jcr/ucw008.

Russell, J. A. (1980). A circumplex model of affect. *Journal of Personality and Social Psychology, 39*(6), 1161–1178. Available from https://doi.org/10.1037/h0077714; http://content.apa.org/journals/psp/39/6/1161.

Samant, S. S., & Seo, H.-S. (2020). Influences of sensory attribute intensity, emotional responses, and non-sensory factors on purchase intent toward mixed-vegetable juice products under informed tasting condition. *Food Research International, 132*, 109095. Available from https://doi.org/10.1016/j.foodres.2020.109095.

3. Digitalization in instrumental, neurological, psychological and behavioural methods:
Current applications and opportunities

Schubert, E., Smith, E., Brydevall, M., Lynch, C., Ringin, E., Dixon, H., Kashima, Y., Wakefield, M., & Bode, S. (2021). General and specific graphic health warning labels reduce willingness to consume sugar-sweetened beverages. *Appetite, 161*, 105141. Available from https://doi.org/10.1016/j.appet.2021.105141.

Seo, H.-S. (2020). Sensory nudges: The influences of environmental contexts on consumers' sensory perception, emotional responses, and behaviors toward foods and beverages. *Foods, 9*(4), 509. Available from https://doi.org/10.3390/foods9040509; https://www.mdpi.com/2304-8158/9/4/509.

Smeets, P. A. M., Dagher, A., Hare, T. A., Kullmann, S., van der Laan, L. N., Poldrack, R. A., Preissl, H., Small, D., Stice, E., & Veldhuizen, M. G. (2019). Good practice in food-related neuroimaging. *The American Journal of Clinical Nutrition, 109*(3), 491–503. Available from https://doi.org/10.1093/ajcn/nqy344; https://academic.oup.com/ajcn/article/109/3/491/5369498.

Smets, E., Rios Velazquez, E., Schiavone, G., Chakroun, I., D'Hondt, E., De Raedt, W., Cornelis, J., Janssens, O., Van Hoecke, S., Claes, S., Van Diest, I., & Van Hoof, C. (2018). Large-scale wearable data reveal digital phenotypes for daily-life stress detection. *NPJ Digital Medicine, 1*(1), 67. Available from https://doi.org/10.1038/s41746-018-0074-9; http://www.nature.com/articles/s41746-018-0074-9.

Smith, E. E., Reznik, S. J., Stewart, J. L., & Allen, J. J. B. (2017). Assessing and conceptualizing frontal EEG asymmetry: An updated primer on recording, processing, analyzing, and interpreting frontal alpha asymmetry. *International Journal of Psychophysiology, 111*, 98–114. Available from https://doi.org/10.1016/j.ijpsycho.2016.11.005.

Sun, W., & Kober, H. (2020). Regulating food craving: From mechanisms to interventions. *Physiology & Behavior, 222*(March). Available from https://doi.org/10.1016/j.physbeh.2020.112878; https://linkinghub.elsevier.com/retrieve/pii/S0031938420301955.

Sweller, J. (1988). Cognitive load during problem solving: Effects on learning. *Cognitive Science, 12*(2), 257–285. Available from https://doi.org/10.1207/s15516709cog1202_4; http://doi.wiley.com/10.1207/s15516709cog1202_4.

Symmank, C. (2019). Extrinsic and intrinsic food product attributes in consumer and sensory research: Literature review and quantification of the findings. *Management Review Quarterly, 69*(1), 39–74. Available from https://doi.org/10.1007/s11301-018-0146-6; http://link.springer.com/10.1007/s11301-018-0146-6.

Tashiro, N., Sugata, H., Ikeda, T., Matsushita, K., Hara, M., Kawakami, K., Kawakami, K., & Fujiki, M. (2019). Effect of individual food preferences on oscillatory brain activity. *Brain and Behavior, 9*(5), 1–7. Available from https://doi.org/10.1002/brb3.1262.

Tseng, A., Bansal, R., Liu, J., Gerber, A. J., Goh, S., Posner, J., Colibazzi, T., Algermissen, M., Chiang, I.-C., Russell, J. A., & Peterson, B. S. (2014). Using the Circumplex Model of Affect to Study Valence and Arousal Ratings of Emotional Faces by Children and Adults with Autism Spectrum Disorders. *Journal of Autism and Developmental Disorders, 44*(6), 1332–1346. Available from https://doi.org/10.1007/s10803-013-1993-6.

Tyler, W. J., Boasso, A. M., Mortimore, H. M., Silva, R. S., Charlesworth, J. D., Marlin, M. A., Aebersold, K., Aven, L., Wetmore, D. Z., & Pal, S. K. (2015). Transdermal neuromodulation of noradrenergic activity suppresses psychophysiological and biochemical stress responses in humans. *Scientific Reports, 5*(1). Available from https://doi.org/10.1038/srep13865; http://www.nature.com/articles/srep13865.

van Bochove, M. E., Ketel, E., Wischnewski, M., Wegman, J., Aarts, E., de Jonge, B., Medendorp, W. P., & Schutter, D. J. L. G. (2016). Posterior resting state EEG asymmetries are associated with hedonic valuation of food. *International Journal of Psychophysiology, 110*, 40–46. Available from https://doi.org/10.1016/j.ijpsycho.2016.10.006.

Verastegui-Tena, L., van Trijp, H., & Piqueras-Fiszman, B. (2019). Heart rate, skin conductance, and explicit responses to juice samples with varying levels of expectation (dis)confirmation. *Food Quality and Preference, 71* (August 2018), 320–331. Available from https://doi.org/10.1016/j.foodqual.2018.08.011.

Vossel, S., Geng, J. J., & Fink, G. R. (2014). Dorsal and ventral attention systems. *The Neuroscientist, 20*(2), 150–159. Available from https://doi.org/10.1177/1073858413494269; http://journals.sagepub.com/doi/10.1177/1073858413494269.

Wang, C.-A., Baird, T., Huang, J., Coutinho, J. D., Brien, D. C., & Munoz, D. P. (2018). Arousal effects on pupil size, heart rate, and skin conductance in an emotional face task. *Frontiers in Neurology, 9*. Available from https://doi.org/10.3389/fneur.2018.01029; https://www.frontiersin.org/article/10.3389/fneur.2018.01029/full.

Wang, Q. J., Barbosa Escobar, F., Alves Da Mota, P., & Velasco, C. (2021). Getting started with virtual reality for sensory and consumer science: Current practices and future perspectives. *Food Research International, 145*, 110410. Available from https://doi.org/10.1016/j.foodres.2021.110410.

Zeinstra, G. G., Koelen, M. A., Colindres, D., Kok, F. J., & de Graaf, C. (2009). Facial expressions in school-aged children are a good indicator of 'dislikes', but not of 'likes'. *Food Quality and Preference, 20*(8), 620–624. Available from https://doi.org/10.1016/j.foodqual.2009.07.002.

Zhang, B., & Seo, H.-S. (2015). Visual attention toward food-item images can vary as a function of background saliency and culture: An eye-tracking study. *Food Quality and Preference, 41,* 172–179. Available from https://doi.org/10.1016/j.foodqual.2014.12.004; https://linkinghub.elsevier.com/retrieve/pii/S0950329314002602.

Zhang, Y., Chen, Y., Bressler, S. L., & Ding, M. (2008). Response preparation and inhibition: The role of the cortical sensorimotor beta rhythm. *Neuroscience, 156*(1), 238–246. Available from https://doi.org/10.1016/j.neuroscience.2008.06.061; https://linkinghub.elsevier.com/retrieve/pii/S0306452208009809.

Zhang, W., Chen, Z., Huang, J., & Wan, X. (2019). The influence of placing orientation on searching for food in a virtual restaurant. *Food Quality and Preference, 78,* 103728. Available from https://doi.org/10.1016/j.foodqual.2019.103728.

Zijlstra, F., Veltman, D. J., Booij, J., van den Brink, W., & Franken, I. H. A. (2009). Neurobiological substrates of cue-elicited craving and anhedonia in recently abstinent opioid-dependent males. *Drug and Alcohol Dependence, 99*(1-3), 183–192. Available from https://doi.org/10.1016/j.drugalcdep.2008.07.012.

Added value of implicit measures in sensory and consumer science

René A. de Wijk[1] and Lucas P.J.J. Noldus[2]

[1]Wageningen Food & Biobased Research, WUR, Wageningen, The Netherlands
[2]Section Neurophysics, Donders Centre for Neuroscience, Radboud University, Nijmegen, The Netherlands

13.1 Introduction

13.1.1 Why implicit measures?

Traditionally, consumer testing in the food and beverages industry relies primarily on so-called explicit testing, where consumers evaluate their conscious food experiences using self-report questionnaires. These tests typically take place in sensory laboratories, central testing locations and other consumption locations such as consumers' own homes. If explicit testing would be successful in the development of food products that do well in the marketplace, then there would be no reason to add other tests to this repertoire. Unfortunately, this is not the case. Of the many new food products that reach the marketplace every year, the majority will not be successful and will disappear within one or two years from the marketplace even though most of these products have successfully passed numerous explicit sensory and consumer tests prior to their market introduction. This suggests that the 'standard' sensory and consumer tests, which typically include explicit sensory analytical profiling, liking and, more recently, emotion tests, have a low predictive validity with respect to commercial product performance. Possibly, consumer food choice outside the laboratory is less based on cognitive information processing and rational reasoning, and more on unarticulated/unconscious motives, emotions and associations. Reasons for likes or dislikes of specific foods are typically difficult to articulate but may determine much of our food choice.

Fitzsimons et al. (2002) reviewed accumulating evidence for the enhanced role of nonconscious influences on consumer responses ranging from perception and memory to affect and choice. Dijksterhuis et al. (2005) further argued for the role of the unconscious in the routine behaviour of consumers and proposed that much of it involves automatic goal pursuit. According to these authors, measures that rely on conscious and thorough information processing are unable to account for a large part of consumer choices, and, in fact, the vast majority of choices are 'not the result of much information processing at all' (Dijksterhuis et al., 2005). Instead, they involve decisions that are contextually or environmentally cue-induced and either engage automatically activated attitudes or are completely devoid of deliberate attitude processing. Unarticulated/unconscious motives, emotions and associations may not be captured very well by traditional explicit tests based on conscious cognitive processes and may be better captured by other measures, for example measures that measure food-related responses implicitly rather than explicitly. In a complex consumption landscape largely determined by nonconscious influences, implicit measures would seem to be potentially useful tools for detecting consumers' 'true' responses. An implicit measure is defined as *'a measurement outcome that is causally produced by the to-be-measured attribute in the absence of certain goals, awareness, substantial cognitive resources, or substantial time'* (p. 350, De Houwer et al., 2009). So far, explicit measures have dominated studies in consumer and sensory research, but implicit measures are becoming more and more popular (Lagast et al., 2017).

13.1.2 Factors related to the use of implicit measures in academic and industrial research

Once a research need for the use implicit measures is identified, it needs to be decided (1) which implicit measure(s) best suit the research goal, (2) which technologies are used for the implicit measure and (3) what expertise is needed to collect and analyse the implicit data.

1. *Which implicit measure(s) best suit the research goal?* The number of implicit measures is virtually limitless but can be categorised into (1) measures that reflect the activity of the brain, such as electroencephalogram (EEG), functional magnetic resonance imaging (fMRI), and functional near-infrared spectroscopy (fNIRS), (2) measures of activity of the autonomic nervous system (ANS) such as skin conductance, heart rate and pupil dilation, 3) expressive measures, such as facial expressions and 4) behavioural measures, such as the way and speed with which food is sampled. A more complete inventory of implicit measures can be found in the review of Smets et al. (2018), which was later adapted by Low et al. (2022) and Kaneko et al. (2018). Just like there is no single explicit measure that captures the full consumer's experience, there is neither one implicit measure that captures the full experience. Some implicit measures like EEG and MRI reflect brain activity, whereas other implicit measures, such as skin conductance and heart rate, reflect actions resulting from the brain activity. Other implicit measures such as facial expressions can, for example, serve to communicate experiences to others: expressions signalling happiness assure the fellow consumer that the food is safe and delicious and encourages the fellow consumer to join the meal. Yet other implicit measures reflect the combined effect of central and ANS activity on behaviour: a slight concern about the food's identity − and resulting doubts about the food's safety and palatability − may result in a more cautious sampling behaviour. Hence, the choice of implicit measure(s) will depend on the research question.

Choosing an appropriate implicit measure is relatively easy for certain research questions but difficult for others. An example of the former type of question is: does my product attract attention on the shelf in the supermarket? This question can be relatively easy answered by using eye tracking techniques with consumers standing on from of the shelf. A much more difficult research question is how to predict long-term consumer' acceptance of new (food) products, one of the industry's holy grails. If long-term acceptance would only depend on the sensory characteristics of the product in combination with an effective marketing strategy, then the current explicit measurements should be sufficient to guarantee success. However, the fact that many new products fail in the marketplace demonstrates that this is not the case, and that additional measures are needed. Which additional (implicit) measures are needed, depends on the factors that are believed to determine long-term acceptance. Some factors may relate to the emotional responses triggered by the new food when this is consumed in specific consumption situations. Other factors may relate to events prior to consumption, such as the placement of the product at the point of sale, the look and feel of the package or the way the food is prepared. This example illustrates that the choice of implicit measures depends heavily on insights in the mechanisms underlying the research question.

2. *Which technologies are used for the implicit measure?* Once the implicit measure(s) have been selected, the appropriate technologies need to be identified. Often, there are multiple technological solutions for a specific implicit measure. For example, skin conductance – or sweat – responses can be measured using surface electrodes attached to the inside of the hand palm, or simply by attaching a sensor with Velcro tape to a finger. Similarly, facial expressions can be measured with sensors attached to the face that measure the activity of specific facial muscles, or with video images of the frontal face that are subsequently analysed with specially developed automated facial expression software. The choice of the technologies will depend on factors such as required level of accuracy, costs and required user expertise.

3. *What expertise is needed to collect and analyse the implicit data?* The required expertise varies considerably with the selected implicit measure(s) and technologies, and with the degree to which they are built into plug-and-play systems. In academic research, technologies are typically selected based on criteria such as reliability and availability of the raw data for (nonstandard) data analysis, and use of these technologies typically requires a high level of expertise. Moreover, technologies are often combined in research, which implies that expertise is required in multiple areas. Furthermore, each measurement technology is often implemented in a separate tool (for eye tracking, facial expression analysis, physiological data acquisition, behavioural scoring), each with its own user interface. This results in a steep learning curve for the researcher and can make multimodal data synchronisation and system integration a daunting task. Luckily, more and more implicit measures are being combined in systems that are offered with a user-friendly front-end and multimodal data analysis software for users with less technical expertise. An example is the automated facial expression analysis software that is based on video recordings of the frontal face. Instead of recording the activities of specific muscles associated with specific facial expressions, which require a considerable expertise, these systems automatically recognise specific facial features

and translate changes in these features into a selected number of facial expressions such as sad and happy. Another example relates to the measurement of physical and mental stress. A reliable stress measurement combines various measurements such as activity, sweat response and heart rate. Nowadays, there are plug-and-play systems that combine all these measurements in a wristband. Sophisticated algorithms combine all measurements into an output that is easy to understand for persons that lack signal analysis and pattern recognition expertise.

13.1.3 Are implicit measures primarily used in academic research?

The complexity of implicit measures may suggest that they would be primarily used in academic research. However, results of a recent small survey conducted by Noldus Information Technology with 58 professionals from industry and academia in the area of sensory and consumer behaviour research suggest otherwise. The results show that more than 50% of the respondents from academia and industry use implicit measures as well as more traditional measures such as self-report/questionnaires.

Reaction times, eye tracking and facial expressions were the implicit measures used most often, and the latter two measures were valued most by their users.

As reasons for using implicit measures, 80% of the respondents indicated that 'the measures answered their research goal', followed by 'the methods are cost and time-efficient' (approximately 30%). Industrial users of implicit measures were found somewhat more often in marketing departments than in R&D departments (41% vs 33%, respectively). The respondents indicated that they would prefer their research to take place on-site as well as in the laboratory, which illustrates the need of industry as well as academia for research in real-life situations.

13.1.4 How do results from implicit measures compare to results from explicit measures?

Studies that combine implicit and explicit measures typically take place in the laboratory and with relatively simple foods and drinks, that is under standardised conditions that do not resemble real-life consumption situations. Under these conditions, the added value of implicit measures is sometimes difficult to identify. For example, in a laboratory study, implicit and explicit responses to a number of commercially available breakfast drinks were compared (De Wijk et al., 2014). Explicit liking ratings showed few significant differences between the breakfast drinks. Implicit heart rate, skin temperature and skin conductance measures of ANS responses did show more significant differences, but these differences were relatively small and difficult to interpret. These results are in line with results from other studies. For example, Samant et al. (2017) failed to demonstrate large contributions of implicit ANS measures to the prediction of acceptance and preference for a number of taste solutions. Similarly, Mojet et al. (2015) tested three different implicit measurement methods (facial expressions, emotive projection and autobiographical congruency) on their effectiveness in measuring the emotional effects of consumption of a number of commercial yoghurts and found that at least two of them were unsuccessful. Danner et al. (2014), who measured skin conductance level, skin temperature, heart rate, pulse volume amplitude, facial expressions and liking during and shortly after samples of commercial juices were tasted, showed no simple relations between self-reported liking and implicit ANS responses or facial expressions. Kaneko et al. (2019), who measured sip size, heart rate, skin conductance level, facial expressions, pupil diameter, EEG frontal alpha asymmetry and valence and arousal for a set of well-liked commercial drinks and one disliked drink (vinegar), found that the implicit measures discriminated between the liked and disliked drinks but not between the liked drinks. Kyto et al. (2019) predicted food purchases from results of implicit (approach-avoidance task, EEG, joystick responses and pupil size responses) and explicit measures in a central location test and concluded that purchase behaviour was stronger associated with explicit measures than with implicit measures (but mentioned the unique contribution of certain implicit measures to the understanding of purchasing behaviour).

Even though some of the lack of results may be caused by artefacts in the measurements (Mojet et al., 2015), these studies *suggest* that under tightly controlled laboratory conditions and with well-liked foods, implicit measures fail to demonstrate clear advantages over explicit measures.

Taken together, the results of implicit responses and facial expressions to foods, and explicit food tests show some similarities and some differences, and implicit tests may not add much to explicit tests. This is especially true in standard sensory food testing where a range of similar tasting foods from the same category of foods are rated blindly on valence under laboratory conditions. This means that these foods are presented without the factors that (also) determine real-life food experiences, such as the foods' packaging, brand name and/or the physical and social context in which these foods are typically consumed. These missing factors relate strongly to the type of factors that are hypothesised to affect implicit measures, such as the relevance of the food to the consumer (e.g. novelty/intrinsic pleasantness/goals), the implications for the consumer (e.g. outcome probability, discrepancy from expectations, urgency), coping (control/power) and normative significance (see Fig. 13.1 for an overview by Coppin & Sanders, 2016).

In standard laboratory tests with blind tasting of similar products, factors such as implications and relevance to the consumer will be very similar across foods, and the outcomes in terms of autonomic physiology, action tendencies and motor expression will therefore also be similar across foods. However, factors such as novelty, implications and normative significance become important in real-life, that is outside the well-controlled conditions of the food laboratory. For example, the average supermarket may offer 30,000 or more products, of which many are food products produced by a large variety of brands. Some of these products and/or brands may be more familiar to the consumer than others, whereas other products may be completely new. Moreover, certain products may be very relevant to the consumer because he is hungry, or because the product is on his/her shopping list. Yet other products may not be especially relevant but may still be appealing because they just look or smell nice, that is intrinsically pleasant, or their brand may fit with the consumer's norms and values. After the foods are purchased, other factors may become important during consumption. For example, does the taste of the food confirm the consumer's expectation based on earlier encounters with the food, or based on the information conveyed by the package? Or does the food go well with a specific environment

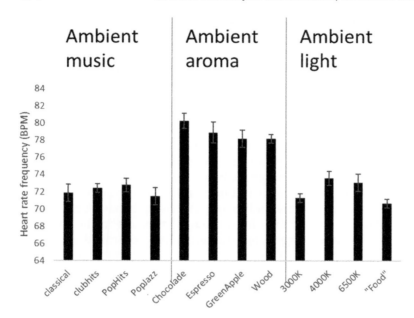

FIGURE 13.1 *Black arrows on top of the figure represent the effects of appraisal criteria on other cognitive processes (e.g. attention, memory). Grey arrows represent the effects of different components (e.g. autonomic physiology, action tendencies) on other cognitive processes. The different components are synchronised during an emotional episode, as shown by the black arrows on the bottom of the figure. Source: From Coppin, G, & Sanders, D. (2016). Theoretical approaches to emotion and its measurement. In H. L. Meiselman (Ed.), Emotion measurements. Elsevier.*

(e.g. three-course dinner served in a canteen), a specific time (cornflakes for lunch instead of breakfast), a specific metabolic state (the three-course dinner when one is already satiated), a specific social context (cheese fondue is more enjoyable with friends) or with one's own believes (a steak presented to a vegetarian). Such factors may not (always) be reflected by traditional explicit tests that assess the food's hedonic and analytical properties, but they do affect food choices and acceptance in the real-world.

In fact, the effects of different locations on implicit measures often seem larger than the differences between foods. Implicit measures are not only sensitive to relatively large differences between locations but also smaller differences. In an unpublished study (De Wijk et al., 2021), heart rate and skin conductance were recorded for participants during a 15-minute stay in a room where aroma, lighting or sound conditions were systematically varied. The results showed not only clear differences in heart rate and skin conductance between aroma, lighting and sound modalities but also within the same modality (Fig. 13.2). Participants also reported their emotions/mood during their stay and the results suggest that these differences in physiological parameters are related to differences in mood triggered by the ambient conditions.

Finally, implicit measures are sensitive to the nutritional needs of the consumer, as demonstrated by a study in which implicit responses were recorded during meal consumption. As the amount of consumed food increased, and the consumer became more satiated, heart rate increased, whereas skin conductance level and skin temperature decreased (He et al., 2017).

We will argue that real-life food experiences (1) are far more complex than food experiences measured in the laboratory, (2) combine the results of conscious and unconscious processing and (3) are best captured by a combination of implicit and explicit measures.

(A)

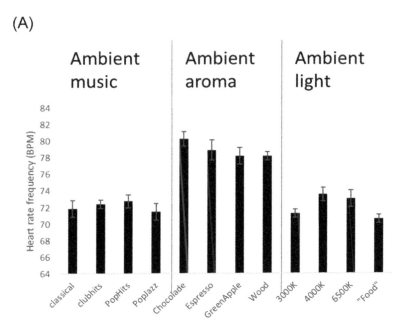

FIGURES 13.2 Averaged heart rate frequency (A) and skin conductance (B) during a 15-min stay in a room with varying ambient music, aromas and lighting conditions. *Source: From de Wijk, R. A., & Noldus, L. P. J. J. (2021). Implicit and explicit measures of food emotions. In Emotion measurement (pp. 169–196). Woodhead Publishing.*

(B)

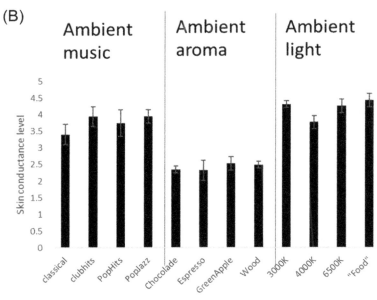

13.1.5 The complexity of real-life food experiences

Bisogni et al. (2007) characterised seven dimensions of eating and drinking episodes: food and drink, time, activities, social setting, mental processes, physical condition and recurrence. Each of these dimensions may affect the way foods are experienced. This means that food experiences reflect not only the perceived properties of the food but also properties of the social and physical environment in which the food is consumed, and physical properties of the consumer (e.g. level of satiation). For example, food can be delicious if served in a nice restaurant on a beautiful plate and consumed in the company of close friends. The same food can be experienced differently when it is served on a plastic plate in a fast-food restaurant in the presence of people that are not-so-close friends. When the food is not entirely in line with what is expected based on the description on the menu, or packaging, the experience will yet again be different. Finally, the experience can again be different when one was not very hungry at the start of the meal.

The fact that food experiences are based on much more than the food's taste is also recognised by marketeers who coined the phrase 'customer journey'. This is defined as the complete sum of experiences that customers go through when interacting with a company and brand. This journey includes not only the food's intrinsic taste properties but also factors such as brand, brand name/image and packaging. Other trends such as increased transparency of a food's origin, ingredients and processing method also acknowledge the multitude of factors that determine the food experience. As a result of this changed perspective, methods are needed that not only monitor consumer reactions to the foods once it is put in the mouth but also reactions to the various points of the customer's journey prior to the moment of consumption. These points may include reactions to, for example the food's origin, packaging, unpacking and the way the food is prepared. The journey may even be extended to postmeal ingestion and waste disposal. An example of the use of implicit measures during part of a customer's food journey is provided by Brouwer et al. (2017), who monitored participants' physiological reactions (EEG, skin conductance, heart rate) during various stages ranging from visual inspection, frying, to consumption, in this case of either mealworms or chicken. The physiological reactions show clear differences between the foods, associated with specific states. Tonic skin conductance responses, a well-known objective measure of arousal, show a gradual increase of arousal between visual inspection ('exposure') and eating for mealworms but not for chickens. Even though this is an extreme example in terms of test foods, the results illustrate that food experiences are not related to a specific phase, such as eating, but may develop prior to eating as a result of expectations.

13.1.6 Testing in real-life

Eating and drinking in real-life situations outside the food laboratories has been primarily the domain of studies using explicit measures. The results often demonstrate effects of context on explicit hedonic and analytical food ratings. Numerous studies, starting with the ground-breaking studies from the US Army's Natick Laboratories (summarised by Hirsch et al., 2005), did find effects of test location on explicit measures (e.g. Weber et al., 2004; King et al., 2004; Boutrolle et al., 2005, 2007; Delarue & Boutrolle, 2010; Meiselman et al., 1988, 2000). These studies typically compare ratings from the laboratory to ratings in real-life situations that differ along many of the dimensions mentioned by Bisogni (2007) and others. Not only does this make identification of the contributions of each of these dimensions to explicit ratings virtually impossible, explicit ratings of liking and emotions may not be the best instruments to capture experiences that involve not only the food itself but also the broader context in which the food is consumed. All experiences are lumped together in hedonic and possible analytical scores making it difficult if not impossible to determine the effects of the food itself and those of all other possible factors. Separation of these effects may be especially important for the food industry in their quest to develop foods that are appreciated by consumers. In this sense, the current trend of testing foods in real-life situations offers a catch-22 situation: we know that food scores from the laboratory are reliable but not very relevant for real-life food appreciation because of its focus on the food rather than the food's interaction with the context. Unfortunately, by introducing the context in real-life testing, it becomes difficult to isolate the effects of the food itself, which can be controlled by a food producer, from the effect of all other factors, which cannot or only to a certain extent be controlled by a food producer.

Application of implicit measures in real-life testing is still relatively difficult because of technical challenges and the relatively low experimental control that is possible in real-life situations. New technological developments in sensory technologies will make real-life applications increasingly possible. Another recent development combines the relevance of real-life testing with the reliability of laboratory testing by recreating real-life contexts in the laboratory. Recent studies have used immersive technologies to recreate the physical context of a consumption situation in a laboratory with visual, auditory and olfactory cues. Initial findings suggest that the use of immersive technologies can improve the predictive validity and reliability of liking scores in consumer testing (e.g. Andersen et al., 2018; Bangcuyo et al., 2015; Delarue, Brasset, Jarrot, & Abiven, 2019; Hannum, Forzley, Popper, & Simons, 2019; Sinesio, Saba, Peparaio, Saggia Civitelli, Paoletti, & Moneta, 2018). The aim of the present study is to compare the effects of recreated contexts on implicit and explicit (e.g. liking) food measures. The data shown here are collected in a study of which the explicit results were recently reported by Van Bergen et al. (2021). Van Bergen et al. repeatedly exposed participants to test foods presented in one of two immersive contexts: a recreated beach and a recreated restaurant. The test foods were either congruent (e.g. sushi in a sushi restaurant or popsicle at the beach) or incongruent with their immersive environment (e.g. sushi at the beach or popsicle in the sushi restaurant). In contrast to some other studies (De Wijk et al., 2019; Zandstra et al., 2020), social interactions during testing were allowed to enhance the degree to which participants felt immersed in the text contexts. The results showed that immersive contexts affected expected liking, that is anticipated liking prior to tasting; incongruent food-context combinations triggered lower expected liking and desire-to-eat scores relative to congruent combinations. During tasting, these group-level differences disappeared, but individual liking scores for incongruent combinations were found to be less consistent over repeated exposures than those of congruent combinations. In the same study, emotions and facial expressions were measured as well. The results of these measurements will be reported here in the second case study.

13.2 The case studies

Two case studies will demonstrate that (1) implicit measures are not restricted to laboratories but can also be used at consumption locations such as the consumers' homes, (2) implicit measures can also be successfully used in recreated consumption locations and (3) implicit measures are sensitive for factors that influence and ultimately determine real-life food experiences such as physical consumption environment. In the first case study, the effect of consumption context on implicit expressive (facial expressions) and physiological responses (heart rate) and explicit food liking responses will be investigated. The results will show that consumption context (laboratory or own home) had relatively little effect on explicit liking scores for test foods, and much larger effects on expressive and physiological responses. In the second case study, explicit emotion scores and facial expressions were investigated for two recreated contexts that were either fitting or nonfitting with the test foods. The results of this case study will show that facial expressions were sensitive to effects of recreated contexts and to differences between test foods, and that these effects were independent from each other.

13.2.1 Case study 1

Effects of real-life and laboratory contexts on explicit and implicit responses to foods (De Wijk et al., 2019):

Background: In food laboratories, consumer' responses to foods are typically measured explicitly using questionnaires, and occasionally, using implicit physiological and behavioural measures. Outside the laboratories, in more natural consumption situations, typically only explicit measures are used. To evaluate the possible use of implicit physiological and behavioural measures in nonlaboratory situations, a study was conducted with 18 Dutch consumers (18–65 years of age) who consumed samples of three test foods ten times alternately in the laboratory and in their own home. Factors such as test procedures, social context and time of day were kept the same in both locations. In addition to standard questionnaires related to the foods' hedonic and sensory properties, video images of the consumers' faces were used to remotely record facial expressions, video-based heart rates (using a technique called remote photoplethysmography, measuring the minute changes of the skin colour resulting from blood pulsation in the hair vessels) and behavioural chewing durations. Screening criteria for participation of consumers in the study were the availability of a laptop with Google Chrome and a webcam, and a stable internet connection. Consumers' own laptops with built-in video cameras were used for collection of the images of the consumer's face during consumption of the test foods, that is no specific software or hardware was provided by the experimenters. These images were automatically uploaded to a server where they were made available to the researchers for further processing. This processing included the automated analysis of facial expressions and heart rate using FaceReaderTM software. Chewing durations were calculated from key presses by the consumers indicating the moments that the food was placed in the mouth and the moment that the food was swallowed.

Results from both test locations were similar with respect to the quality of the images, which allowed a direct comparison of the effects of location on physiological, behavioural and sensory responses to foods. Interestingly, the results showed very few effects of location on explicit sensory evaluations, suggesting that the test foods were perceived similarly in the laboratory and in the consumers' own homes. However, the physiological and behavioural responses paint a very different picture. Compared to consumption in the laboratory, consumption at home triggered more intense facial expressions of happiness, contempt, disgust and boredom (see Fig. 13.3), which were accompanied by significantly higher heart rates (72.6 ± 0.9 vs 71.3 ± 0.9 BPM, F(1,2967) = 10.5, P = .001). Consumption at home was faster than consumption in the lab (29.0 vs 30.1 s, F(1,1245) = 4.2, P = .04), and consumption became faster with replication (average duration 33.3 s for rep 1 to 26.7 s for rep 5) (F(4,1244) = 15.4, P < .001). Presumably, the faster eating rate at home was accompanied by increased muscle activity, which in turn resulted in higher heart rates.

Thus the fact that participants were tested at home instead of the laboratory had clear effects on their physiological and behavioural responses to the test foods, without corresponding effects on their explicit hedonic and sensory judgements. It has to be kept in mind that in this study, everything except for the location was kept constant, that is the same participants consumed the same foods at the same time using the same plates and utensils sitting alone in from of a webcam. The physical location of testing was the only dimension that was varied. It is under those circumstances that implicit measures proved to be more sensitive than explicit measures. Since then, we replicated this finding in another study in which the test location varied between a laboratory, a simulated grand café and an actual grand café (Zandstra et al., 2020). Again, all other variables were kept constant, and the results of the explicit measures again showed no systematic effects of location. Other studies did demonstrate effects of variables such as location and plating on explicit measures. However, in these studies, not only the physical context was varied but also other variables such as the social context and time of day were varied too. This makes it difficult to identify the specific variables that are responsible for the explicit effects.

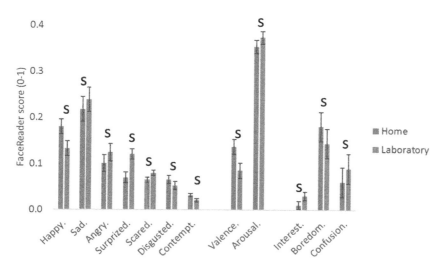

FIGURE 13.3 Facial expressions to test foods consumed at home and in the laboratory. 'S' indicate significant differences between locations at P < .05. *Source: From De Wijk, R. A., Kaneko, D., Dijksterhuis, G. B., Zoggel, M, van Schiona, I., Visalli, M., & Zandstra, E. H. (2019). Food perception and emotion measured over time in-lab and in-home. Food Quality and Preference, 75, 170−178.*

The results of this study also demonstrates that it is possible to monitor implicit food reactions at locations that are typically not accessible for these types of measurements. Moreover, these measurements were made with instruments that are already available in most households. Obviously, there are privacy issues associated with home testing using the internet, but these can be solved easily with systems that allow anonymous storage of privacy-sensitive information on servers.

13.2.2 Case study 2: effects of recreated consumption contexts on facial expressions

Background: The first case study demonstrated the effects of consumption location on food experiences, especially when they are measured implicitly. In this first case study, real-life consumption locations were tested. In the second case study, effects of recreated consumption locations will be tested implicitly. Explicit results of this study reported previously by van Bergen et al. (2021), with recreated beach and sushi restaurant contexts, demonstrated small effects of context on explicit ratings for test products. Moreover, these effects were limited to anticipated taste properties and not to actual taste properties. The context effects were specific for the food/context combination. Anticipated taste for fitting combinations (e.g. sushi tasted in the sushi restaurant context) differed from nonfitting combinations (e.g. sushi taste at the beach context).

Thirty-five participants were exposed repeatedly in seven sessions in one of two recreated contexts (sushi restaurant and beach, see Fig. 13.4) to foods that were either fitting (sushi in the sushi restaurant, popsicle at the beach and iced tea) or nonfitting (sushi at the beach and popsicle in the sushi restaurant) with the recreated context. Contexts were switched in the eighth and final session. Facial expressions to test foods were recorded in sessions 2, 7 and 8. Details of the study can be found in De Wijk et al. (2022).

Facial expression profiles varied between foods [F(18,2270) = 3.70, $P < .001$] (Fig. 13.5). Significant food product differences were found for expressions of sadness, disgust, as well as overall valence and neutrality (all $P < .05$). Facial expressions for sushi were sadder and more negatively valenced than expressions for the other foods. Facial expressions for iced tea were more disgusted.

FIGURES 13.4   Impressions of the immersive beach (left) and sushi restaurant (left) contexts.

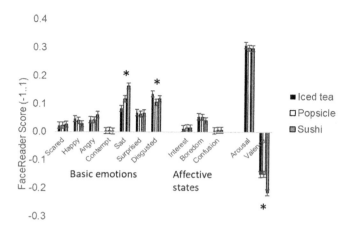

FIGURE 13.5 Effect of foods on facial expressions. Facial expressions are averaged across repeated exposures (sessions 1–7) and immersive contexts. *Source: From De Wijk, R. A., Kaneko, D., Dijksterhuis, G. B., van Bergen, G., Vingerhoeds, M. H., Visalli, M., & Zandstra, E. H. (2022). A preliminary investigation on the effect of immersive consumption contexts on food-evoked emotions using facial expressions and subjective ratings. Food Quality and Preference, 99, 104572.*

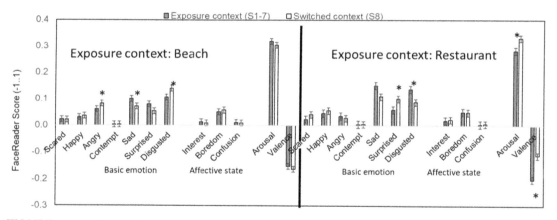

FIGURE 13.6 Effect of immersive context on facial expressions. Facial expression scores are averaged across repeated exposures (sessions 1–7) and foods. Significant effects are indicated by *. *Source: From De Wijk, R. A., Kaneko, D., Dijksterhuis, G. B., van Bergen, G., Vingerhoeds, M. H., Visalli, M., & Zandstra, E. H. (2022). A preliminary investigation on the effect of immersive consumption contexts on food-evoked emotions using facial expressions and subjective ratings. Food Quality and Preference, 99, 104572.*

Profiles varied significantly with context (interaction: $F(9,2270) = 4.0$, $P < .001$): foods presented in the recreated beach context triggered more neutral expressions, fewer expressions of sadness and disgust and were, in general, less negatively valenced than the same foods presented in the recreated restaurant context (all $P < .05$) (Fig. 13.6). The effect of context was not food specific, as indicated by the lack of significant interactions between foods and context over time ($F(18,2270) = 0.38$, n.s.). Facial expressions did not change systematically after repeated exposure (no significant main and/or interaction effects).

Facial expression profiles after the switch in session 8 varied significantly with the previous context (sessions 1 and 7) (interaction: $F(9,2270) = 3.2$, $P < .01$) and with the new context (session 8) (interaction: $F(9,2270) = 3.7$, $P < .001$), that is expressions did not only vary with the context at the moment of testing but also varied with the context in which the foods had previously been presented. Participants that had previously been exposed to the foods in the beach context showed more expressions of anger and disgust, and fewer expressions of sadness than the participants that had previously been exposed to the restaurant context ($P < .05$).

Previously, it was found that explicit liking ratings primarily reflect properties of the foods, and that effects of consumption context were limited to anticipated liking of fitting- and nonfitting food-context combinations (van Bergen et al., 2021). In contrast, facial expressions during actual consumption reported here reflected properties of the food and of the context. The effects of context and test foods were independent, that is the effects of contexts were similar for fitting- and nonfitting food-context combinations.

The effects of context are probably mediated by longer-lasting effects of context on emotions (or mood). Compared to the restaurant context, participants felt more interested, less calm and more surprised in the beach context not only during periods of consumption but also before and after. Similarly, facial expressions during consumption in the beach context were more positively valenced compared to the restaurant context. Or in other words, people feel better in certain contexts compared to other contexts, irrespectively whether they eat sushi, lick a popsicle or drink iced tea. This result is probably not a surprise for the hospitality industry where restaurants are designed to optimise customers' experiences. Somewhat surprising is that the results of this study suggest that it may not be necessary to make the type of food consistent with the theme of the restaurant. In the extreme case, this may imply that good Chinese food is experienced equally positive in a nice Chinese restaurant and in a nice Italian restaurant. Obviously, additional research is needed to explore possible boundaries of the (lack of) congruency effects.

The immersive contexts used in this study clearly affected the way participants felt as evidenced by facial expressions. Participants exposed to the immersive beach context report that they feel more interested, surprised and calm than participants exposed to the immersive restaurant context. This was corroborated by the facial expressions that were more positively valenced in the immersive beach context compared to the restaurant context.

13.3 What are the learnings from these case studies?

Both case studies show larger sensitivity of implicit measures/facial expressions compared to explicit measures for factors such as consumption context, irrespective whether this context is a real-life context or a recreated context. Implicit measures (heart rate, consumption duration in the first use case study) not only showed effects of context on reactions to the foods but also on the way foods were actually consumed. In the second case study, facial expressions not only demonstrated effects of contexts but also demonstrated that these effects were independent of the specific foods.

There are various possible reasons for the higher sensitivity of implicit measures compared to explicit measures observed in the case studies described above. For example, explicit measures reflect by definition consumer experiences that are accessible via introspection, whereas implicit measures reflect also experiences that are inaccessible, for example because they are too fast to reach the consumer's consciousness. Another possible reason is that implicit measures may reflect processes that serve other functions than those reflected by explicit measures. Some implicit measures of ANS responses are indicative for very basic reactions of fight and flight, that is they are meant to avoid harm to the consumer. In contrast, explicit responses typically reflect whether the consumer likes the food, or whether this food tastes sweeter than other foods. Moreover, most explicit measurements are rather static and reflect the end result of numerous mental processes, some of which are conscious, and others are unconscious. These mental processes are typically collapsed — or integrated — over time, resulting in one explicit score. To understand real-life food experiences, that reflect factors such as food, time, activities, social setting, mental processes, physical condition and recurrence as well as their interactions, explicit scores may not be sensitive enough. In contrast, implicit measures are dynamic and represent a temporal window on processes, some of which are unconscious or subconscious and fast while others are conscious and slower. Moreover, different implicit measures reflect different types of conscious and unconscious reactions, some of which are fast whereas others are slow. The wide range of implicit measures and their ability to continuously monitor fast and slower reactions to food stimuli makes them well-suited to study the richness of food experiences.

The bottom line: Combined, these results suggest that implicit measures in food studies reflect reactions to the sensory properties of the food itself, just like explicit measures. In addition, implicit measures also reflect reactions to the social and physical context in which the food is consumed, to expectations raised by cues such as food packaging and to the internal state of the food consumer (hungry/satiated). Thus in contrast to explicit measures, implicit measures seem to reflect the broader food experience, rather than food-specific experiences.

This suggests that the real potential contributions of implicit measures to food sciences may be in real-life applications, that is in situations where foods are normally consumed and where factors such as social context, branding, novelty and expectations play a large role. So far, implicit measures have been primarily limited to laboratory conditions. However, as demonstrated by the second case study, recreated consumption situations in the laboratory may provide a useful alternative for real-life studies, at least for the present generation of implicit measurement techniques. In fact, new studies in our laboratory not only recreate the consumption context but also other phases in which consumers interact with food products, such as point of sale, ingredient specifications, as well as food preparation. 'Consumers' join a virtual journey that takes them from the supermarket via the kitchen to the product that is actually tasted at the end of the journey. Implicit measures such as heart rate, skin conductance and eye tracking are continuously monitored during the various phases of the journey. Effects of experimental variations during each of the phases, such as a different product placement in the supermarket or a different ingredient specification, can be measured instantaneously and during consumption of the product (to verify long-lasting effects of the variations on the actual food experiences).

13.4 The road towards user-friendly implicit measures that can be used at any place and at any time and by everybody

Ideally, implicit measures would be plug-and-play with regard to the data recordings as well as their processing, analysis and interpretation. Currently, this is not yet possible but encouraging steps have been made, especially with regard to the data recordings. The first development is the miniaturisation of sensors and devices for heart rate, skin conductance, eye tracking and EEG which makes them better suited for real-life applications. These techniques can be combined with video recordings of the face allowing for the simultaneous recording of facial expressions and even implicit measures such as heart rate.

Real-life consumption situations such as supermarkets and restaurants are not yet the natural habitat of implicit measures because they typically require a high degree of control over the food stimulus, the consumer and the situation, all of which are lacking in real-life situations. However, we believe that current technological developments will facilitate the use of implicit measures in real-life situations. The first development is the miniaturisation of sensors and devices for heart rate, skin conductance, eye tracking and EEG, which makes them better suited for real-life applications. These techniques can be combined with video recordings of the face allowing for the simultaneous recording of facial expressions and even implicit measures such as heart rate. Software developments will facilitate analyses of the results, for example by automatically linking the responses of the consumer to the location of the consumer in the food environment (e.g. location in the supermarket). Another promising technical development is the use of virtual reality that allows consumers to be exposed to food environments via VR glasses while they are sitting in the laboratory. This would combine the increased control of the laboratory with the increased relevance of a real-life situation. Yet another promising development is the so-called immersive technologies where real-life consumption situations are recreated in the laboratory using combinations of image projectors, sound systems and sometimes even odour dispensers (Bangcuyo et al., 2015; Holthuysen et al., 2017; Delarue et al., 2019; Hannum et al., 2019; Zandstra et al., 2020). The result is a simulated situation with the look and feel of the real situation, but with the added benefits of the increased control associated with a laboratory.

Implicit measures cannot only be used for specific studies carried out in specific contexts (either real-life or VR) but could in principle also be used for 24/7 monitoring of consumer' experiences. Mobile devices and wearable sensors have become so powerful and nonintrusive that the combination of a smartwatch (as a body-worn sensor) and a smartphone (as a communication and display device) offers a wealth of new possibilities for continuous measurement of implicit and explicit responses to food and drinks. Several models are available nowadays that offer heart rate and heart rate variability (through photoplethysmography), galvanic skin conductance, skin temperature, activity (through 3D accelerometer and gyroscope) and a battery life exceeding 24 h. In a study with continuous data collection through a wearable sensor and a smartphone, and more than 1000 participants, Smets et al. (2018) were able to establish distinct 'digital phenotypes' related to stress. For personalised feedback and to capture consumer experience in real-time, a programmable display, response buttons and the possibility to extend the firmware with new algorithms are desirable. This is still a rare feature among commercially available study watches. Increasing functionality and ongoing cost reduction of wearable sensors, such as research-grade smartwatches, will stimulate widespread adoption of these tools for implicit measurement of emotions by the consumer science community.

What will future food experience studies look like? A three-week trial could include 24/7 measurement of basic activity and physiological parameters via the smartwatch, while the smartphone prompts the subject to fill in a short questionnaire during each meal, while his/her facial expression and gaze direction is captured via the phone camera with built-in eye tracker. Data from smartwatch and smartphone are then stored in a single, secure time-series database for integrated analysis. This perspective is appealing for researchers because wrist-worn watches allow for continuous, scalable, unobtrusive and ecologically valid data collection of behavioural and physiological data.

13.5 In conclusion and future perspectives

It is argued that implicit measures of food experiences should not be regarded as a more complex substitute of established explicit measures. Instead, each type of measure provides complementary information. Whereas explicit measures capture especially the sensory aspects of the food itself, implicit measures capture especially the total food experience from pre- to postconsumption, which not only relates to the food itself but also to factors such as the food's package and branding or the physical and social context in which foods are consumed in real-life. This requires that implicit measures are applied outside the conventional laboratory habitat. Fortunately, this becomes increasingly possible with current technical developments.

References

Andersen, I. N. S. K., Kraus, A. A., Ritz, C., & Bredie, J. (2018). Desires for beverages and liking of skin care product odors in imaginative and immersive virtual reality beach contexts. *Food Research International, 117*, 10−18.

Boutrolle, I., Arranz, D., Rogeaux, M., & Delarue, J. (2005). Comparing central location test and home use test results: Application of a new criterion. *Food Quality and Preference, 16*(8), 704−713.

Boutrolle, I., Delarue, J., Arranz, D., Rogeaux, M., & Köster, E. P. (2007). Central location test vs. home use test: Contrasting results depending on product type. *Food Quality and Preference, 18*(3), 490−499.

Bisogni, C., Falk, L., Madore, E., Blake, C., Jastran, M., Sobal, J., & Devine, C. (2007). Dimensions of everyday eating and drinking episodes. *Appetite, 48*, 218−231.

Brouwer, A.-M., Hogervorst, M. A., Grootjen, M., van Erp, J. B. F., & Zandstra, E. H. (2017). Neurophysiological responses during cooking food associated with different emotions. *Food Quality and Preference, 62*, 307−316.

Bangcuyo, R. G., Smith, K. J., Zumach, J. L., Pierce, A. M., & Guttman, G. A. (2015). The use of immersive technologies to improve consumer testing: The role of ecological validity, context and engagement in evaluating coffee. *Food Quality & Preference, 41*, 84−95.

Coppin, G., & Sanders, D. (2016). Theoretical approaches to emotion and its measurement. In H. L. Meiselman (Ed.), *Emotion measurements*. Elsevier.

Delarue, J., & Boutrolle, I. (2010). The effects of context on liking: Implications for hedonic measurements in new product development. In S. R. Jaeger, & H. Macfie (Eds.), *Consumer-driven innovation in food and personal care products*. Great Abington, UK: Woodhead.

Delarue, J., Brasset, A. C., Jarrot, F., & Abiven, F. (2019). Taking control of product testing context thanks to a multi-sensory immersive room. A case study on alcohol-free beer. *Food Quality and Preference, 75*, 78−86.

De Wijk, R. A., He, W., Mensink, M. M., Verhoeven, R., & de Graaf, C. (2014). ANS responses and facial expressions differentiate between repeated exposures of commercial breakfast drinks. *PLoS One, 9*(4).

Danner, L., Joechl, M., Duerrschmid, K., & Haindl, S. (2014). Facial expressions and autonomous nervous system responses elicited by tasting different juices. *Food Research International, 64*, 81−90.

De Wijk, R. A., Kaneko, D., Dijksterhuis, G. B., Zoggel, M., van; Schiona, I., Visalli, M., & Zandstra, E. H. (2019). Food perception and emotion measured over time in-lab and in-home. *Food Quality and Preference, 75*, 170−178.

De Wijk, R. A., Kaneko, D., Dijksterhuis, G. B., van Bergen, G., Vingerhoeds, M. H., Visalli, M., & Zandstra, E. H. (2022). A preliminary investigation on the effect of immersive consumption contexts on food-evoked emotions using facial expressions and subjective ratings. *Food Quality and Preference, 99*, 104572.

de Wijk, R. A., & Noldus, L. P. J. J. (2021). Implicit and explicit measures of food emotions. In *Emotion measurement* (pp. 169−196). Woodhead Publishing.

De Houwer, J., Teige-Mocigemba, S., Spruyt, A., & Moors, A. (2009). Implicit measures: A normative analysis and review. *Psychological Bulletin, 135*, 347−368. Available from https://doi.org/10.1037/a0014211.

De Wijk, R. A., Ushiama, S., Ummels, M., Zimmerman, P., Kaneko, D., & Vingerhoeds, M. H. (2021). Effect of branding and familiarity of soy sauces on valence and arousal using facial expressions, physiological measures, emojis, and ratings. *Frontiers in Neuroergonomics*, section Consumer Neuroergonomics, 2, 651682.

Dijksterhuis, A., Smith, P. K., & van Baaren, R. B. (2005). The unconscious consumer: Effects of environment on consumer behavior. *Journal of Consumer Psychology, 15*(3), 193−202.

Fitzsimons, G. J., Hutchinson, J. W., Williams, P., Alba, J. W., Chartrand, T. L., Huber, J., & Tavassoli, N. T. (2002). Non-conscious influences on consumer choice. *Marketing Letters, 13*(3), 269, 269.

He, W., Boesveldt, S., Delplanque, S., de Graaf, C., & de Wijk, R. A. (2017). New insights into sensory-specific satiety from autonomic nervous system responses and facial expressions. *Physiology and Behavior, 170*, 12−18.

Hannum, M., Forzley, S., Popper, R., & Simons, C. T. (2019). Does environment matter? Assessments of wine in traditional booths compared to an immersive and actual wine bar. *Food Quality & Preference, 76*, 100−108.

Hirsch, E., Kramer, M., & Meiselman, H. L. (2005). Effects of food attributes and feeding environment on acceptance, consumption and body weight: Lessons learned in a twenty year program of military ration research. *Appetite, 44*(1), 33−45.

Holthuysen, N. T. E., Vrijhof, M. N., De Wijk, R. A., & Kremer, S. (2017). "Welcome on board": Overall liking and just-about-right ratings of airplane meals in three different consumption contexts - laboratory, re-created airplane, and actual airplane. *Journal of Sensory Studies, 32.*

Kyto, E., Bult, H., Aarts, E., Wegman, J., Ruijschop, R. M. A. J., & Mustonen, S. (2019). Comparison of explicit vs. implicit measurements in predicting food purchases. *Food Quality and Preference, 78*103733.

Kaneko, D., Hogervorst, M., Toet, A., van Erp, J. B. F., Kallen, V., & Brouwer, A.-M. (2019). Explicit and implicit responses to tasting drinks associated with different tasting experiences. *Sensors, 19*, 4397.

Kaneko, D., Toet, A., Brouwer, A.-M., Kallen, V., & van Erp, J. B. F. (2018). Methods for evaluating emotions evoked by food experiences: A literature review. *Frontiers in Psychology, 9*, 911. Available from https://doi.org/10.3389/fpsyg.2018.00911, 911.

King, S. C., Weber, A. J., Meiselman, H. L., & Lv, N. (2004). The effect of meal situation, social interaction, physical environment and choice on food acceptability. *Food Quality and Preference, 15*(7), 645−653.

Lagast, S., Gellynck, X., Schouteten, J. J., De Herdt, V., & De Steur, H. (2017). Consumers' emotions elicited by food: A systematic review of explicit and implicit methods. *Trends in Food Science & Technology: Part A, 69*, 172−189.

Low, Y. Q., Janin, N., Traill, R. M., & Hort, J. (2022). The who, what, where, when, why and how of measuring emotional response to food. A systematic review. *Food Quality and Preference, 100*, 104607. Available from https://doi.org/10.1016/j.foodqual.2022.104607, 104607.

Mojet, J., Dürrschmid, K., Danner, L., Joöchl, M., Heiniö, R. L., Holthuysen, N., & Köster, E. (2015). Are implicit emotion measurements evoked by food unrelated to liking? *Food Research International, 76*(P2), 224−232.

Meiselman, H. L., Hirsch, E. S., & Popper, R. D. (1988). Sensory, hedonic, and situational factors in food acceptance and consumption. In D. M. H. Thomson (Ed.), *Food acceptability*. Barking, UK: Elsevier Science Publishers.

Meiselman, H. L., Johnson, J. L., Reeve, W., & Crouch, J. E. (2000). Demonstrations of the influence of the eating environment on food acceptance. *Appetite, 35*, 231−237.

Samant, S. S., Chapko, M. J., & Seo, H.-S. (2017). Predicting consumer liking and preference based on emotional responses and sensory perception: A study with basic taste solutions. *Food Research International: Part 1* (100), 325−334.

Sinesio, F., Saba, A., Peparaio, M., Saggia Civitelli, E., Paoletti, F., & Moneta, E. (2018). Capturing consumer perception of vegetable freshness in a simulated real-life taste situation. *Food Research International, 105*, 764−771. Available from https://doi.org/10.1016/j.foodres.2017.11.073.

Smets, E., Rios Velazquez, E., Schiavone, G., Chakroun, I., D'Hondt, E., De Raedt, W., Cornelis, J., Janssens, O., Van Hoecke, S., Claes, S., Van Diest, I., & Van Hoof, C. (2018). Large-scale wearable data reveal digital phenotypes for daily life stress detection. *NPJ Digital Medicine, 1,* 67.

van Bergen, G., Zandstra, E. H., Kaneko, D., Dijksterhuis, G. B., & de Wijk, R. A. (2021). Sushi at the beach: Effects of congruent and incongruent immersive contexts on food evaluations. *Food Quality and Preference, 91.* Available from https://edepot.wur.nl/542276.

Weber, A., King, S., & Meiselman, H. L. (2004). The effect of social interaction, physical environment and food choice freedom on consumption in a meal-testing environment. *Appetite, 42*(1), 115–118.

Zandstra, E. H., Kaneko, D., Dijksterhuis, G. B., Vennik, E., & de Wijk, R. A. (2020). Implementing immersive technologies in consumer testing: Liking and Just-About-Right ratings in a laboratory, immersive simulated café and real café. *Food Quality and Preference, 84.*

Immersion technologies, context and sensory perception

14

Using virtual reality as a context-enhancing technology in sensory science

Emily Crofton[1] and Cristina Botinestean[2]

[1]Food Quality and Sensory Science Department, Teagasc Food Research Centre, Ashtown, Dublin, Ireland [2]Food Industry Development Department, Teagasc Food Research Centre, Ashtown, Dublin, Ireland

14.1 Introduction

14.1.1 The role of context in the sensory evaluation of food

Imagine it is 11:00 a.m. on a Wednesday morning and you are participating in a consumer taste trial involving chocolate. You are sitting in a partitioned booth in a typical sensory testing facility that is carefully controlled in terms of noise, temperature and lighting and is stripped of any decorative features. Would you consider this a normal consumption situation? Most people would think probably not. Instead, we tend to eat either alone or with family and friends, in our home or at a restaurant or café. While sensory testing facilities are purposefully designed to be bland in an attempt to minimise the influence of external cues on sensory perception, the setting does not represent how consumers interact with food products in the real world (Low et al., 2021a; Stelick & Dando, 2018). The sensory sensations we perceive while eating are multisensory in nature and are influenced not only by the intrinsic sensory attributes contained within the food (e.g. colour, aroma, taste, texture) but also by the surrounding environment or context in which we consume it. Context can be defined as the specific physical, social and situational conditions in which food and beverages are consumed. As such, the location or setting, the temperature and humidity, and the type of music, lighting or sound are all examples of extrinsic contextual cues that may influence consumer's sensory and hedonic perception, purchase intent and other food-related behaviours (Bangcuyo et al., 2015; Torrico et al., 2021).

Food companies all over the world conduct consumer sensory studies in an attempt to predict how newly launched products will perform in the marketplace. However, it is clear that sensory data collected in these studies cannot reliably predict the success of a product, and it is not uncommon for products scoring well in consumer trials to quickly fail in the marketplace (de Wijk et al., 2012; Köster, 2009). Although numerous factors can contribute to new product failures, sensory scientists are increasingly questioning whether consumer consumption behaviour would be better predicted from sensory data generated in a more realistic context (Bangcuyo et al., 2015; Crofton et al., 2021; Giezenaar & Hort, 2021; Low et al., 2021a; Torrico et al., 2021). To create context in sensory evaluation studies, product assessments are typically conducted in an immersive environment. An immersive environment (in terms of consumer testing) has been recently described by Schöniger (2022) as using a physical or virtual approach to evoke context in an attempt to give a food product a more complete meaning and create a higher degree of internal and external validity. They note that an immersive environment should have a high sensory input level, a high immersion level and an aim to achieve a high level of presence in the evoked situation.

Several approaches have been used to create a more immersive environment in which the impact of consumption context on consumer sensory perception can be studied. These include using written or imagined scenarios (Hein et al., 2010; Nijman et al., 2019), or physically creating an immersive room using a combination of visual, auditory, olfactory and/or tactile stimuli in the form of a coffeehouse (Bangcuyo et al., 2015; Lichters et al., 2021), bar (Hannum et al., 2020; Sester et al., 2013), beach (van Bergen et al., 2021) or holiday farm (Sinesio et al., 2019). Immersive rooms typically consist of projecting a video with sound onto the walls of the room and using various props to mimic a real world environment. Studies have also been conducted in actual real life settings such as a bar (Hannum et al., 2020; Nijman et al., 2019; Sinesio et al., 2019), café (Low et al., 2021a), coffeehouse (Lichters et al., 2021) and aeroplane (Holthuysen et al., 2017). In general, results from studies investigating the influence of context on sensory data have been encouraging, with consumers reporting higher levels of engagement and, in some cases, providing more discriminative product data in an immersive setting, compared to traditional sensory booths (Bangcuyo et al., 2015; Hathaway & Simons, 2017; Holthuysen et al., 2017; Sester et al., 2013). However, in the long term, it is not realistic or feasible in terms of time and cost in having to create physical immersive rooms or gain access to real world settings, to run consumer sensory trials that are more ecologically valid. As the digital world continues to evolve at a rapid pace, researchers are considering the use of virtual reality (VR) technology as a means to design immersive environments to reflect a more realistic eating context. The goal of this chapter is to investigate VR technology as a digital tool for creating immersive environments in consumer sensory testing. Different VR systems will be introduced and the outcomes from recent studies using VR as a context-enhancing technology will be summarised, in conjunction with learnings from case studies involving beef and chocolate. The potential opportunities and limitations of measuring consumer sensory response in a VR environment will also be discussed.

14.1.2 Immersive virtual reality technologies

VR technology is gaining momentum for its ability to create realistic immersive environments in which external contextual cues can be fully controlled. Due to continuous advances in VR technologies and the conflicting meaning of the words 'virtual' and 'reality', the term virtual reality is notoriously difficult to define and countless definitions exist in the literature as a result (Aukstakalnis, 2017). Nonetheless, VR is generally described as an immersive human—computer interaction which enables a person to explore and interact with a three-dimensional world as if the surrounding environment was real. A fully immersive VR experience is usually accomplished through the use of a head-mounted display (HMD), which completely replaces the person's view of the real world with an interactive computer-generated environment. To date, consumer sensory responses collected in immersive VR environments (involving some form of HMD device) have been studied for a range of food products including chocolate (Crofton et al., 2021; Torrico et al., 2021), coffee (Wang et al., 2021), cheese (Stelick & Dando, 2018), tea-break snacks (Low et al., 2021a), beef (Crofton et al., 2021) and alcoholic beverages (Picket & Dando, 2019; Sinesio et al., 2019; Yang et al., 2022). A summary of recent studies utilising VR technology is presented in Table 14.1.

Exposure to an immersive VR environment can induce an intense feeling of presence, whereby the user genuinely feels removed from the real world. As the feeling of presence is subjective, people immersed in the same VR system may experience different levels of presence, although in general, a more immersive system induces a more intense feeling of presence, bridging the gap between the real and virtual world (Schöniger, 2022; Wang et al., 2021). VR systems can provide different levels of immersion depending on the design. Currently, HMDs can be broadly classified as either tethered, standalone or mobile-powered devices. Tethered HMDs (e.g. Oculus Rift and HTC Vive) are high-end devices that need to be physically connected with a cable (e.g. HDMI/USB) to another computing device, such as a PC computer or video game console. Tethered devices provide the highest quality and most immersive experience compared to any other type of VR; however, they require the person wearing the HMD to be physically connected to a computer with significant processing power, thus limiting the person's physical ability to move around. As a result, research involving tethered HMDs tends to be limited to laboratory-based studies. In contrast, standalone VR HMDs (e.g. Oculus Quest and HTC Vive Focus) have built-in processing power and battery, so they do not require a connection to any sort of computing device. Generally speaking, the quality and capabilities of standalone HMDs are lower than tethered ones, but since they are wireless, the person is not confined to a specific room or area. Nonetheless, it can be argued whether the mobility capability of a HMD device impacts sensory evaluation studies, as consumers tend to be seated during the assessment of food (Wang et al., 2021). Tethered and standalone VR systems often include motion-sensing handheld controllers enabling the person to manipulate virtual objects and interact with the computer simulation, enhancing the feeling of immersion (Siegrist et al., 2019). In contrast, mobile-powered HMD (e.g. Samsung Gear) generally consists of a simple headset with custom lenses in which the person inserts a compatible smartphone. The quality of the VR experience depends on the quality of the smartphone being used, but in general, mobile-powered devices are on the lower end of the VR quality spectrum. In addition, while tethered and standalone HMDs have software to show both 360-degree content and a computer-generated three-dimensional (3D) VR environment, mobile-powered VR only supports viewing 360-degree videos, meaning the person can look in every direction around them (up and down, side to side, forwards and backwards), but cannot interact or move around the virtual space.

TABLE 14.1 Summary of studies using virtual reality (VR) technology (via some form of head-mounted display device) to measure sensory perception of food.

References	Sample size	Product	Testing environment	Headset	Studied factors	Main findings
Sinesio et al. (2019)	513	Beer	Real pub and three immersive rooms: 1. Immersive room projecting the situation on flat walls with videos at 180 degrees and with a pub set-up; 2. VR headset with the projection of the situation from a 360-degree video; 3. VR headset with the projection of the situation from 3D modelling and 360-degree photos.	Trust Urban VR (360-degree video) Oculus Rift (3D Model)	Overall liking and intention to re-taste beer samples. Consumer's emotional response and perceived engagement was also measured.	More similarity of the results with the real pub was obtained for the immersive room and VR 3D modelling. Less discrimination for assessments in the real environment and in immersive approaches. Higher consumer engagement was observed for the more immersive techniques.
Oliver and Hollis (2021)	15	Pizza rolls	VR restaurant and a VR table in an empty room.	HTC Vive	Hunger, fullness, desire to eat, amount eaten, presence, palatability, saltiness, texture and acceptability.	Participants reported a greater sense of presence, and their heart rate and skin temperature were higher in the restaurant condition. However, there were minimal changes in food intake, masticatory parameters and the sensory evaluation of the pizza bites.
Ammann et al. (2020)	50	Chocolate	Two virtual rooms: The first VR room (control) consisted of a piece of chocolate on a table. The second room (disgust) saw a dog that walked across the table and stopped halfway to produce dog feces that looked like a piece of chocolate.	HTC Vive	Food digest and willingness to eat the chocolate.	Participants in the disgust condition were more likely to refuse consumption than those in the control condition. Physical presence mediated the relationship between participants' food disgust sensitivity and willingness to eat the chocolate in the disgust condition.

Study	N	Food	Environment	VR device	Measures	Results
Ammann et al. (2020)	102	Juice and cake slices	Virtual environment and real environment.	HTC Vive	Colour and flavour identification.	The result patterns for the real conditions and VR conditions were comparable. However, the effects tended to be smaller in the VR condition than in the real environment.
Alba-Martínez et al. (2022)	110	Variety of cakes	Traditional sensory booth and virtualised sensory booth.	Oculus Rift	Visual expectations about appearance, serving size, deliciousness and liking under both real and virtual environments.	The real or virtual environments did not have a significant influence on participants' expectations of visual appearance, serving size, deliciousness or expected liking for the variety of cakes analysed.
Crofton et al. (2021)	30	Beef and chocolate	Beef: Traditional sensory booth and a VR restaurant. Chocolate: Traditional sensory booth and two VR contexts including an Irish countryside and a busy city.	Oculus Go	Chocolate was evaluated for liking in terms of smell, flavour, sweetness, texture, smoothness and overall liking. Beef was evaluated for liking in terms of smell, tenderness, juiciness, beef flavour and overall liking.	Beef was rated as significantly higher in terms of liking for all sensory attributes when consumed in the VR restaurant, compared to the sensory booths. While for chocolate, the VR countryside context generated significantly higher hedonic scores for flavour and overall liking in comparison to the sensory booth. The VR conditions resulted in a more memorable testing experience for both products.
Torrico et al. (2021)	50	Chocolate	Traditional sensory booth and two VR environments: positive-VR (aesthetically open-field forest) and negative-VR (closed-space old room).	DELL visor mixed reality headset	Acceptability of chocolate in terms of sweetness, bitterness, texture, mouth-coating, aftertaste and overall liking was assessed. Emotional response was also measured.	Chocolate type and VR did not affect the liking of attributes. However, full-sugar samples had higher sweetness intensity than no-sugar samples for positive-VR.

(Continued)

TABLE 14.1 (Continued)

References	Sample size	Product	Testing environment	Headset	Studied factors	Main findings
Torrico et al. (2021)	53	Wine	Traditional sensory booth, two real restaurant environments and two VR restaurants: bright-VR restaurant with bright lights, and dark-VR restaurant with dimly lit candles.	Oculus Go	Acceptability of aroma, sweetness, acidity, astringency, mouthfeel, aftertaste, and overall liking, and intensities of sweetness, acidity and astringency.	Results showed that context (booths, real or VR) affected the perception of the wine's floral aroma. Liking of the sensory attributes did not change under different environmental conditions.
Kong et al. (2020)	67	Chocolate	Traditional sensory booth and two VR contexts: a pleasant sightseeing tour and a live music concert.	Oculus VR	Chocolate acceptability was rated for flavour, sweetness, bitterness, cocoa flavour, dairy flavour, texture, hardness, smoothness, aftertaste and overall liking. Sweetness, bitterness, cocoa flavour, dairy flavour and overall texture were also evaluated by a just-about-right-scale (JAR) in terms of both intensity and acceptability. Emotional response was also measured.	No significant effects of context type on the tasting experience were found; however, there were significant effects of chocolate type. Milk and white chocolates were preferred over dark chocolate irrespective of the context type. Additionally, more positive emotions were elicited for the dark chocolate in the 'virtual live concert' environment.
Low et al. (2021a)	120	Tea-break snacks	Sensory booth, real café and an evoked mixed reality café.	HoloLens device	Overall liking and liking of six sensory attributes (chocolaty, chewy, sweet, crumbly, buttery and dry aftertaste) were assessed. Emotional response was also measured.	No significant differences for most affective ratings between data obtained from the evoked mixed reality café and real café. Affective response in sensory booths was not representative of real-context response.
Wang et al. (2021)	32	Coffee	Two virtually presented coffee colours (light brown and dark brown).	HTC Vive	Overall liking and liking of sweetness and creaminess.	Coffee colour as viewed in VR significantly influenced perceived creaminess, with the light-brown coffee rated to be creamier than dark-brown coffee. However, coffee colour did not influence perceived sweetness or overall liking.

Study	N	Product	Context	VR Device	Measures	Results
Yang et al. (2022)	27	Beer	Evoked context (picture slideshow and sound recording to simulate the exact same bar context) and VR bar.	HTC Vive	Liking and emotional response.	In general, no significant differences were found for liking and emotional responses of beer between contexts. VR was shown to generate higher participant engagement than the evoked context.
Picket and Dando (2019)	59	Sparkling wine and beer	Two virtual reality environments: a typical college bar and a tasting room at an expensive local winery.	Samsung Gear VR	Samples were rated for overall liking, and intensity of sweetness, bitterness and carbonation.	Panellists had significantly higher hedonic ratings for sparkling wine consumed in the winery context compared to the bar. Liking for beer was higher, although not significantly, in the bar context.
Stelick and Dando (2018)	50	Blue cheese	Sensory booth and two VR environments: a park bench in a natural open setting and a cow barn.	Samsung Gear VR	Samples were rated in terms of overall liking, and intensity of pungency and saltiness.	No differences in liking of cheese was observed across environments. However, intensity of pungency was significantly higher in barn context compared to the sensory booth and VR park bench.

14.2 Creating context through an immersive virtual reality environment: case studies with beef and chocolate

In the following section, we present and discuss findings from a research study (Crofton et al., 2021) which investigated how consumers' hedonic ratings of two different food products (beef steaks and milk chocolate) were influenced by different consumption contexts created using immersive VR digital technology, in comparison to a standard sensory booth. Aspects relating to the level of consumer engagement were also explored, with literature indicating that consumers feel more engaged in a digital immersive environment compared to a standard sensory booth (Giezenaar & Hort, 2021). Research has shown that certain products are better suited to specific situations of consumption (Piqueras-Fiszman & Jaeger, 2014). For example, beef is usually eaten in the context of a meal while chocolate can be considered a snack suited to a range of eating occasions. Therefore a virtual environment relevant to each type of food product was created to ensure consumers assessed products in an appropriate eating context.

14.2.1 Beef trial

Consumers (n = 30, 15 female, 15 male) assessed the meat from the same beef steak in two different environments: a standard sensory booth (Fig. 14.1) and an immersive VR restaurant (Fig. 14.2). They were not aware of the fact that they were tasting the same sample in both conditions. The VR restaurant environment was captured using a Samsung Gear 360 4K Ultra High Definition camera at a restaurant in Dublin, Ireland. To ensure that the atmosphere created was as close as possible to real-life conditions, the restaurant was open to customers during the recording. Audio recordings consisting of background noises and indistinguishable conversations were also recorded. Consumers viewed the video through a HMD (Oculus Go). Beef steak (striploin cut) was assessed for liking in terms of smell, tenderness, juiciness, beef flavour and overall liking on a 9-point hedonic scale, where 1 = dislike extremely and 9 = like extremely.

FIGURE 14.1 Contextual settings for the sensory evaluation of beef samples: standard sensory booth context.

FIGURE 14.2 Contextual settings for the sensory evaluation of beef samples: immersive virtual reality restaurant context.

Consumers assessed samples while wearing the HMD device and answered questions verbally in all conditions in an effort to retain the level of immersion in the VR condition. Prior to starting the trial, participants took part in a short training session in which they were familiarised with the testing protocol and how to use the VR headset.

The results showed that the environmental condition had a significant effect of liking of beef attributes smell ($P = .012$), tenderness ($P = .002$), juiciness ($P = .002$) and flavour ($P = .003$), with beef consumed in the VR restaurant environment setting receiving higher mean liking scores, compared to the sensory booths. Consumer overall liking scores for beef were also significantly greater ($P = 0.004$) when consumed in the VR context, suggesting that the VR restaurant had a positive influence on participants' hedonic ratings of beef steaks. The descriptive statistics are illustrated in Fig. 14.3.

In addition to hedonic response, consumers were asked questions regarding the memorability of the experience, effort to assess the samples, level of distraction and purchase intent. While there was no difference in the effort required to perform the assessment between the two environments, the VR restaurant was found to be more distracting ($P = .013$). However, the VR restaurant resulted in a more memorable testing experience ($P < .0001$) and consumers were more willing to purchase the beef when consumed in the VR restaurant than the sensory booth ($P < .05$).

14.2.2 Chocolate trial

In a separate study, consumers ($n = 30$, 18 females, 12 males) assessed three identical square pieces of chocolate in three different environmental contexts: a standard sensory booth, a VR busy city (Fig. 14.4) and a VR Irish countryside (Fig. 14.5). The VR environments were 360-degree videos, captured using a Garmin VIRB 360 camera and presented to consumers through a HMD (Oculus Go). The busy city 360-degree video was recorded in Dublin City, Ireland, and depicted the International Financial Services Centre, public transport and people walking through the streets. The sounds of a busy city environment were included in the video. The Irish countryside 360-degree video was recorded in a

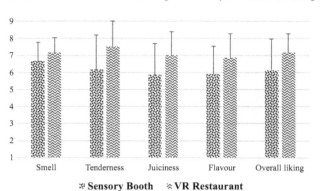

FIGURE 14.3 Mean scores and standard deviations for liking of beef in each testing condition. Responses recorded on a 9-point hedonic scale where 1 = dislike extremely and 9 = like extremely.

FIGURE 14.4 Contextual settings for the sensory evaluation of chocolate: virtual reality busy city (Dublin, Ireland).

FIGURE 14.5 Contextual settings for the sensory evaluation of chocolate: virtual reality Irish countryside.

rural part of Meath, Ireland, and provided a view of a typical Irish countryside setting with bright sunlight, the movement of the grass and the sound of the wind in the background. Consumers assessed the chocolate for liking in terms of smell, flavour, sweetness, texture, smoothness and overall liking on a 9-point hedonic scale, where 1 = dislike extremely and 9 = like extremely. Similar to the approach taken in the beef trial, participants provided answers to the questions verbally in all conditions and took part in a training session before beginning the trial.

In terms of results, a significant difference was found in liking of flavour ($P = .019$) and overall liking ($P = .005$), depending which environment the chocolate was consumed. The flavour and overall liking of chocolate was significantly more liked when consumed in the VR countryside setting, compared to the standard sensory booth. The surrounding environment did not influence hedonic ratings of other chocolate attributes. In general, the VR countryside setting produced higher liking scores for chocolate compared to the VR busy city and sensory booth environment (Fig. 14.6).

Similar to beef, the degree of memorability, effort, distraction and purchase intent was assessed during the chocolate trial. No difference was observed in the effort required to perform the trial between environments. Difference in the level of distraction between the three environments was approaching significance ($P = 0.055$), with consumers indicating that the VR busy city environment was more distracting than the sensory booth. Both VR settings were perceived as more memorable compared to the sensory booth ($P < 0.0001$). No significant differences were observed for purchase intent of chocolate between the three environmental conditions.

14.3 Discussion, conclusion and future perspectives

In our case studies, the VR restaurant (beef) and the VR Irish countryside (chocolate) produced significantly higher liking scores for all attributes (beef) and in terms of flavour and overall liking (chocolate), compared to the sensory booth. No significant differences in liking were observed between the two VR environments in the chocolate study, although the VR busy city setting tended to produce lower hedonic ratings. While only a limited number of studies have investigated how immersive VR technologies (involving some form of HMD device) impact hedonic liking, the results have been conflicting to date. Some studies have observed similar findings to ours. For instance, consumers had a significantly higher liking for sparkling wine than for beer in a VR winery context, while beer consumed in a VR bar was scored slightly higher for liking compared to wine (Picket & Dando, 2019). Similarly, Yang et al. (2022) showed that beer consumed in a VR bar context was liked more than in an evoked session. In addition, overall liking scores for snacks

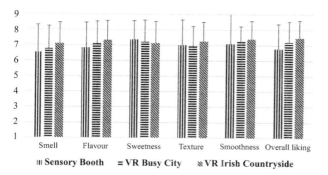

FIGURE 14.6 Mean scores and standard deviations for liking of chocolate in each testing condition. Responses recorded on a 9-point hedonic scale where 1 = dislike extremely and 9 = like extremely.

(caramel slice and chocolate digestive biscuit) were slightly higher in a café context created using an augmented mixed reality device (Microsoft Hololens) compared to sensory booths (Low et al., 2021a). The authors also note that choice of snacks was more influenced by liking (overall, sensory attributes) in an immersive testing context.

In contrast, however, other studies observed no changes in liking ratings between foods and beverages consumed in different immersive VR setting and/or sensory booths. A VR environment (sightseeing tour or a live music concert) had no influence on consumer liking ratings for chocolate in a study by Kong et al. (2020). Similarly Torrico et al. (2021) showed that both positive (open-field forest) and negative (closed-space room) VR environmental contexts did not affect the liking of any chocolate attributes; however, intensity of sweetness was higher in the positive VR setting, and no-sugar chocolate was associated with overall liking in the positive VR. Other studies also found no differences in consumer liking scores between immersive VR settings and traditional sensory booths for cheese (Stelick & Dando, 2018) or beer (Sinesio et al., 2019). However, pungency intensity of cheese was higher in the VR context (cow barn) compared to a sensory booth and a VR park bench in a natural open setting (Stelick et al., 2018). In addition, viewing a simulated coffee colour in VR was found to significantly influence perceived creaminess but had no impact on sweetness or liking (Wang et al., 2020). As noted by Giezenaar and Hort (2021), immersive environments created using digital technologies may influence consumer intensity ratings of specific sensory attributes, but not necessarily liking of sensory attributes, although more research is needed before any conclusions can be made.

There are a number of factors that could influence findings from sensory studies using immersive digital VR technologies to create context. For instance, the context-sensitivity of the food product being tested is likely to play a role. Recently Lichters et al. (2021) demonstrated that differences in the test environments impacted context-sensitive products more strongly than context-insensitive products. In their study, beer was considered context-sensitive while cappuccino was context-insensitive. Low et al. (2021a) also reported that consumers liking response was dependent on the appropriateness of the product-context. Main dishes tend to be context-sensitive while snacks are likely to be rather context-insensitive (Schöniger, 2022). The products' context-sensitivity may explain the findings from our case studies, in which beef (which could be defined as context-sensitive as it is usually eaten as part of a meal) received significantly higher liking ratings for all attributes in the VR restaurant context compared to the sensory booth. In contrast, liking ratings for chocolate (which could be defined as context-insensitive as it is suited to a range of eating occasions) were only significantly higher for flavour and overall liking in one VR environment. Also, as noted above, the surrounding context had no impact of liking ratings of chocolate in studies by Torrico et al. (2021) and Kong et al. (2020). Thus immersive environments could be more relevant for context-sensitive products (Lichters et al., 2021), and digital VR technologies could prove particularly useful for creating context for these types of products.

In addition to the specific context-sensitivity of a product, Nijman et al. (2019) reported that consumers can be segmented according to liking based on their individual sensitivity to context. For instance, using cluster analysis, three distinct consumer clusters of liking beer were identified, one of which was more influenced by context and showed a more discriminating response in a bar context compared to a sensory booth. However, the immersive environment used in this study was an actual real bar, and thus the possibility of classifying consumers according to their context-sensitivity while immersed in a digital VR world is currently unknown and requires further investigation (Giezenaar & Hort, 2021). Nonetheless, the importance of segmentation analysis was also highlighted by Lichters et al. (2021), who showed that consumer segments derived from an immersive consumption setting overlap more strongly with those from a natural consumption environment in comparison to those from a sensory lab. In addition, consumers have been segmented according to their emotional response to tea-break snacks across different contexts including the use of mixed reality technology (Low et al., 2021b). Taken together, data analysis should consider the segmentation of consumers to gain deeper insights into product response in immersive contexts (Lichters et al., 2021).

Another important factor to consider is how VR technology is being applied in sensory studies. Our study created immersive VR environments by displaying custom-recorded 360-degree VR videos and their corresponding sounds through an HMD, similar to others (Kong et al., 2020; Picket & Dando, 2019; Stelick et al., 2018; Torrico et al., 2021). In contrast, other researchers have taken a more technically challenging approach and created computer-simulated 3D models of products such as coffee (Wang et al., 2021), chocolate (Ammann et al., 2020) and cake and juice (Ammann et al., 2020a). While the technology enabling the creation of 3D food in VR is advancing, consumers' sensory perception of foods in a completely virtual environment is likely to be dependent on the visual realism of the food image depicted. For instance, fruits and vegetables could be perceived as less appealing in VR than in the real world (Siegrist et al., 2019). Regardless of the approach, participants must be given enough time to feel immersed in the surrounding VR environment, without having to assess the food product, which may enhance their feeling of presence in the evoked context (Schöniger, 2022).

As noted above, an HMD fully replaces the users' view of the real world, restricting their ability to interact visually with the 'real' food product. In order to measure consumers' hedonic responses while fully immersed in the VR environment, participants in our case studies, like others (Wang et al., 2020) provided answers to the questions verbally, which were subsequently recorded by the researcher. In other studies, participants were instructed to simply remove their VR headsets after tasting and answer questions related to the samples (Kong et al., 2020; Sinesio et al., 2019; Torrico et al., 2021), while others displayed the questionnaire and scales in the VR environment, which naturally enhanced the technical complexity of the design (Picket & Dando, 2019; Stelick et al., 2018; Yang et al., 2022). While each approach has clear advantages and disadvantages, future studies could investigate which approach can best improve the ecological validity of consumer sensory data, while continuing to retain participant engagement in the task. Nonetheless, this challenge could be overcome by using an augmented mixed reality device, such as Microsoft HoloLens, which integrates specific elements from the 'real-world' within the surrounding VR space, enabling the user to interact with the real food product while simultaneously answering questions in a controlled environment. Mixed reality has been used to understand consumer response to tea-break snacks and was shown to evoke ecologically valid data comparable to a real-life context (Low et al., 2021b).

It is abundantly clear that digital technologies such as VR will be a part of our future. While research on the use of VR as a context-enhancing technology is still in its infancy, it is expected further research will confirm its confident application in improving the ecological validity of consumer sensory evaluations. Researchers should continue to investigate the role context-sensitivity plays in immersive digital environments, both in terms of the product category tested and its relevance to consumers' product experience. Future research should also focus on integrating olfactory and tactile cues to enhance multisensory digital experiences (Crofton et al., 2019). In the next few years, the technical ability of HMD devices is expected to advance, providing additional opportunities to gain even deeper insights into consumer response to sensory attributes and, more accurately, predict product perception and performance in the marketplace. The enabling technologies have been developed; the next step is creating the innovative applications for sensory science.

References

Alba-Martínez, J., Sousa, P. M., Alcañiz, M., Cunha, L. M., Martínez-Monzó, J., & García-Segovia, P. (2022). Impact of context in visual evaluation of design pastry: Comparison of real and virtual. *Food Quality and Preference, 97*. Available from https://doi.org/10.1016/j.foodqual.2021.104472; https://www.journals.elsevier.com/food-quality-and-preference.

Ammann, J., Hartmann, C., Peterhans, V., Ropelato, S., & Siegrist, M. (2020). The relationship between disgust sensitivity and behaviour: A virtual reality study on food disgust. *Food Quality and Preference, 80*. Available from https://doi.org/10.1016/j.foodqual.2019.103833, Article 103833.

Ammann, J., Stucki, M., & Siegrist, M. (2020a). True colours: Advantages and challenges of virtual reality in a sensory science experiment on the influence of colour on flavour identification. *Food Quality and Preference, 86*, 103998. Available from https://doi.org/10.1016/j.foodqual.2020.103998.

Aukstakalnis, S. (2017). Applications of augmented and virtual reality in science and engineering. *Practical augmented reality: A guide to the technologies, applications and human factors for AR and VR*. America: Pearson Education.

Bangcuyo, R. G., Smith, K. J., Zumach, J. L., Pierce, A. M., Guttman, G. A., & Simons, C. T. (2015). The use of immersive technologies to improve consumer testing: The role of ecological validity, context and engagement in evaluating coffee. *Food Quality and Preference, 41*, 84–95. Available from https://doi.org/10.1016/j.foodqual.2014.11.017.

Crofton, E. C., Botinestean, C., Fenelon, M., & Gallagher, E. (2019). Potential applications for virtual and augmented reality technologies in sensory science. *Innovative Food Science and Emerging Technologies, 56*, 102178. Available from https://doi.org/10.1016/j.ifset.2019.102178.

Crofton, E., Murray, N., & Botinestean, C. (2021). Exploring the effects of immersive virtual reality environments on sensory perception of beef steaks and chocolate. *Foods, 10*(6), 1154. Available from https://doi.org/10.3390/foods10061154.

de Wijk, R. A., Kooijman, V., Verhoeven, R. H. G., Holthuysen, N. T. E., & de Graaf, C. (2012). Autonomic nervous system responses on and facial expressions to the sight, smell, and taste of liked and disliked foods. *Food Quality and Preference, 26*(2), 196–203. Available from https://doi.org/10.1016/j.foodqual.2012.04.015.

Giezenaar, C., & Hort, J. (2021). A narrative review of the impact of digital immersive technology on affective and sensory responses during product testing in digital eating contexts. *Food Research International, 150*, 110804. Available from https://doi.org/10.1016/j.foodres.2021.110804.

Hannum, M. E., Forzley, S., Popper, R., & Simons, C. T. (2020). Further validation of the engagement questionnaire (EQ): Do immersive technologies actually increase consumer engagement during wine evaluations? *Food Quality and Preference, 85*. Available from https://doi.org/10.1016/j.foodqual.2020.103966; https://www.journals.elsevier.com/food-quality-and-preference.

Hathaway, D., & Simons, C. T. (2017). The impact of multiple immersion levels on data quality and panelist engagement for the evaluation of cookies under a preparation-based scenario. *Food Quality and Preference, 57*, 114–125. Available from https://doi.org/10.1016/j.foodqual.2016.12.009.

Hein, K. A., Hamid, N., Jaeger, S. A., & Delahunty, C. M. (2010). Application of a written scenario to evoke a consumption context in a laboratory setting: Effects on hedonic ratings. *Food Quality and Preference, 21*, 410–416. Available from https://doi.org/10.1016/j.foodqual.2009.10.003.

Holthuysen, N. T. E., Vrijhof, M. N., de Wijk, R. A., & Kremer, S. (2017). "Welcome on board": Overall liking and just-about-right ratings of airplane meals in three different consumption contexts—Laboratory, re-created airplane, and actual airplane. *Journal of Sensory Studies, 32*(2). Available from https://doi.org/10.1111/joss.12254; http://www.blackwellpublishing.com/journal.asp?ref = 0887-8250&site = 1.

Kong, Y., Sharma, C., Kanala, M., Thakur, M., Li, L., Xu, D., Harrison, R., & Torrico, D. D. (2020). Virtual reality and immersive environments on sensory perception of chocolate products: A preliminary study. *Foods, 9*(4), 515. Available from https://doi.org/10.3390/foods9040515.

Köster, E. P. (2009). Diversity in the determinants of food choice: A psychological perspective. *Food Quality and Preference, 20*(2), 70–82. Available from https://doi.org/10.1016/j.foodqual.2007.11.002.

Lichters, M., Möslein, R., Sarstedt, M., & Scharf, A. (2021). Segmenting consumers based on sensory acceptance tests in sensory labs, immersive environments, and natural consumption settings. *Food Quality and Preference, 89*, 104138. Available from https://doi.org/10.1016/j.foodqual.2020.104138.

Low, J. Y. Q., Diako, C., Lin, V. H. F., Jun Yeon, L., & Hort, J. (2021a). Investigating the relative merits of using a mixed reality context for measuring affective response and predicting tea break snack choice. *Food Research International, 150*, 110718. Available from https://doi.org/10.1016/j.foodres.2021.110718; http://www.elsevier.com/inca/publications/store/4/2/2/9/7/0.

Low, J. Y. Q., Lin, V. H. F., Jun Yeon, L., & Hort, J. (2021b). Considering the application of a mixed reality context and consumer segmentation when evaluating emotional response to tea break snacks. *Food Quality and Preference, 88*. Available from https://doi.org/10.1016/j.foodqual.2020.104113; https://www.journals.elsevier.com/food-quality-and-preference.

Nijman, M., James, S., Dehrmann, F., Smart, K., Ford, R., & Hort, J. (2019). The effect of consumption context on consumer hedonics, emotional response and beer choice. *Food Quality and Preference, 74*, 59–71. Available from https://doi.org/10.1016/j.foodqual.2019.01.011.

Oliver, J. H., & Hollis, J. H. (2021). Virtual reality as a tool to study the influence of the eating environment on eating behavior: A feasibility study. *Foods, 10*(1). Available from https://doi.org/10.3390/foods10010089; https://www.mdpi.com/2304-8158/10/1/89/pdf.

Picket, B., & Dando, R. (2019). Environmental immersion's influence on hedonics, perceived appropriateness, and willingness to pay in alcoholic beverages. *Foods, 8*(2), 42. Available from https://doi.org/10.3390/foods8020042.

Piqueras-Fiszman, B., & Jaeger, S. R. (2014). The impact of evoked consumption contexts and appropriateness on emotion responses. *Food Quality and Preference, 32*, 277–288. Available from https://doi.org/10.1016/j.foodqual.2013.09.002.

Schöniger, M. K. (2022). The role of immersive environments in the assessment of consumer perceptions and product acceptance: A systematic literature review. *Food Quality and Preference, 99*, 104490. Available from https://doi.org/10.1016/j.foodqual.2021.104490.

Sester, C., Deroy, O., Sutan, A., Galia, F., Desmarchelier, J. F., Valentin, D., & Dacremont, C. (2013). Having a drink in a bar: An immersive approach to explore the effects of context on drink choice. *Food Quality and Preference, 28*(1), 23–31. Available from https://doi.org/10.1016/j.foodqual.2012.07.006.

Siegrist, M., Ung, C. Y., Zank, M., Marinello, M., Kunz, A., Hartmann, C., & Menozzi, M. (2019). Consumers' food selection behaviors in three-dimensional (3D) virtual reality. *Food Research International, 117*, 50–59. Available from https://doi.org/10.1016/j.foodres.2018.02.033; http://www.elsevier.com/inca/publications/store/4/2/2/9/7/0.

Sinesio, F., Moneta, E., Porcherot, C., Abbá, S., Dreyfuss, L., Guillamet, K., Bruyninckx, S., Laporte, C., Henneberg, S., & McEwan, J. A. (2019). Do immersive techniques help to capture consumer reality. *Food Quality and Preference, 77*, 123–134. Available from https://doi.org/10.1016/j.foodqual.2019.05.004.

Stelick, A., & Dando, R. (2018). Thinking outside the booth—The eating environment, context and ecological validity in sensory and consumer research. *Current Opinion in Food Science, 21*, 26–31. Available from https://doi.org/10.1016/j.cofs.2018.05.005; http://www.journals.elsevier.com/current-opinion-in-food-science/.

Stelick, A., Penano, A. G., Riak, A. C., & Dando, R. (2018). Dynamic context sensory testing—A proof of concept study bringing virtual reality to the sensory booth. *Journal of Food Science, 83*(8), 2047–2051. Available from https://doi.org/10.1111/1750-3841.14275; http://onlinelibrary.wiley.com/journal/10.1111/(ISSN)1750-3841.

Torrico, D. D., Sharma, C., Dong, W., Fuentes, S., Gonzalez Viejo, C., & Dunshea, F. R. (2021). Virtual reality environments on the sensory acceptability and emotional responses of no- and full-sugar chocolate. *LWT, 137*. Available from https://doi.org/10.1016/j.lwt.2020.110383; http://www.elsevier.com/inca/publications/store/6/2/2/9/1/0/index.

van Bergen, G., Zandstra, E. H., Kaneko, D., Dijksterhuis, G. B., & de Wijk, R. A. (2021). Sushi at the beach: Effects of congruent and incongruent immersive contexts on food evaluations. *Food Quality and Preference, 91*, 104193. Available from https://doi.org/10.1016/j.foodqual.2021.104193.

Wang, Q. J., Barbosa Escobar, F., Alves Da Mota, P., & Velasco, C. (2021). Getting started with virtual reality for sensory and consumer science: Current practices and future perspectives. *Food Research International, 145*. Available from https://doi.org/10.1016/j.foodres.2021.110410; http://www.elsevier.com/inca/publications/store/4/2/2/9/7/0.

Wang, Q. J., Meyer, R., Waters, S., & Zendle, D. (2020). A dash of virtual milk: Altering product color in virtual reality influences flavor perception of cold-brew coffee. *Frontiers in Psychology, 11*, 3491. Available from https://doi.org/10.3389/fpsyg.2020.595788.

Yang, Q., Nijman, M., Flintham, M., Tennet, P., Hidrio, C., & Ford, R. (2022). Improving simulated consumption context with virtual reality: A focus on participant experience. *Food Quality and Preference, 98*, 104531. Available from https://doi.org/10.1016/j.foodqual.2022.104531.

Next-generation sensory and consumer science: data collection tools using digital technologies

Rebecca Ford, Imogen Ramsey and Qian Yang

Sensory Science Centre, Division of Food, Nutrition and Dietetics, School of Biosciences, University of Nottingham, Sutton Bonington Campus, Loughborough, Leicestershire, United Kingdom

15.1 Introduction

Sensory methods provide objective data, whereas consumer methods generally provide subjective information about products. Looking back at the history of sensory and consumer science, the first named sensory method was published in the 1930s and was known as the paired-eating method (Howgate, 2015). Since then, sensory and consumer science research has evolved significantly, with not only new methods, but also new ways to collect and analyse data.

Many of us will remember a time when the majority of sensory and consumer data was collected by pen and paper ballots. Although these were highly mobile, they had numerous disadvantages including the time-consuming nature of data collection, data processing and analysis, difficulties of recognising hand-writing, issues with hand-designing a balanced order of presentation, as well as the risk of sheets being lost or sections being missed by assessors (Kemp, 2009; Moskowitz, 2003). In addition, there was also the environmental damage of using considerable quantities of paper. Any last-minute changes to questions on the form were difficult to rectify once the study was in place and questions were limited to specific formats. A data entry operator was needed to collate all the responses by hand, making the whole process costly and time-consuming. Furthermore, specific methods, such as continuous time intensity (Cliff & Heymann, 1993; Lee & Pangborn, 1986), required graph paper to continuously record intensity perception of each assessor, sample and replicate, and manual data analysis.

Thankfully, in the late 1980s, the first sensory software companies such as Compusense (Canada) and Fizz Biosystems (France) were founded, and the development of computerised forms, for example Fizz Forms, tackled some of these difficulties. Further developments saw networked desktop computers installed in sensory booths, which were commonly managed by a local network server on a master computer, allowing digital data collection and facilitating faster data collection, analysis and even standardised reporting (Stone et al., 2012). Additional data or procedures could also be collected and implemented, such as time stamps, as well as enforced breaks for palate cleansing. This also allowed an increased number of sensory and consumer methodologies to be developed, such as temporal dominance of sensations (Pineau & Schilch, 2015; Pineau et al., 2003; Pineau et al., 2009), temporal check all that apply (Ares et al., 2016; Castura et al., 2016) and temporal order of sensations (Pecore et al., 2011), as well as polarised sensory positioning (Teillet et al., 2010, 2015). These enhanced softwares also allowed immediate feedback of results, such as panel performance measures during or after data collection to both panel leader and sensory panellists to increase panel training efficiency (Findlay et al., 2003).

Networked desktop computers, however, had some drawbacks, which included the lack of flexibility in collecting data in various locations. Therefore with the software revolution came cloud platforms to store and manage data from different data collection devices. This enabled the use of handheld touchscreen devices, allowing sensory and consumer data to be collected outside the sensory laboratory, facilitating 'on-the-go' and 'in-context' consumer evaluations (Stone et al., 2012), in a bar or in a restaurant, for example. Waterproof handheld devices could also collect real-time sensory data of personal-care products such as shampoo, shower gel and shaving products. However, as these new digital data collection techniques were developed, research was needed to understand how these affect the results collected. A recent study did just this, by exploring the impact of handheld devices on emotional responses of consumers using continuous line scales (Nijman 2019). No significant impact of device was found when comparing handheld touchscreens to desktop computers, but when the data were analysed separately, there was an indication that desktop computers may be more sensitive at discriminating between products for emotional response (Nijman, 2019). This could be due to the larger representation of the line scale on the desktop computer screen or due to the accuracy of using a mouse (desktop computer) over a finger (handheld device) to mark the line scale. However, the difference was subtle, indicating the data collection device has limited impact on consumers and it is therefore likely that sensory assessors would be even less affected due to their training on scale-use.

Digital developments have allowed this evolution of sensory data from consumers in more realistic consumption environments (as opposed to controlled laboratory environments required for objective sensory evaluations) (Kemp, 2009; Lawless, 1999). Traditionally, when consumer evaluations were conducted in sensory labs, it was to reduce the effects of psychological factors and physical conditions, as well as minimise distractions and gain experimental control (Meiselman, 2013). However, these testing laboratories are far from normal consumption environments. Eating is a multisensory experience (Spence, 2013) and we are influenced by the context and environment in which we consume products (Bell et al., 1994; Edwards et al., 2016). For example, beer is not consumed in a sterile environment, without social interaction, and deodorant is not applied in a precise way with a sensory panel leader watching. Thus consumer testing in laboratory environments has little predictive power for how consumers experience products in the real world and therefore limits the extent to which the testing environment is representative of the context of interest (Galiñanes Plaza et al., 2019), or is ecologically valid (de Graaf et al., 2005; Köster, 2003). It has also been discussed that this is one of the reasons for high failure rates of new product launches, as participants have stated lack of engagement and boredom in central location tests (Bangcuyo et al., 2015). Therefore there has been increased interest in research 'in-context' to counteract this (Galiñanes Plaza et al., 2019; Jaeger, Hort, et al., 2017).

Context has been defined as 'a set of events and experiences that are not part of the reference event that one is interested in but have some relationship to it' (Rozin & Tuorila, 1993). This was further adapted by Meiselman (1996), to define the reference event as the inclusion of the sensory, physiological and behavioural responses to the reference food, whilst the context included those factors which surround and influence the reference event. Previous research has shown that context can impact both food choice and acceptability, with liking scores found to vary between sensory laboratories and home use tests (HUT) (Boutrolle et al., 2007), restaurants (King et al., 2007; Meiselman et al., 2000) and cafeterias (Meiselman et al., 2000). Consumer choice behaviour can also be manipulated by the decoration in a restaurant (Bell et al., 1994) or the ambience of a bar (Sester et al., 2013). Different strategies using context to improve the validity of sensory consumer tests, such as virtual reality (VR), augmented reality (AR) and mixed reality (MR) are discussed in the case study in Section 15.2.

Another advancement is the adaption of measuring liking alone, to also include emotional response measurements, as these have been found to provide additional discriminatory information. There are many different ways to measure emotional response, including self-reported emotional response questionnaires collected either through line scale or check-all-that-apply (King & Meiselman, 2010; Nestrud et al., 2016), self-reported visual measures using images, or animations instead of words to express emotions, such as the product emotion measurement instrument (PreEmo) (Desmet, 2018), mood portraits (Churchill & Behan, 2010) and use of emoji/emoticons (Ares et al., 2016; Jaeger, Lee, et al., 2017; Rini et al., 2022). What is important with all of these methods is that the data are captured as close to the reference event as possible, which is aided by technology. Other implicit measures that rely on digital technology include those monitoring changes in physiological response [e.g. electromyography (EMG), electrodermal activity (EDA), electrocardiogram (ECG), functional magnetic resonance imaging (fMRI)], as well as the development of facial expression recognition software. Smart wearables present considerable opportunities both for sensory and consumer data collection in the future, as they also benefit from the integration of various applications providing data on location, heart rate (HR) and blood oxygen content. However, these tools are still at their infancy in relation to collecting sensory and consumer data and so will be discussed in Section 15.3.

15.2 Case study on virtual reality/augmented reality to capture sensory data

A combination of different stimuli including visual, auditory, tactile and olfactory can be used as important resources to create contextual information to stimulate natural testing environments. A realisation of the importance of context has led to the use of different approaches to bring relevant contexts to traditional controlled test settings, as an attempt to optimise both experimental control and ecological validity. Different strategies include written scenarios (Dorado et al., 2016; Hein et al., 2010), exposure to images (Andersen, Kraus, et al., 2019; Hersleth et al., 2015), video or sound recordings (Hathaway & Simons, 2017; Liu et al., 2019), decorating rooms to be more similar to real-life environments (Holthuysen et al., 2017) and video projection (Sester et al., 2013; Sinesio et al., 2018). In recent years, affordable technological advances have allowed more immersive consumer testing approaches using video walls (Bangcuyo et al., 2015; Hannum et al., 2019; Hathaway & Simons, 2017; Worch et al., 2020; Zandstra et al., 2020) and VR (Andersen, Kraus, et al., 2019; Barbosa Escobar et al., 2021; Sinesio et al., 2019; Stelick et al., 2018; Torrico et al., 2021; Wang et al., 2020; Worch et al., 2020).

VR immerses participants within a computer-generated virtual environment, whereas AR enhances views of the real world by overlaying what participants see with computer-generated information. This is achieved by taking images from a camera and processing it back into the virtual world in real-time. Augmented virtuality (AV) involves adding elements from the real world into the virtual world. Generally speaking, MR is the merging of both virtual and real worlds. There are a number of practical considerations that should be taken into account when designing food tasting experiences in VR/AR/MR environments. Firstly, when participants wear a VR headset, their eyes are covered, blocking physical reality and making food and drink consumption a challenging task. Previous studies have used object tracking techniques that attach to the drinking vessel and match to a 3D modelled object in VR to allow the consumer to view and achieve a drinking experience in the VR environment (Wang et al., 2020; Yang et al., 2022). However, when considering eating food in a bowl or on a plate, different techniques need to be considered, as object tracking is only capable of tracking the plate or bowl and not the food upon or within it. Thus MR is required to bring the real and digital world together. In MR, participants can interact with and manipulate both virtual and physical environments, allowing them to view and interact with a food product without taking off their VR headsets.

Another consideration is data collection tools, as the idea of creating an immersive environment is to create the relevant context for consumers. However, some established VR experiences require participants to remove the headset to answer questions on a tablet or paper (Ammann et al., 2020; Barbosa Escobar et al., 2021; Torrico et al., 2021), whilst others ask participants to describe their answers verbally (Nivedhan et al., 2020; Wang et al., 2020) which disturbs the flow of the study and would ultimately take participants out of the immersive environment. So far, there are only a handful of studies which have embedded the questionnaire in the virtual environment. Stelick et al. (2018) used a controller and Picket and Dando (2019) asked participants to turn their head to manoeuvre the cursor to position their responses. Both of these require a technician to check responses in real-time and input responses into the system manually. Building upon this, Sinesio et al. (2019) used a joystick to answer questions in a 3D modelled bar and Yang et al. (2022) developed a unique feature to augment a physical pen with a tracker to collect responses, therefore allowing more natural responses within the VR environment, which is believed to be more similar to reality. A drawback of this, however, is that the embedded questionnaire needs additional programming, offering reduced flexibility in future testing. In addition, these features would be highly dependent on the model of VR headset that is selected.

Previous VR studies involving food evaluation have only evaluated appearance or smell of products (Barbosa Escobar et al., 2021), requiring tasting after VR environment exposure (Wen & Leung, 2021) or having participants hold the sample (e.g. chocolate) in their hands to enable the eating experience (Torrico et al., 2021). Interestingly, some VR studies have used more advanced technologies — object tracking to enable drinking/eating experience in the VR environment that explored colour—taste interaction of drink with no relevant context (Ammann et al., 2020; Wang et al., 2020) and context effect for beer evaluation (Yang et al., 2022) and pizza eating (Oliver & Hollis, 2021). However, for some eating experience, it requires a much more complex hand involvement than drinking a liquid or eating a solid food such as snack product. Therefore it is essential to be able to see physical hands and food in a real-time manner, for example a meal or semisolid food (e.g. breakfast cereal or yogurt). Thus MR could be considered to tackle some of these challenges. The case study below considers the innovative approach of an embedded camera on a VR headset to merge participant's real-life view to the virtual world (e.g. a modelled kitchen), with the added complexity of embedding questionnaire in the virtual context. This design offers considerable flexibility to change product categories and questionnaires used.

15.2.1 Case study — evaluating breakfast in a mixed reality kitchen environment

Objective: Investigate the feasibility of evaluating a semisolid breakfast product in a designed MR kitchen environment.

15.2.1.1 Materials and method

Participants: 39 consumers of the semisolid breakfast, who consume the product at least once a month and are non-rejectors of the flavour under test.

Testing environment: Participants wore a VR headset (HTC Vice, HTC Corporation, Taiwan) which was connected to a gaming-specification computer running a custom application built using the Unity game engine (Unity Technology, California, USA). The VR headset also had integrated headphones to allow sound application in the VR experience. The MR environment immersed a virtually modelled kitchen with a neutral colour scheme as the background, and the built-in camera in the VR headset was enabled to allow real-time tracking of hand and body movements when participants were interacting with objects and foods in the foreground. To blend the real and virtual content, a white opaque table was used in the real world that connected with the virtual white table smoothly in the MR environment. Virtual objects including toast, coffee cafetière and product packaging were designed on the table, to add a realistic visual breakfast environment to the experience. Sound stimuli including a bird chirping and the sound of a kettle were also embedded in the VR experience to help simulate an 'at home' kitchen environment in the morning. Finally, a TV attached to one of the virtual kitchen walls allowed projection of the questionnaire on the TV screen and participants used a wireless mouse to submit their responses.

Sample preparation and questions: Three commercially available products of the same flavour with known differences in sensory properties were selected. Each sample was prepared following instructions given on the pack, and named as Product 1 (P1), Product 2 (P2) and Product 3 (P3).

To assess the ability of consumers to interact with different scale types in the MR environment, participants were asked to rate a range of hedonic responses for the products and the optimal attribute intensity for the given products collected by just about right (JAR) scale. Consumer opinions regarding the virtual experiences were also captured by questionnaire after product evaluation.

15.2.1.2 Results and discussion

Consumer opinions regarding the virtual experience

In general, participants were positive about the MR kitchen environment, stating that it enabled an immersive, relaxing, interesting, engaging and realistic environment that most participants liked and related to. Participants found the reality of the visual kitchen setting and the sound of the bird chirping helped with immersion. All participants managed to eat the product without spilling, indicating that the MR kitchen design allowed smooth hand-to-mouth interactions and satisfactory real-time imagery to be fed back into the headset enabling complex eating experiences. Although some participants indicated limitations, with the headset being mentioned as chunky and heavy thus increasing the difficulty of getting the spoon into their mouth. The innovative approach of embedding the questionnaire on the TV screen also received positive feedback, as the TV screen acted as a data collection media, thus offering flexibility regarding the type of sensory and consumer data that could be collected. In general, positive feedback was received regarding using the mouse to answer the questionnaire. Data collected on the sensory properties of the product (JAR) indicated that the technology impacted the perceived consistency and colour of the product rather than any flavour attributes. Therefore the use of MR technologies may not yet be suited to tests comparing the appearance of products.

Furthermore, another limitation of the technology used was the resolution of the in-built camera in the HTC Vive headset resulting in a less clear field of view and some blurring which affected product appearance and the clarity of the edge of the questionnaire. It is hoped that future technologies will seek to improve this resolution allowing a much clearer view, as well as lighter headsets which would ultimately improve perceived realism.

15.2.2 Conclusion

This study explored an innovative solution allowing participants to consume a semi-solid food in a context-relevant virtual experience using MR technology. This study shows that MR can successfully create such a virtual experience, allowing the participant's to complete complex hand movements during eating. The MR kitchen design was perceived by participants as an immersive, relaxing and interesting environment, with additional stimuli (e.g. 3D kitchen model, sound, objects placed on the virtual table) having increased the level of realism. Embedding the questionnaire on the TV screen on a virtual kitchen wall, connected to the web browser offered flexibility regarding data collection and received positive feedback. The benefit of the MR kitchen design is that it offers considerable flexibility of altering the context or the questionnaire needed for different studies, therefore, once the virtual environment has been developed, it can be used to test different product categories or for exploring different questions without any additional programming. However, there were limitations to the technologies, with the field of view described as less clear and the headset discussed as being chunky and heavy thus impacting eating. Products and questionnaires on screen were also described as a bit blurry. It is hoped that technological improvements could see development of a lighter version of a headset with a higher quality built-in camera to increase consumer experience.

15.3 Future perspectives

15.3.1 Digital tools currently used by sensory and consumer scientists

Sensory and consumer science is an evolving field with increased interest in incorporating technologies from different disciplines to collect sensory and consumer information. As shown in the case study, VR/MR techniques can now successfully simulate a relevant context. But there are many other ways in which digital technologies can be used to acquire sensory and consumer data, including technologies to measure physiological reactions, emotional response via facial expressions and taste and aroma data using the e-nose and e-tongue. These technological tools allow (in some part) the collection of data from assessors without the need to directly ask questions.

15.3.1.1 Physiological measures

When consuming a food or drink product, there are many factors, such as intrinsic sensory properties, extrinsic packaging cues and environmental factors which can influence a response. Traditional sensory methods involve self-reported responses which could have cognitive bias, whereas physiological reactions help to collect unconscious responses. As discussed in Chapter 12, Emerging biometric methodologies for human behaviour measurement in applied sensory and consumer science, Danni Peng-Li, Qian Janice Wang, and Derek Victor Byrne, these consumer neuroscience tools can measure changes in the autonomic nervous system, such as facial expression, HR and skin conductance, as well as changes in brain activity and emotional response (Beyts et al., 2017; Songsamoe et al., 2019). These measurements provide advantages over more traditional methods as they measure processes that consumers themselves are not aware of and which may not be captured in questionnaires, thus tapping into their subconscious responses. The technologies available to measure peripheral nervous systems responses include EDA, ECG, EMG, with central nervous system responses using electroencephalography (EEG) and fMRI. Chapter 12 introduces the strengths and weaknesses of these neuroscience tools, but here these will be discussed in detail.

15.3.1.2 Peripheral

EDA is a measurement of skin conductance response, which measures the change in sweat emitted by eccrine glands (Bell et al., 2018). This physiological response has been used to measure consumer reactions and predict their decisions with advertising stimuli (Ohme et al., 2009), hedonistic and utilitarian products (Guerreiro et al., 2015) and product packaging (Vila-López & Küster-Boluda, 2019). Researchers have found that this is a useful technique in assessing consumer's reactions and indicates preference without the need for self-report measures (Bell et al., 2018). The disadvantages of this method, however, are that it is susceptible to many environmental factors, such as room temperature and humidity, caffeine, medication, nicotine and exercise (Figner & Murphy, 2011) and thus may have limited application in sensory research (Bell et al., 2018).

ECG is used to measure HR and consists of two electrodes on the skin, one on the right collar bone and the other on the lower left rib (Brouwer et al., 2017). Numerous studies have measured HR to understand the influence of images and videos (Verastegui-Tena et al., 2017; Walsh et al., 2017), aromas (Beyts et al., 2017; He et al., 2014), consumption of food and drink (de Wijk et al., 2012, 2014; Gunaratne et al., 2019; Verastegui-Tena et al., 2019) and cooking of foods (Brouwer et al., 2017, 2019), however, there have been inconclusive results so far. Researchers in one study found that HR increased with more disliked foods (de Wijk et al., 2012), decreased for more liked smells (He et al., 2014) and increased for more liked foods (de Wijk et al., 2014). Other studies found no significant correlation between HR and liking (Beyts et al., 2017; Danner et al., 2014). The authors discussed that the differences amongst studies may be due to the fact the products were too similar. Whereas other studies where significant differences were found used products from different categories (Beyts et al., 2017), limiting the method to compare competitor products against each other. The sensors used within these studies have also been shown to be invasive, increasing stress levels as participants are aware they are being monitored (Gonzalez Viejo et al., 2018).

EMG is a measure of facial muscle activity, with placement of electrodes near the eyebrows and cheeks to record changes in electrical impulses generated by contracting muscles (Bell et al., 2018). Previous studies have measured the effect of food images on facial muscle activity (Nath et al., 2019, 2020), as well as on food and drink products (Horio, 2003; Korb, Götzendorfer, et al., 2020; Korb, Massaccesi, et al., 2020; Sato, Minemoto, et al., 2020), whilst others have explored the effect of nutritional information (Sato, Yoshikawa, et al., 2020). Reports suggested that EMG signals were associated with subjective hedonic ratings such as liking and wanting of both images and products (Nath et al., 2019, 2020; Sato, Minemoto, et al., 2020). Other studies have used EMG to measure differences in eating behaviours, linking this to sensory data to understand differences in overall liking and sensory attributes of cheese (Jack et al., 1993), apples (Ioannides et al., 2007) and chocolate (Carvalho-da-Silva et al., 2011). Results discovered that sensory attributes can be linked to EMG data, with the descriptor 'mouth-coating' correlated to number of chews, swallowing events and residence time in mouth (Carvalho-da-Silva et al., 2011). There are a number of disadvantages with this technique, however, which include; misplacement of sensors leading to lower accuracy of measurement, electrodes picking up other movements such as speech and body, and individual variation between eating behaviours (Bolls et al., 2001; Carvalho-da-Silva et al., 2011; Huang et al., 2004).

15.3.1.3 *Central measures*

EEG is an electrophysiological measurement used to record the electrical activity of the brain, and is the most commonly used method for measuring brain waves with high temporal resolution (Songsamoe et al., 2019). Multiple electrodes are attached to a cap, which the participants places on the scalp and forehead. Participants are often asked to conduct a task during evaluation and signals are linked to the task during data analysis. However, EEG lacks spatial resolution meaning that whilst a signal can be linked to a sensory event/task (such as eating), the specific part of the brain responding to that event cannot. EEG technology has been used by neuroscientists to study consumer behaviour (Lagast et al., 2017), the effect of appearance (Toepel et al., 2009; Walsh et al., 2017) and flavour, taste and texture (Andersen, Kring, et al., 2019; Hashida et al., 2005; Horská et al., 2016; Labbe et al., 2011) on emotional and behavioural responses of food products, as well as the effect of food consumption on human brain functioning (Allen et al., 2014; De Pauw et al., 2017). Studies have shown a correlation between EEG results and hedonic attitudes measured through a questionnaire, with increased activation of the left hemisphere of the brain when the participant has a more positive attitude towards the food (van Bochove et al., 2016). Although a useful technique in measuring unconscious processes, there are some limitations in addition to the low spatial resolution, which include the need for specialist equipment and experienced data analysts (Songsamoe et al., 2019).

fMRI, on the other hand, has high spatial resolution but at the expense of temporal resolution. However, careful experimental designs allow neuronal activity to specific parts of the brain to be measured by quantifying blood flow. It has been used to study neural pathways of flavour perception (Eldeghaidy, Marciani, Pfeiffer, et al., 2011), taste—aroma interactions (Verhagen & Engelen, 2006) and taste—tactile interactions, as well as to understand the impact of taste phenotypes on cortical responses (Eldeghaidy, Marciani, McGlone, et al., 2011; Hort et al., 2016). However, limitations include only being able to consume a very small quality of sample, restricted movement and difficulty swallowing whilst laying down/still, specialist equipment and high costs.

Numerous studies have combined measurements from different physiological responses to further understand sensory changes (Beyts et al., 2017; Danner et al., 2014; de Wijk et al., 2012, 2014; Verastegui-Tena et al., 2019), with research suggesting that these measures could be used to discriminate between samples that did not differ in terms of liking scores (Danner et al., 2014; de Wijk et al., 2012, 2014). Future research could use advancements in technologies within smart wearables to combine outputs from different sensors to give an overall view on physiological responses, revealing differences in liking and sensory properties in a more realistic consumption environment.

15.3.1.4 *Facial expression recognition*

Methods used to measure changes in emotional response include facial expression through the use of software such as FaceReader (Noldus Information Technology, Wageningen, the Netherlands). This software is able to record fast-changing emotions by analysing facial expressions, which represent the six basic emotions (angry, happy, disgusted, sad, scared and surprised) and taps into the subconscious experience whilst consuming a product (van Bommel et al., 2020). Researchers within the field aim to provide deeper insights and discriminative abilities compared to liking scores alone (Beyts et al., 2017), but research so far has concluded that facial expression analyses are more suitable to differentiate disliked products compared to liked products (Danner et al., 2014; van Bommel et al., 2020; Zeinstra et al., 2009). Advantages of using these techniques are that they report multidimensional aspects, such as intensity and unfolding of emotions during consumption, which are not found by traditional self-report methods, such as questionnaires (Danner et al., 2014). Nevertheless, there are many disadvantages to this technique, which include data losses due to technical failures or coverage of the face, as well as large data sets which are time-consuming and difficult to analyse. Furthermore, the quality of results are highly product dependent due to oral processing changes during consumption (van Bommel et al., 2020).

15.3.1.5 E-nose and e-tongue

The development of the e-nose and e-tongue occurred due to the need to address certain problems with traditional sensory methods, including the fact they are time-consuming, expensive and sensory panellists can be influenced by fatigue, product carry-over or illness (e.g. COVID-19) or assessors might not be available (e.g. during COVID-19 lockdowns and isolations) (Ross, 2021). These multisensory systems are comprised of gas (e-nose) or chemical (e-tongue) sensors which send data to a processing system for analysis using pattern recognition. Previous research has shown that the e-nose can be used to measure quality and processing-related factors of food production, including coca/coffee bean roasting, cocoa/tea fermentation and the freshness/spoilage of meat (Tan & Xu, 2020). Originally the e-tongue was applied to discriminate the five basic tastes of sweet, sour, bitter, salty and umami, however, it is now being used to measure food quality, adulteration and authenticity (Tan & Xu, 2020). In addition, sensors have also been included for metallic, spicy, astringency and aftertaste attributes, which are typically difficult to assess using sensory methodology due to physiological or carryover issues (Ross, 2021). There are still, however, some limitations, which include the reduced specificity of sensors to target compounds and the fact that they measure static rather than dynamic in vivo release of compounds during eating. In addition, specialised operators and data analysts are needed to collect reliable outputs. For more detailed information regarding the current trends and future needs of the e-nose and e-tongue, please refer to Chapter 13, Added value of implicit measures in sensory and consumer science by René A. de Wijk and Lucas P.J.J. Noldus.

15.3.2 What does the future hold?

The future of sensory science is exciting and even though there are still challenges, many of these can be fulfilled by the advancements in digital technologies, which are not only able to assist with data collection, but are also able to store and analyse in real-time. These technologies include remote sensors, smart wearables, the application of artificial intelligence and digital simulations.

15.3.2.1 Remote sensing

One of the drawbacks of the measurement of physiological responses is that these are normally measured using physical sensors/electrodes, which mean that participants are aware that they are being monitored. Therefore suggestions for future research are to use remote sensing techniques, such as computer vision analysis, to assess physiological changes. Recently researchers developed and validated a remote sensing technique which used an integrated camera system, controlled by an app, to assess HR, blood pressure and facial expressions, showing that this was a reliable and accurate technique compared to the more traditional methods (Gonzalez Viejo et al., 2018). In addition, this technique was found to require less processing times than other methods, however, suggestions for future work were to implement an automatic algorithm to help track specific facial regions (Gonzalez Viejo et al., 2018).

15.3.2.2 Smart wearables

In more recent years, wearable devices such as smart watches, smart glasses and smart jewellery have been developed by technology companies. Several hand/wrist worn wearable devices have been proposed as a promising technology to collect consumer data, with the use of built in sensors such as gyroscopes, accelerometers, GPS and microphones. In earlier studies, model devices were developed to monitor behaviour, with the use of the built in sensors, such as accelerometers and gyroscopes to detect and record food intake (Dong & Biswas, 2017; Dong et al., 2012), whilst commercial products were used in other studies to record eating episodes and identify the mode of eating (e.g. using cutlery, chopsticks or bare hands) (Sen et al., 2015). Other authors took a different approach, and used a smartwatch to record acoustic signals in the identification of bites and swallows (Kalantarian & Sarrafzadeh, 2015). Furthermore, ownership of fitness bands and smart watches has increased (Mintel, 2020), indicating higher user acceptance as health-monitoring devices (Kalantarian & Sarrafzadeh, 2015). It is therefore clear that these devices are one of the most desirable and natural options for collecting data (Farooq & Sazonov, 2016), with research documenting the reliability and validity of these tools (Bell et al., 2018). They could also be used for data collection, with questionnaires developed and designed within apps and programmed to appear on smartwatch/smartphone screens when a specific situation (e.g. product interaction) is detected by the in-built sensors.

Another wearable technology capable of recording food intake are smart spectacles/glasses. Glasses with piezoelectric sensors can monitor muscle and jaw movements whilst on-the-go (Farooq & Sazonov, 2016) whilst others use multimodal sensors (gyroscopes located around the ear to monitor chewing, an accelerometer to monitor vibrations from the throat related to swallowing, a proximity sensor to record hand-to-mouth gestures and a camera to capture consumed foods) to capture food consumption in real-world settings (Bedri et al., 2020). The design of multisensor devices benefit from being noninvasive and unobtrusive (Farooq & Sazonov, 2016) but have been found to lack accuracy in classifying differences between eating and drinking episodes (Farooq & Sazonov, 2016). Further concerns include privacy and social acceptability due to capturing images or audio in public settings (Mintel, 2020), making them the least favourable option of the wearable devices for monitoring food intake (Kalantarian & Sarrafzadeh, 2015). However, large companies are reinvesting in these products, with Google investing in AR smart glasses firm North and Facebook launching a produced in partnership with EssilorLuxottica (Mintel, 2020).

Research has also explored the development of smart jewellery to collect consumer information, with one study developing a smart necklace consisting of a piezoelectric sensor to monitor jaw motion, a proximity sensor for detection of hand-to-mouth gestures and an accelerometer to monitor body motions (Fontana et al., 2014). Further research with the same smart necklace monitored duration of eating episodes, and was found to be more accurate than traditional food diaries (Alshurafa et al., 2015; Doulah et al., 2017). Advantages to this technique are that the necklace is nonintrusive, can be comfortably worn and is able to collect continuous data in more real-life situations (Alshurafa et al., 2015). However, limitations were that reported accuracy values only corresponded to solid foods, with further studies needed to determine feasibility in detecting liquid intake in real-life scenarios (Fontana et al., 2014). In recent years, commercial smart jewellery has also shown some interesting developments, with the launch of the Amazon Echo Loop (a smart ring that includes Alexa), however, no current research has used these products to monitor consumer eating behaviour or to assist in sensory data collection from consumers.

Overall, it has been reported that the combination of outputs from all sensors embedded into wearable devices outperforms individual ones (Bedri et al., 2020), with a successful example of this being shown in the case study in Chapter 12, Emerging biometric methodologies for human behaviuor measurement in applied sensory and consumer science, Danni Peng-Li, Qian Janice Wang, and Derek Victor Byrne. However, one of the limitations with this is that copious amounts of data are produced. This can cause problems with storage and data analysis, and thus artificial intelligence (AI), machine learning (ML) and deep learning (DL) techniques are recommended for future exploration.

15.3.2.3 *Artificial intelligence techniques/machine learning*

Trained sensory panel evaluations are costly and time-consuming, consisting of recruitment of participants, training on sensory methodology and product-specific references, final evaluation sessions, data handling and statistical analysis. In addition, traditional statistical techniques used to analyse large data sets, such as principal component analysis, partial least squares regression and preference mapping, can sometimes not provide enough accuracy and extendibility, especially when applying to larger volumes of data (Jiménez-Carvelo et al., 2019). As sensory science progresses, collection of continuous data will become the new norm and therefore new techniques will need to be employed. AI, ML and DL techniques have, over recent years, been discussed within the sensory science field. The main purpose of these techniques, however, is not to remove the need for trained sensory panels or consumers, but to incorporate these into models which are able to predict outcomes and allow real-time changes to products (Fuentes et al., 2021). AI is a technique which is able to incorporate human intelligence into machines, whilst ML and DL are subsets of this, which enable the machines to learn by themselves using data to make accurate decisions. These techniques have been used in the food industry to predict quality attributes (García-Esteban et al., 2018; Gonzalez Viejo & Fuentes, 2020) and consumer preference (Bi et al., 2020). One example was the development of predictive models in a computer system called ICatador, which used trained sensory panel scores of cheeses to train the system, with the additional input of instrumental measures. The system then automatically predicted the sensory attributes of the measured products through supervised machine learning (García-Esteban et al., 2018). More information on AI techniques can be found in Chapter 18, Shining through the smog: how to tell powerful, purposeful stories about sensory science in a world of digital overload by Sam Knowles. These techniques are still at their infancy, however, and research and development with these practices need to be conducted to fill the gaps in our understanding.

15.4 Conclusions

Over the last decade, digital technology has evolved rapidly allowing developments in sensory data collection. For example, the use of VR and immersive rooms in consumer studies to create relevant consumption environments for sensory product evaluation. The use of MR is now being considered as a solution to the challenges of uninterrupted and 'natural' drinking and eating experiences in virtual environments as presented in the case study. However, further research is needed to explore the most important features to create a realistic context for the consumer.

There is an increased interest in capturing data from consumers about sensory properties of products without the need to directly ask questions due to reduced bias and an ability to tap into system 1 thinking (implicit). Established technologies used in the field of sensory and consumer science capable of this include physiological measurement, fMRI and facial expression. However, data analysis from these methods can be complex, time-consuming and expensive. The more advanced technologies that are not widely used in the field but could shape future perspectives include remote sensing, AI technologies and smart wearables. These technologies enable data collection in real-life scenarios and utilising ML to investigate sensory properties and consumer responses. Although there are limitations and the data analysis required to understand outputs is complicated, the future potentials of these novel technologies as sensory data collection tools are full of opportunities. However, it does highlight the need for multidisciplinary collaborations across sensory and consumer science with data science and computer science to develop these exciting opportunities together.

References

Allen, A. P., Jacob, T. J. C., & Smith, A. P. (2014). Effects and after-effects of chewing gum on vigilance, heart rate, EEG and mood. *Physiology and Behavior, 133*, 244−251. Available from https://doi.org/10.1016/j.physbeh.2014.05.009.

Alshurafa, N., Kalantarian, H., Pourhomayoun, M., Liu, J. J., Sarin, S., Shahbazi, B., & Sarrafzadeh, M. (2015). Recognition of nutrition intake using time-frequency decomposition in a wearable necklace using a piezoelectric sensor. *IEEE Sensors Journal, 15*(7), 3909−3916. Available from https://doi.org/10.1109/JSEN.2015.2402652.

Ammann, J., Stucki, M., & Siegrist, M. (2020). True colours: Advantages and challenges of virtual reality in a sensory science experiment on the influence of colour on flavour identification. *Food Quality and Preference, 86*, 103998. Available from https://doi.org/10.1016/j.foodqual.2020.103998.

Andersen, C. A., Kring, M. L., Andersen, R. H., Larsen, O. N., Kjaer, T. W., Kidmose, U., ... Kidmose, P. (2019). EEG discrimination of perceptually similar tastes. *Journal of Neuroscience Research, 97*(3), 241−252. Available from https://doi.org/10.1002/jnr.24281.

Andersen, I. N. S. K., Kraus, A. A., Ritz, C., & Bredie, W. L. P. (2019). Desires for beverages and liking of skin care product odors in imaginative and immersive virtual reality beach contexts. *Food Research International*, *117*, 10−18. Available from https://doi.org/10.1016/j.foodres.2018.01.027.

Ares, G., Castura, J., Antúnez, L., Vidal, L., Giménez, A., Coste, E., ... Jaeger, S. (2016). Comparison of two TCATA variants for dynamic sensory characterization of food products. *Food Quality and Preference*, *54*. Available from https://doi.org/10.1016/j.foodqual.2016.07.006.

Bangcuyo, R. G., Smith, K. J., Zumach, J. L., Pierce, A. M., Guttman, G. A., & Simons, C. T. (2015). The use of immersive technologies to improve consumer testing: The role of ecological validity, context and engagement in evaluating coffee. *Food Quality and Preference*, *41*, 84−95. Available from https://doi.org/10.1016/j.foodqual.2014.11.017.

Barbosa Escobar, F., Petit, O., & Velasco, C. (2021). Virtual terroir and the premium coffee experience [original research]. *Frontiers in Psychology*, *12*(560). Available from https://doi.org/10.3389/fpsyg.2021.586983.

Bedri, A., Li, D., Khurana, R., Bhuwalka, K., & Goel, M. (2020). FitByte: Automatic diet monitoring in unconstrained situations using multimodal sensing on eyeglasses. In: *Proceedings of the 2020 CHI conference on human factors in computing systems*.

Bell, L., Vogt, J., Willemse, C., Routledge, T., Butler, L. T., & Sakaki, M. (2018). Beyond self-report: A review of physiological and neuroscientific methods to investigate consumer behavior. *Frontiers in Psychology*, *9*. Available from https://doi.org/10.3389/fpsyg.2018.01655, Article 1655.

Bell, R., Meiselman, H. L., Pierson, B. J., & Reeve, W. G. (1994). Effects of adding an Italian theme to a restaurant on the perceived ethnicity, acceptability, and selection of foods. *Appetite*, *22*(1), 11−24. Available from https://doi.org/10.1006/appe.1994.1002.

Beyts, C., Chaya, C., Dehrmann, F., James, S., Smart, K., & Hort, J. (2017). A comparison of self-reported emotional and implicit responses to aromas in beer. *Food Quality and Preference*, *59*, 68−80. Available from https://doi.org/10.1016/j.foodqual.2017.02.006.

Bi, K., Qiu, T., & Huang, Y. (2020). A deep learning method for yogurt preferences prediction using sensory attributes. *Processes*, *8*(5), 518. Available from https://www.mdpi.com/2227-9717/8/5/518.

Bolls, P., Lang, A., & Potter, R. (2001). The effects of message valence and listener arousal on attention, memory, and facial muscular responses to radio advertisements. *Communication Research*, *28*(5), 627−651. Available from https://doi.org/10.1177/009365001028005003.

Boutrolle, I., Delarue, J., Arranz, D., Rogeaux, M., & Köster, E. P. (2007). Central location test vs. home use test: Contrasting results depending on product type. *Food Quality and Preference*, *18*(3), 490−499. Available from https://doi.org/10.1016/j.foodqual.2006.06.003.

Brouwer, A. M., Hogervorst, M. A., Grootjen, M., van Erp, J. B. F., & Zandstra, E. H. (2017). Neurophysiological responses during cooking food associated with different emotions. *Food Quality and Preference*, *62*, 307−316. Available from https://doi.org/10.1016/j.foodqual.2017.03.005.

Brouwer, A.-M., Hogervorst, M. A., van Erp, J. B. F., Grootjen, M., van Dam, E., & Zandstra, E. H. (2019). Measuring cooking experience implicitly and explicitly: Physiology, facial expression and subjective ratings. *Food Quality and Preference*, *78*, 103726. Available from https://doi.org/10.1016/j.foodqual.2019.103726.

Carvalho-da-Silva, A. M., Van Damme, I., Wolf, B., & Hort, J. (2011). Characterisation of chocolate eating behaviour. *Physiology and Behavior*, *104*(5), 929−933. Available from https://doi.org/10.1016/j.physbeh.2011.06.001.

Castura, J. C., Antúnez, L., Giménez, A., & Ares, G. (2016). Temporal check-all-that-apply (TCATA): A novel dynamic method for characterizing products. *Food Quality and Preference*, *47*, 79−90. Available from https://doi.org/10.1016/j.foodqual.2015.06.017.

Churchill, A., & Behan, J. (2010). Comparison of methods used to study consumer emotions associated with fragrance. *Food Quality and Preference*, *21*(8), 1108−1113. Available from https://doi.org/10.1016/j.foodqual.2010.07.006.

Cliff, M., & Heymann, H. (1993). Development and use of time-intensity methodology for sensory evaluation: A review. *Food Research International*, *26*(5), 375−385. Available from https://doi.org/10.1016/0963-9969(93)90081-S.

Danner, L., Sidorkina, L., Joechl, M., & Duerrschmid, K. (2014). Make a face! Implicit and explicit measurement of facial expressions elicited by orange juices using face reading technology. *Food Quality and Preference*, *32*, 167−172. Available from https://doi.org/10.1016/j.foodqual.2013.01.004.

de Graaf, C., Cardello, A. V., Matthew Kramer, F., Lesher, L. L., Meiselman, H. L., & Schutz, H. G. (2005). A comparison between liking ratings obtained under laboratory and field conditions: The role of choice. *Appetite*, *44*(1), 15−22. Available from https://doi.org/10.1016/j.appet.2003.06.002.

De Pauw, K., Roelands, B., Van Cutsem, J., Marusic, U., Torbeyns, T., & Meeusen, R. (2017). Electro-physiological changes in the brain induced by caffeine or glucose nasal spray. *Psychopharmacology, 234*(1), 53—62. Available from https://doi.org/10.1007/s00213-016-4435-2.

de Wijk, R. A., He, W., Mensink, M. G. J., Verhoeven, R. H. G., & de Graaf, C. (2014). ANS responses and facial expressions differentiate between the taste of commercial breakfast drinks. *PLoS One, 9*(4), e93823. Available from https://doi.org/10.1371/journal.pone.0093823.

de Wijk, R. A., Kooijman, V., Verhoeven, R. H. G., Holthuysen, N. T. E., & de Graaf, C. (2012). Autonomic nervous system responses on and facial expressions to the sight, smell, and taste of liked and disliked foods. *Food Quality and Preference, 26*(2), 196—203. Available from https://doi.org/10.1016/j.foodqual.2012.04.015.

Desmet, P. (2018). Measuring emotion: Development and application of an instrument to measure emotional responses to products. In M. Blythe, & A. Monk (Eds.), *Funology 2: From usability to enjoyment* (pp. 391—404). Springer International Publishing. Available from https://doi.org/10.1007/978-3-319-68213-6_25.

Dong, B., & Biswas, S. (2017). Meal-time and duration monitoring using wearable sensors. *Biomedical Signal Processing and Control, 32*, 97—109. Available from https://doi.org/10.1016/j.bspc.2016.09.018.

Dong, Y., Hoover, A., Scisco, J., & Muth, E. (2012). A new method for measuring meal intake in humans via automated wrist motion tracking. *Applied Psychophysiology and Biofeedback, 37*(3), 205—215. Available from https://doi.org/10.1007/s10484-012-9194-1.

Dorado, R., Chaya, C., Tarrega, A., & Hort, J. (2016). The impact of using a written scenario when measuring emotional response to beer. *Food Quality and Preference, 50*, 38—47. Available from https://doi.org/10.1016/j.foodqual.2016.01.004.

Doulah, A., Farooq, M., Yang, X., Parton, J., McCrory, M. A., Higgins, J. A., & Sazonov, E. (2017). Meal microstructure characterization from sensor-based food intake detection [original research]. *Frontiers in Nutrition, 4*(31). Available from https://doi.org/10.3389/fnut.2017.00031.

Edwards, J., Hartwell, H., & Giboreau, A. (2016). *15 — Emotions studied in context: The role of the eating environment* (pp. 377—403). Woodhead Publishing. https://doi.org/10.1016/B978-0-08-100508-8.00015-1.

Eldeghaidy, S., Marciani, L., McGlone, F., Hollowood, T., Hort, J., Head, K., ... Francis, S. T. (2011). The cortical response to the oral perception of fat emulsions and the effect of taster status. *Journal of Neurophysiology, 105*(5), 2572—2581. Available from https://doi.org/10.1152/jn.00927.2010.

Eldeghaidy, S., Marciani, L., Pfeiffer, J. C., Hort, J., Head, K., Taylor, A. J., ... Francis, S. (2011). Use of an immediate swallow protocol to assess taste and aroma integration in fMRI studies. *Chemosensory Perception, 4*(4), 163—174. Available from https://doi.org/10.1007/s12078-011-9094-4.

Farooq, M., & Sazonov, E. (2016). A novel wearable device for food intake and physical activity recognition. *Sensors, 16*(7), 1067. Available from https://doi.org/10.3390/s16071067.

Figner, B., & Murphy, R. O. (2011). Using skin conductance in judgment and decision making research. *A handbook of process tracing methods for decision research: A critical review and 'user's guide* (pp. 163—184). Psychology Press.

Findlay, C. J., Castura, J. C., & Lesschaeve, I. (2003). Feedback calibration: A training method for descriptive panels. In: *5th Pangborn sensory science symposium*, Boston, MA, USA.

Fontana, J. M., Farooq, M., & Sazonov, E. (2014). Automatic ingestion monitor: A novel wearable device for monitoring of ingestive behavior. *IEEE Transactions on Bio-Medical Engineering, 61*(6), 1772—1779. Available from https://doi.org/10.1109/tbme.2014.2306773.

Fuentes, S., Tongson, E., & Gonzalez Viejo, C. (2021). Novel digital technologies implemented in sensory science and consumer perception. *Current Opinion in Food Science, 41*, 99—106. Available from https://doi.org/10.1016/j.cofs.2021.03.014.

Galiñanes Plaza, A., Delarue, J., & Saulais, L. (2019). The pursuit of ecological validity through contextual methodologies. *Food Quality and Preference, 73*, 226—247. Available from https://doi.org/10.1016/j.foodqual.2018.11.004.

García-Esteban, J. A., Curto, B., Moreno, V., González-Martín, I., Revilla, I., & Vivar-Quintana, A. (2018, 18—20 July 2018). A digitalization strategy for quality control in food industry based on Artificial Intelligence techniques. In: *2018 IEEE 16th international conference on industrial informatics* (INDIN).

Gonzalez Viejo, C., & Fuentes, S. (2020). Low-cost methods to assess beer quality using artificial intelligence involving robotics, an electronic nose, and machine learning. *Fermentation, 6*(4), 104. Available from https://www.mdpi.com/2311-5637/6/4/104.

Gonzalez Viejo, C., Fuentes, S., Torrico, D. D., & Dunshea, F. R. (2018). Non-contact heart rate and blood pressure estimations from video analysis and machine learning modelling applied to food sensory responses: A case study for chocolate. *Sensors, 18*(6), 1802. Available from https://www.mdpi.com/1424-8220/18/6/1802.

Guerreiro, J., Rita, P., & Trigueiros, D. (2015). Attention, emotions and cause-related marketing effectiveness. *European Journal of Marketing, 49*, 1728−1750. Available from https://doi.org/10.1108/EJM-09-2014-0543.

Gunaratne, T. M., Fuentes, S., Gunaratne, N. M., Torrico, D. D., Gonzalez Viejo, C., & Dunshea, F. R. (2019). Physiological responses to basic tastes for sensory evaluation of chocolate using biometric techniques. *Foods, 8* (7), 243. Available from https://www.mdpi.com/2304-8158/8/7/243.

Hannum, M., Forzley, S., Popper, R., & Simons, C. T. (2019). Does environment matter? Assessments of wine in traditional booths compared to an immersive and actual wine bar. *Food Quality and Preference, 76*, 100−108. Available from https://doi.org/10.1016/j.foodqual.2019.04.007.

Hashida, J. C., Carolina de Sousa Silva, A., Souto, S., & Costa, E. J. X. (2005). EEG pattern discrimination between salty and sweet taste using adaptive Gabor transform. *Neurocomputing, 68*, 251−257. Available from https://doi.org/10.1016/j.neucom.2005.04.004.

Hathaway, D., & Simons, C. T. (2017). The impact of multiple immersion levels on data quality and panelist engagement for the evaluation of cookies under a preparation-based scenario. *Food Quality and Preference, 57*, 114−125. Available from https://doi.org/10.1016/j.foodqual.2016.12.009.

He, W., Boesveldt, S., de Graaf, C., & de Wijk, R. (2014). Dynamics of autonomic nervous system responses and facial expressions to odors [original research]. *Frontiers in Psychology, 5*(110). Available from https://doi.org/10.3389/fpsyg.2014.00110.

Hein, K. A., Hamid, N., Jaeger, S. R., & Delahunty, C. M. (2010). Application of a written scenario to evoke a consumption context in a laboratory setting: Effects on hedonic ratings. *Food Quality and Preference, 21*(4), 410−416. Available from https://doi.org/10.1016/j.foodqual.2009.10.003.

Hersleth, M., Monteleone, E., Segtnan, A., & Naes, T. (2015). Effects of evoked meal contexts on consumers' responses to intrinsic and extrinsic product attributes in dry-cured ham [article]. *Food Quality and Preference, 40*, 191−198. Available from https://doi.org/10.1016/j.foodqual.2014.10.002.

Holthuysen, N. T. E., Vrijhof, M. N., de Wijk, R. A., & Kremer, S. (2017). "Welcome on board": Overall liking and just-about-right ratings of airplane meals in three different consumption contexts—laboratory, re-created airplane, and actual airplane. *Journal of Sensory Studies, 32*(2), e12254. Available from https://doi.org/10.1111/joss.12254.

Horio, T. (2003). EMG activities of facial and chewing muscles of human adults in response to taste stimuli. *Perceptual and Motor Skills, 97*(1), 289−298. Available from https://doi.org/10.2466/pms.2003.97.1.289.

Horská, E., Besík, J., Krasnodebski, A., Matysik-Pejas, R., & Bakayova, H. (2016). Innovative approaches to examining consumer preferences when choosing wines. *Agricultural Economics-zemedelska Ekonomika, 62*, 124−133.

Hort, J., Ford, R. A., Eldeghaidy, S., & Francis, S. T. (2016). Thermal taster status: Evidence of cross-modal integration. *Human Brain Mapping, 37*(6), 2263−2275. Available from https://doi.org/10.1002/hbm.23171.

Howgate, P. (2015). A history of the development of sensory methods for the evaluation of freshness of fish. *Journal of Aquatic Food Product Technology, 24*(5), 516−532. Available from https://doi.org/10.1080/10498850.2013.783897.

Huang, C.-N., Chen, C.-H., & Chung, H.-Y. (2004). The review of applications and measurements in facial electromyography. *Journal of Medical and Biological Engineering, 25*(1), 15−20.

Ioannides, Y., Howarth, M. S., Raithatha, C., Defernez, M., Kemsley, E. K., & Smith, A. C. (2007). Texture analysis of Red Delicious fruit: Towards multiple measurements on individual fruit. *Food Quality and Preference, 18*(6), 825−833. Available from https://doi.org/10.1016/j.foodqual.2005.09.012.

Jack, F. R., Piggott, J. R., & Paterson, A. (1993). Relationships between electromyography, sensory and instrumental measures of cheddar cheese texture. *Journal of Food Science, 58*(6), 1313−1317. Available from https://doi.org/10.1111/j.1365-2621.1993.tb06173.x.

Jaeger, S. R., Hort, J., Porcherot, C., Ares, G., Pecore, S., & MacFie, H. J. H. (2017). Future directions in sensory and consumer science: Four perspectives and audience voting. *Food Quality and Preference, 56*, 301−309. Available from https://doi.org/10.1016/j.foodqual.2016.03.006.

Jaeger, S. R., Lee, S. M., Kim, K.-O., Chheang, S. L., Jin, D., & Ares, G. (2017). Measurement of product emotions using emoji surveys: Case studies with tasted foods and beverages. *Food Quality and Preference, 62*, 46−59. Available from https://doi.org/10.1016/j.foodqual.2017.05.016.

Jiménez-Carvelo, A. M., González-Casado, A., Bagur-González, M. G., & Cuadros-Rodríguez, L. (2019). Alternative data mining/machine learning methods for the analytical evaluation of food quality and authenticity - a review. *Food Research International, 122*, 25−39. Available from https://doi.org/10.1016/j.foodres.2019.03.063.

Kalantarian, H., & Sarrafzadeh, M. (2015). Audio-based detection and evaluation of eating behavior using the smartwatch platform. *Computers in Biology and Medicine, 65*, 1−9. Available from https://doi.org/10.1016/j.compbiomed.2015.07.013.

Kemp, S. E. (2009). *Sensory evaluation: A practical handbook/Sarah E. Kemp, Tracey Hollowood, Joanne Hort*. Chichester: Wiley-Blackwell.

King, S. C., & Meiselman, H. L. (2010). Development of a method to measure consumer emotions associated with foods. *Food Quality and Preference, 21*(2), 168−177. Available from https://doi.org/10.1016/j.foodqual.2009.02.005.

King, S. C., Meiselman, H. L., Hottenstein, A. W., Work, T. M., & Cronk, V. (2007). The effects of contextual variables on food acceptability: A confirmatory study. *Food Quality and Preference, 18*(1), 58−65. Available from https://doi.org/10.1016/j.foodqual.2005.07.014.

Korb, S., Götzendorfer, S. J., Massaccesi, C., Sezen, P., Graf, I., Willeit, M., . . . Silani, G. (2020). Dopaminergic and opioidergic regulation during anticipation and consumption of social and nonsocial rewards. *eLife, 9*, e55797. Available from https://doi.org/10.7554/eLife.55797.

Korb, S., Massaccesi, C., Gartus, A., Lundström, J. N., Rumiati, R., Eisenegger, C., & Silani, G. (2020). Facial responses of adult humans during the anticipation and consumption of touch and food rewards. *Cognition, 194*, 104044. Available from https://doi.org/10.1016/j.cognition.2019.104044.

Köster, E. P. (2003). The psychology of food choice: Some often encountered fallacies. *Food Quality and Preference, 14*(5), 359−373. Available from https://doi.org/10.1016/S0950-3293(03)00017-X.

Labbe, D., Martin, N., Le Coutre, J., & Hudry, J. (2011). Impact of refreshing perception on mood, cognitive performance and brain oscillations: An exploratory study. *Food Quality and Preference, 22*(1), 92−100. Available from https://doi.org/10.1016/j.foodqual.2010.08.002.

Lagast, S., Gellynck, X., Schouteten, J. J., De Herdt, V., & De Steur, H. (2017). Consumers' emotions elicited by food: A systematic review of explicit and implicit methods. *Trends in Food Science and Technology, 69*, 172−189. Available from https://doi.org/10.1016/j.tifs.2017.09.006.

Lawless, H. T. (1999). *Sensory evaluation of food: Principles and practices/Harry T. Lawless, Hildegarde Heymann*. New York London: Kluwer Academic/Plenum Publishers.

Lee, W., III, & Pangborn, M. (1986). Time-intensity: The temporal aspects of sensory perception. *Food Technology*.

Liu, R., Hannum, M., & Simons, C. T. (2019). Using immersive technologies to explore the effects of congruent and incongruent contextual cues on context recall, product evaluation time, and preference and liking during consumer hedonic testing. *Food Research International, 117*, 19−29. Available from https://doi.org/10.1016/j.foodres.2018.04.024.

Meiselman, H. L. (1996). The contextual basis for food acceptance, food choice and food intake: The food, the situation and the individual. In H. L. Meiselman, & H. J. H. MacFie (Eds.), *Food choice, acceptance and consumption* (pp. 239−263). US: Springer. Available from https://doi.org/10.1007/978-1-4613-1221-5_6.

Meiselman, H. L. (2013). The future in sensory/consumer research: Evolving to a better science. *Food Quality and Preference, 27*(2), 208−214. Available from https://doi.org/10.1016/j.foodqual.2012.03.002.

Meiselman, H. L., Johnson, J. L., Reeve, W., & Crouch, J. E. (2000). Demonstrations of the influence of the eating environment on food acceptance. *Appetite, 35*(3), 231−237. Available from https://doi.org/10.1006/appe.2000.0360.

Emmanuel, Z. (2020). *Wearable technology: Inc Impact of Covid-19 - UK - November 2020. Mintel [online]*. https://reports.mintel.com/display/990136/.

Moskowitz, H. R. (2003). In H. R. Moskowitz, A. M. Muñoz, & M. C. Gacula (Eds.), *Viewpoints and controversies in sensory science and consumer product testing*. Trumbull, Conn: Food & Nutrition Press.

Nath, E. C., Cannon, P. R., & Philipp, M. C. (2019). An unfamiliar social presence reduces facial disgust responses to food stimuli. *Food Research International, 126*, 108662. Available from https://doi.org/10.1016/j.foodres.2019.108662.

Nath, E. C., Cannon, P. R., & Philipp, M. C. (2020). Co-acting strangers but not friends influence subjective liking and facial affective responses to food stimuli. *Food Quality and Preference, 82*, 103865. Available from https://doi.org/10.1016/j.foodqual.2019.103865.

Nestrud, M. A., Meiselman, H. L., King, S. C., Lesher, L. L., & Cardello, A. V. (2016). Development of EsSense25, a shorter version of the EsSense Profile®. *Food Quality and Preference, 48*(Part A), 107−117. Available from https://doi.org/10.1016/j.foodqual.2015.08.005.

Nijman, M. (2019). *Measuring emotional response to sensory attributes: Context effects*. Nottingham: University of Nottingham.

Nivedhan, A., Mielby, L. A., & Wang, Q. J. (2020). The influence of emotion-oriented extrinsic visual and auditory cues on coffee perception: A virtual reality experiment. In: *Companion publication of the 2020 international conference on multimodal interaction*, Virtual Event, the Netherlands. https://doi-org.nottingham.idm.oclc.org/10.1145/3395035.3425646.

Ohme, R., Reykowska, D., Wiener, D., & Choromanska, A. (2009). Analysis of neurophysiological reactions to advertising stimuli by means of EEG and galvanic skin response measures. *Journal of Neuroscience, Psychology, and Economics*, 2(1), 21–31. Available from https://doi.org/10.1037/a0015462.

Oliver, J. H., & Hollis, J. H. (2021). Virtual reality as a tool to study the influence of the eating environment on eating behavior: A feasibility study. *Foods*, 10(1), 89. Available from https://www.mdpi.com/2304-8158/10/1/89.

Pecore, S., Rathjen-Nowak, C., & Tamminen, T. (2011). Temporal order of sensations. In: *9th Pangborn sensory science symposium*, Toronto, Canada.

Picket, B., & Dando, R. (2019). Enviromental 'immersion's influence on hedonics, perceived appropriateness, and willingness to pay in alcoholic beverages. *Foods*, 8(2), 42.

Pineau, N., & Schilch, P. (2015). Temporal dominance of sensations (TDS) as a sensory profiling technique. In J. Delarue, J. B. Lawlor, & M. Rogeaux (Eds.), *Rapid sensory profiling techniques* (pp. 269–306). Woodhead Publishing. Available from https://doi.org/10.1533/9781782422587.2.269.

Pineau, N., Cordelle, S., & Schlich, P. (2003). Temporal dominance of sensations: A new technique to record several sensory attributes simultaneously over time. In: *5th Pangborn symposium*, 35. Journées de statistique.

Pineau, N., Schlich, P., Cordelle, S., Mathonnière, C., Issanchou, S., Imbert, A., … Köster, E. (2009). Temporal dominance of sensations: Construction of the TDS curves and comparison with time–intensity. *Food Quality and Preference*, 20(6), 450–455. Available from https://doi.org/10.1016/j.foodqual.2009.04.005.

Rini, L., Lagast, S., Schouteten, J. J., Gellynck, X., & De Steur, H. (2022). Impact of emotional state on consumers' emotional conceptualizations of dark chocolate using an emoji-based questionnaire. *Food Quality and Preference*, 99, 104547. Available from https://doi.org/10.1016/j.foodqual.2022.104547.

Ross, C. F. (2021). Considerations of the use of the electronic tongue in sensory science. *Current Opinion in Food Science*, 40, 87–93. Available from https://doi.org/10.1016/j.cofs.2021.01.011.

Rozin, P., & Tuorila, H. (1993). Simultaneous and temporal contextual influences on food acceptance. *Food Quality and Preference*, 4(1), 11–20. Available from https://doi.org/10.1016/0950-3293(93)90309-T.

Sato, W., Minemoto, K., Ikegami, A., Nakauma, M., Funami, T., & Fushiki, T. (2020). Facial EMG correlates of subjective hedonic responses during food consumption. *Nutrients*, 12(4), 1174. Available from https://doi.org/10.3390/nu12041174.

Sato, W., Yoshikawa, S., & Fushiki, T. (2020). Facial EMG activity is associated with hedonic experiences but not nutritional values while viewing food images. *Nutrients*, 13(1), 11. Available from https://doi.org/10.3390/nu13010011.

Sen, S., Subbaraju, V., Misra, A., Balan, R., & Lee, Y. (2015). The case for smartwatch-based diet monitoring. In: *2015 IEEE international conference on pervasive computing and communication workshops (PerCom Workshops)*, 585–590.

Sester, C., Deroy, O., Sutan, A., Galia, F., Desmarchelier, J.-F., Valentin, D., & Dacremont, C. (2013). "Having a drink in a bar": An immersive approach to explore the effects of context on drink choice. *Food Quality and Preference*, 28(1), 23–31. Available from https://doi.org/10.1016/j.foodqual.2012.07.006.

Sinesio, F., Moneta, E., Porcherot, C., Abbà, S., Dreyfuss, L., Guillamet, K., … McEwan, J. A. (2019). Do immersive techniques help to capture consumer reality? *Food Quality and Preference*, 77, 123–134. Available from https://doi.org/10.1016/j.foodqual.2019.05.004.

Sinesio, F., Saba, A., Peparaio, M., Saggia Civitelli, E., Paoletti, F., & Moneta, E. (2018). Capturing consumer perception of vegetable freshness in a simulated real-life taste situation. *Food Research International*, 105, 764–771. Available from https://doi.org/10.1016/j.foodres.2017.11.073.

Songsamoe, S., Saengwong-ngam, R., Koomhin, P., & Matan, N. (2019). Understanding consumer physiological and emotional responses to food products using electroencephalography (EEG). *Trends in Food Science and Technology*, 93, 167–173. Available from https://doi.org/10.1016/j.tifs.2019.09.018.

Spence, C. (2013). Multisensory flavour perception. *Current Biology*, 23(9), R365–R369. Available from https://doi.org/10.1016/j.cub.2013.01.028.

Stelick, A., Penano, A. G., Riak, A. C., & Dando, R. (2018). Dynamic context sensory testing—a proof of concept study bringing virtual reality to the sensory booth. *Journal of Food Science, 83*(8), 2047–2051. Available from https://doi.org/10.1111/1750-3841.14275.

Stone, H., Bleibaum, R. N., & Thomas, H. A. (2012). Introduction to sensory evaluation. In H. Stone, R. N. Bleibaum, & H. A. Thomas (Eds.), *Sensory evaluation practices (fourth edition)* (pp. 1–21). Academic Press. Available from https://doi.org/10.1016/B978-0-12-382086-0.00001-7.

Tan, J., & Xu, J. (2020). Applications of electronic nose (e-nose) and electronic tongue (e-tongue) in food quality-related properties determination: A review. *Artificial Intelligence in Agriculture, 4*, 104–115. Available from https://doi.org/10.1016/j.aiia.2020.06.003.

Teillet, E. (2015). Polarized sensory positioning (PSP) as a sensory profiling technique. In J. Delarue, J. B. Lawlor, & M. Rogeaux (Eds.), *Rapid sensory profiling techniques* (pp. 215–225). Woodhead Publishing. Available from https://doi.org/10.1533/9781782422587.2.215.

Teillet, E., Schlich, P., Urbano, C., Cordelle, S., & Guichard, E. (2010). Sensory methodologies and the taste of water. *Food Quality and Preference, 21*(8), 967–976. Available from https://doi.org/10.1016/j.foodqual.2010.04.012.

Toepel, U., Knebel, J.-F., Hudry, J., Le Coutre, J., & Murray, M. M. (2009). The brain tracks the energetic value in food images. *Neuroimage, 44*(3), 967–974. Available from https://doi.org/10.1016/j.neuroimage.2008.10.005.

Torrico, D. D., Sharma, C., Dong, W., Fuentes, S., Gonzalez Viejo, C., & Dunshea, F. R. (2021). Virtual reality environments on the sensory acceptability and emotional responses of no- and full-sugar chocolate. *LWT, 137*, 110383. Available from https://doi.org/10.1016/j.lwt.2020.110383.

van Bochove, M. E., Ketel, E., Wischnewski, M., Wegman, J., Aarts, E., de Jonge, B., . . . Schutter, D. J. L. G. (2016). Posterior resting state EEG asymmetries are associated with hedonic valuation of food. *International Journal of Psychophysiology, 110*, 40–46. Available from https://doi.org/10.1016/j.ijpsycho.2016.10.006.

van Bommel, R., Stieger, M., Visalli, M., de Wijk, R., & Jager, G. (2020). Does the face show what the mind tells? A comparison between dynamic emotions obtained from facial expressions and temporal dominance of emotions (TDE). *Food Quality and Preference, 85*, 103976. Available from https://doi.org/10.1016/j.foodqual.2020.103976.

Verastegui-Tena, L., Schulte-Holierhoek, A., van Trijp, H., & Piqueras-Fiszman, B. (2017). Beyond expectations: The responses of the autonomic nervous system to visual food cues. *Physiology and Behavior, 179*, 478–486. Available from https://doi.org/10.1016/j.physbeh.2017.07.025.

Verastegui-Tena, L., van Trijp, H., & Piqueras-Fiszman, B. (2019). Heart rate, skin conductance, and explicit responses to juice samples with varying levels of expectation (dis)confirmation. *Food Quality and Preference, 71*, 320–331. Available from https://doi.org/10.1016/j.foodqual.2018.08.011.

Verhagen, J. V., & Engelen, L. (2006). The neurocognitive bases of human multimodal food perception: Sensory integration. *Neuroscience and Biobehavioral Reviews, 30*(5), 613–650. Available from https://doi.org/10.1016/j.neubiorev.2005.11.003.

Vila-López, N., & Küster-Boluda, I. (2019). Consumers' physiological and verbal responses towards product packages: Could these responses anticipate product choices? *Physiology and Behavior, 200*, 166–173. Available from https://doi.org/10.1016/j.physbeh.2018.03.003.

Walsh, A. M., Duncan, S. E., Bell, M. A., O'Keefe, S. F., & Gallagher, D. L. (2017). Integrating implicit and explicit emotional assessment of food quality and safety concerns. *Food Quality and Preference, 56*, 212–224. Available from https://doi.org/10.1016/j.foodqual.2016.11.002.

Wang, Q. J., Meyer, R., Waters, S., & Zendle, D. (2020). A dash of virtual milk: Altering product color in virtual reality influences flavor perception of cold-brew coffee [original research]. *Frontiers in Psychology, 11*(3491). Available from https://doi.org/10.3389/fpsyg.2020.595788.

Wen, H., & Leung, X. Y. (2021). Virtual wine tours and wine tasting: The influence of offline and online embodiment integration on wine purchase decisions. *Tourism Management, 83*, 104250. Available from https://doi.org/10.1016/j.tourman.2020.104250.

Worch, T., Sinesio, F., Moneta, E., Abbà, S., Dreyfuss, L., McEwan, J. A., & Porcherot-Lassallette, C. (2020). Influence of different test conditions on the emotional responses elicited by beers. *Food Quality and Preference, 83*, 103895. Available from https://doi.org/10.1016/j.foodqual.2020.103895.

Yang, Q., Nijman, M., Flintham, M., Tennent, P., Hidrio, C., & Ford, R. (2022). Improving simulated consumption context with virtual reality: A focus on participant experience. *Food Quality and Preference, 98*, 104531. Available from https://doi.org/10.1016/j.foodqual.2022.104531.

Zandstra, E. H., Kaneko, D., Dijksterhuis, G. B., Vennik, E., & De Wijk, R. A. (2020). Implementing immersive technologies in consumer testing: Liking and just-about-right ratings in a laboratory, immersive simulated café and real café. *Food Quality and Preference*, *84*, 103934. Available from https://doi.org/10.1016/j.foodqual.2020.103934.

Zeinstra, G. G., Koelen, M. A., Colindres, D., Kok, F. J., & de Graaf, C. (2009). Facial expressions in school-aged children are a good indicator of 'dislikes', but not of 'likes'. *Food Quality and Preference*, *20*(8), 620–624. Available from https://doi.org/10.1016/j.foodqual.2009.07.002.

16

Multisensory immersive rooms: a mixed reality solution to overcome the limits of contexts studies

Adriana Galiñanes-Plaza[1,2], Agnès Giboreau[3] and Jacques-Henry Pinhas[2]

[1]Reperes Insights, Paris, France [2]The Lab in the Bag, Paris, France [3]Research & Innovation Centre, Institut Lyfe (ex Institut Paul Bocuse), Lyon, France

16.1 Introduction

Consumer experience takes place within a multisensory environment where different stimuli interact and modulate perception and behaviour (Velasco & Obrist, 2020). Hence, the ecological validity of sensory and consumer studies conducted in laboratory settings has been heavily questioned; nevertheless, studies in natural contexts are difficult to manage and the gain in realism is obtained to the detriment of control over contextual variables, bringing the reproducibility and transferability of the results into question (Galiñanes Plaza et al., 2019). Immersive technologies have been developed in order to overcome the challenge of ecological validity and bring the extrinsic factors of the consumption experience to the laboratory (Crofton et al., 2019).

Immersive technologies are defined as 'methods and devices that induce targeted behaviour in individuals by creating an effect of identification with immersive media through sensory stimuli' (Hehn et al., 2019); and they are characterised by four main features: lack of awareness of time, loss of awareness of the real world, involvement and a sense of being in the task environment (Jennett et al., 2008).

Within these technologies, virtual reality (VR) and multisensory immersive rooms have started to be used in the sensory and consumer field showing promising results. VR allows to create realistic and very immersive environments that increase participants' engagement. However, aspects related to distorted feelings of presence, motion sickness and the difficulties to interact with real food and people need to be further address (see also Chapter 14, Using virtual reality as a context enhancing technology in sensory science by Emily Crofton and Cristina Botinestean for a detailed description on VR).

Multisensory immersive rooms are characterised by the use of audiovisual scenarios (HD, 360 degrees or interactive videos) in combination with olfactory and haptic stimulation (odours, temperature, lighting, breeze, etc.) providing solutions for realistic consumer experience (Bangcuyo et al., 2015; Sester et al., 2013). These platforms are also called mixed reality (Hartmann & Siegrist, 2019), and according to Hehn et al. (2019), 'they seem to be the best trade-off for food and beverage testing'.

The first study applying this type of approach was conducted by Sester et al. (2013). These researchers aimed to evaluate the effect of bar ambience (warm vs cold interior) on self-reported drink choices from a range of drinks by using physical means (wood vs blue furniture) and clips with visual and music stimuli. Their results showed that drink choices differ according to perceptual, semantic or cognitive associations between drinks and visuals. This type of approach was also applied by Bangcuyo et al. (2015) who found coffee evaluation data to be more discriminating and reliable when obtained in an immersive virtual coffeehouse — audiovisual presentation of sights and sounds recorded from a real coffee house was displayed on high-definition monitors and a subtle cinnamon roll aroma was dispersed — than when collected in laboratory conditions. Besides, participants reported to be more engaged in the multisensory immersive coffeehouse than in controlled conditions, which could contribute to improved data quality. Similar results were obtained by Hathaway and Simons (2017), who further showed that data discriminability and reliability improves when increasing the level of immersion. However, Sinesio et al. (2018) found that freshness discrimination between stored and unstored vegetables was lower in the immersive room than in the lab condition, even if the magnitude of liking was higher in the immersive environment.

Certain products can be more or less context-specific and the salience of contextual information can have more or less impact on consumers' evaluation. Multisensory immersive rooms help to better understand the prioritisation and processing of sensory information in food consumption scenarios for context-sensitive products (Lichters et al., 2021). Liu et al. (2019) manipulated the congruency of visual, auditory and olfactory cues using immersive technologies, and assessed their impact on context recall, evaluation time, preference and liking for cold-brewed coffee. The authors found that olfactory information had a lesser saliency than visual or auditory information in immersive environments when recalling information streams. In order to investigate the congruency of products-specificity and contexts, van Bergen et al. (2021) studied how repeated exposure to foods in congruent and incongruent immersive environments influenced hedonic perception over time. Different products (sushi, popsicle and iced tea) were tested in an immersive beach or sushi restaurant for 7 days. On the eighth day, participants tested the same products but switched among immersive environments. The researchers found that expected liking and desire to eat were higher and consistent in the congruent food-context environment. Besides, food consumption and consistency of the liking scores over time were higher in the typical consumption context such as the sushi restaurant. This type of approach also highlights the possibility that immersive rooms offer to replicate a study, even in different locations by keeping the environmental variables or sensory stimuli under control.

Delarue et al. (2019) found that immersive rooms could overcome the discrepancies between time testing sessions found in laboratory settings and propose more homogeneous conditions. In addition, the authors tested the flexibility offered by this tool to compare the situational appropriateness scores of nonalcoholic beers in reproduced beach or nightclub environments. Consumer liking results differed depending on the immersive environment, which allowed identifying the best product-context couple.

Those studies have shown the potential of multisensory immersive rooms to create unlimited environments that can be used at different stages of product development and sensory evaluation. However, it should be underlined that further research need to be conducted to overcome technological constraints and improve the quality of immersion.

16.2 General principles of the multisensory immersive rooms

Immersion is 'a psychological state characterised by perceiving oneself to be enveloped by, included in, and interacting with an environment that provides a continuous stream of stimuli and experiences' (Witmer & Singer, 1998), giving the effect of 'being' in a particular environment (an effect called *presence* (Cummings & Bailenson, 2016)).

To design an immersive multisensory room, we must first avoid referring to existing techniques. The needs come naturally from research teams and clients who are looking to recreate worlds and environments that are 'as natural as possible' while being able to manage and control all the sensory stimuli.

Controlling means mastering the sensory parameters in terms of different factors as presented in Table 16.1.

In R&D projects, it is key to be able to reproduce the same context and the same parameters at any moment. For example, a scenario on a tropical beach with the same light intensity (or the same variation of a passing cloud) at the same time, the same waves and the same sounds of passing birds, etc. With the multisensory immersive approach, it is possible to compare experiments that are identical in terms of parameters or by varying them in a well-controlled manner. Thus scenarios can be endlessly developed, played and replayed along the project life, as far as more data are needed to evaluate same (or different) products in same (or different) consumptions environments.

In marketing projects, breaking the codes and improving creativity is key for brainstorming and inventiveness (Wang et al., 2021). Thus various scenarios may boost

TABLE 16.1   Variables that need to be controlled in multisensory immersive rooms.

Variables	
Spatiality	• Being able to diffuse or simulate diffusion from specific points (visual scenes, odours, sounds …)
	• *Example: a bee flying moves from one point to another and the sound circles in the room accordingly, then disappears; the noise of a car passing behind a person from right to left*
Temporality	• Being able to start and to stop a stimulus at any precise moment, at various spaces
	• *Example: being in the forest and after a minute the sun will appear for 2 min and then stop giving place to the rain. This can be reproduced and set several times along within the same scenario*
	• Being able to interact with all those stimuli at any precise moment and activate new stimuli
	• *Example: a person takes a glass of wine and this is going to activate the diffusion of a cherry tree odour*
Intensity	• Being able to control the strength of stimulus
	• *Example: the level of heat of a fireplace, of a simple heater*

creativity in different ways having the potential to change the way of thinking and generating ideas or concepts. Immersive scenarios are a great help in brainstorming or design thinking sessions. For instance, one session could be based on one walking in the countryside and suddenly perceiving a food smell and bird or tractor sounds. In equipped multisensory rooms, it is possible to naturally move in order to keep all the spontaneity as possible avoiding one of the major limits when working with virtual reality helmets or any equipment that may involve potential biases.

Moreover, interactivity is easy to set-up in immersive rooms. Interactivity multiplies two-folds:

- **With the environment:** immersive environments that implement interactive art or technical responses (Jeon et al., 2019).
- **With the stakeholders:** immersive environments that recreate different scenarios where participants interact with colleagues (Delarue et al., 2019).

16.3 Technical description of a multisensory immersive room

Several parameters should be considered when designing a multisensory immersive room: **Physical parameters:**

- Able to create spaces that can be easily installed by nonspecialists, of all sizes — from a few square metres up to more than 100 square metres (Fig. 16.1) — often transportable, autonomous, scalable and able to be controlled and supported remotely.

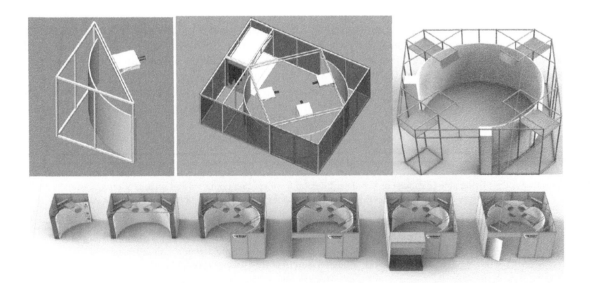

FIGURE 16.1   Different multisensory immersive rooms designs. *Source: The Lab in The Bag.*

- Able to instantly activate any stimulus in space, time and intensity (see above), see examples of spatial designs in Fig. 16.1.

This means cohabiting very diverse processes: strong and weak electrical power, 240 V/110 V to 5 V, wired, Bluetooth and Wi-Fi signals as well as air, water and fragrance flows, etc.

Software parameters:

- Able to control stimuli by ergonomic and powerful apps, user-friendly in all their components with standard devices (tablet, smartphone).
- Able to capture data (events, triggers, records, etc.) and to transcript into standard files (such as comma-separated values file that allow a posteriori analyses.
- Able to load and replace easily any stimulus: video, sound, odour.
- Able to introduce interactivity between stakeholders (inside and outside of the space) and scenarios. For example, moving inside of a house, entering in the kitchen and smelling coffee.

Fig. 16.2 shows how software parameters are integrated within a multisensory immersive room.

Integration system:

- Able to deliver an integrated solution using all criteria hereabove (Fig. 16.2).
- To keep as much flexibility as possible.

16.4 Case study: on soft drinks in multisensory immersive contexts

The Institut Paul Bocuse opened a 80-sqm multisensory immersive room, MIXH, Multisensory Immersive Experience Hall (Fig. 16.3). It allows to explore and measure what makes consumer differently perceive and interact with products in relation to the consumption environment. This room allows creating, parameterising and studying repeatable scenarios, integrating the individual effects and/or the combination of several stimuli at a time. Moreover, the social interaction within guests allows overcoming one of

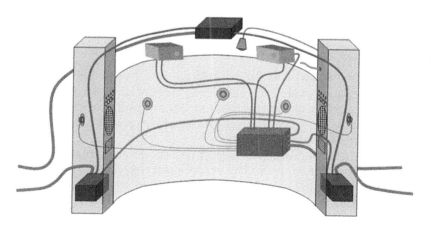

FIGURE 16.2 Software integration schema of a multisensory immersive room. *Source: The Lab in The Bag.*

FIGURE 16.3   Two restaurant settings of the MIXH, the Multisensory Immersive Experience Hall. *Source: Institut Paul Bocuse.*

the limits of laboratory settings and bridging the gap with natural in-context and field studies (see examples of environments in Fig. 16.3).

The immersive room is modular and the combination of technology and equipment allows the simulation of very various types of usage contexts. For instance, a hair-dressing salon in a virtual forest has been set-up to study the effect of the environment on emotional status with consumers having their hair washed (Fig. 16.4). The room was equipped with furniture of a hair salon and a 360-degree immersive video showing a forest with a cascade and relaxing sounds (water, forest and birds sounds) was played in the immersive room.

A case study on beverages was conducted to investigate opportunities and limitations of using multisensory immersive rooms in sensory and consumer research.

Context conveys different emotions inducing changes in the way consumers perceive a product (Danner et al., 2016). The objective of this study was to evaluate the added-value of a multisensory environment on the consumer experience.

As previously described, there is still a need to better understand the integration of visual, auditory, olfactory and tactile stimuli to further engage the senses and to optimise the use of this approach (Zandstra et al., 2020). Hence, we decided to look at the role of the different sensory stimuli by working with different levels of immersion.

We assessed the effect of multisensory experiences on consumers' spontaneous reactions about the environment, and measured their liking, sensory perception and satisfaction. Three different environments (neutral, café and park environment) and two level of immersive conditions (visual and full immersion with audio-visual and olfactory cues) were recreated inside of an immersive room (Fig. 16.5).

In the café environment the same visual was used for visual and full immersions (see Fig. 16.6B). Besides a café background sound and a sweet odour were diffused during the full immersion. In the case of the park environment, both visual and full immersion used the same scenario (Fig. 16.6C); however, only the full immersion included a park background sound and a vegetal odour diffusion.

Iced-tea was chosen for this study because it is a popular product, often consumed more than once a day, and in a variety of contexts. The preparation was identical in all the five experimental conditions. A total of 182 consumers were recruited to participate to an iced-tea and pastry break with a companion.

FIGURE 16.4 A hair-dresser in a virtual forest setting of the MIXH, the Multisensory Immersive Experience Hall. *Source: Institut Paul Bocuse.*

STIMULI	NEUTRAL	CAFE VISUAL ENVIRONMENT	FULL CAFE ENVIRONMENT	PARK VISUAL ENVIRONMENT	FULL PARK ENVIRONMENT
👁		X	X	X	X
👂			X		X
👃			X		X

FIGURE 16.5 Three different environments (neutral, cafe and park environment) and two level of immersive conditions (visual and full immersion with audio-visual and olfactory cues).

For each condition, consumers were first asked to give three words to quote their spontaneous reactions towards the environment. They were then presented with two iced-tea formula. After quoting their spontaneous reactions using three words and scoring their liking on a 7-point hedonic scale (1 = dislike very much; 7 = like very much), consumers were offered the drink of their choice together with a chocolate cake, similar for all participants. Ratings of the environment satisfaction and overall experience were finally collected on a 7-point scale (1 = did not appreciated it at all; 7 = very much appreciated) (see the protocol on Fig. 16.7).

The spontaneous reactions were analysed using the R3m Calculation integrating emotional and linguistics considerations. The collected verbal corpus is computed using different criteria (rank, valence, grammatical nature, subject positioning (judgement, description, etc.) and the consensus between subjects) as the respondent level and sample level. Results obtained regarding the environment showed that the neutral environment was quite segmenting with participants appreciating its calm and mysterious side, whereas others criticised its emptiness. Adding a visual environment greatly reduced this perception, obtaining a more consensual emotional activation in the park environment compared to the café environment.

The enrichment with sound and odours in the full immersion condition shifted from the calm, pleasant and relaxed perception to evocations directly related to background sounds like noisy for the full café environment but also friendly or odours evocations like gourmet for the full café environment and floral or earthy for the full park environment. This enrichment of the atmosphere makes it possible to approach the more natural consumer experiences even if a natural-context experiment should confirm such results.

FIGURE 16.6　(A) Neutral environment; (B) indoor-coffee place environment and (C) outdoor-coffee place environment.

FIGURE 16.7　Description of the four phases of the study protocol.

Besides these results, the impact of the environments on participants' satisfaction is confirmed by the satisfaction scores showing that the two full environments (with visual, sound and odours) obtained significantly higher ratings compared to the neutral environment (Kruskal–Wallis test with Bonferroni correction at 0.01: full café environment ($P = .001$) and full park environment ($P < .0001$)); while only visual environments obtained intermediate scores, not statistically higher, confirming the higher impact of a multisensory immersion that may have enhanced the 'sense' of presence in those environments (Sinesio et al., 2019).

When looking at the differences between the two iced-teas within each individual condition, the neutral environment ($p = 0.005$) and the full café environment ($p = 0.003$) were the only ones in which the liking scores between the two iced-teas significantly differed.

16.5　Conclusions and future perspectives

In the last decade, the use of multisensory immersive rooms have increased, and academia and industry are looking at the unlimited possibilities this approach offers to overcome the limits of contexts studies in sensory and consumer evaluation.

The case study presented here illustrates the use of multisensory rooms to underline the effect of context on consumers' emotional responses, this effect being potentially stronger on emotions than the tested product itself. The importance of going 'beyond liking' and working on more ecologically valid conditions have raised in order to better predict products positioning in the market and consumers' behaviours towards those products. Multisensory immersive rooms can be used in sensory evaluation as a tool to improve ecological validity compare to controlled conditions. Testing products in different contexts and being able to parameter and control different stimuli (odours, sounds or even temperatures), may help to better understand changes on consumers' product perception. Besides, due to this control on the different sensory stimuli, this device can be also combined with implicit measures (such as skin conductance or heart rate) to improve the quality of the explicit collected data while testing different environments at a time (see Chapter 13, Added value of implicit measures in sensory and consumer science by René A. de Wijk and Lucas P.J.J. Noldus).

Besides research projects bringing a better understanding of consumers, immersive rooms represent a flexible tool for concept definition and product positioning. On the one hand, the ability to easily and quickly set various environments is very useful in design thinking projects. The creative team is thus exposed to very different atmospherics and is able to generate ideas and concepts in response to the various experienced situations. On the other hand, more mature projects also benefit from these modular multisensory environments to study consumer acceptance of new formula in a given usage situation. Similarly, several consumption cases can be tested to identify the one better fitting with a given prototype.

More generally, multisensory immersive rooms seem to us to be a fantastic playground for sensory and consumer research. These devices will continue to be equipped with new capabilities, particularly in the field of augmented reality, which should make immersion even more effective. A large field of research is therefore opening up in order to calibrate these tools and explore their full potential.

Acknowledgements

The authors would like to thank the following individuals for their contributions to the case study: Estelle Petit (Institut Paul Bocuse Research R&D Manager); Laura Zerbini (Repères, Quantitative research executive); Chloe Nuvoli, Claire Gabet and Benoit Matthieu (Institut Paul Bocuse Research fellows); Bernard Ricolleau (Maître d'hôtel Service et Arts de la table Institut Paul Bocuse); Spencer Dubreuil (Institut Paul Bocuse student), and, all of the consumers and others who dedicated their time, but are not mentioned by name, who contributed to these works. The authors also thank L'Oréal for their support of the immersive hair-dressing salon experiment.

References

Bangcuyo, R. G., Smith, K. J., Zumach, J. L., Pierce, A. M., Guttman, G. A., & Simons, C. T. (2015). The use of immersive technologies to improve consumer testing: The role of ecological validity, context and engagement in evaluating coffee. *Food Quality and Preference*, *41*, 84−95. Available from https://doi.org/10.1016/j.foodqual.2014.11.017.

van Bergen, G., Zandstra, E. H., Kaneko, D., Dijksterhuis, G. B., & de Wijk, R. A. (2021). Sushi at the beach: Effects of congruent and incongruent immersive contexts on food evaluations. *Food Quality and Preference*, *91*, 104193. Available from https://doi.org/10.1016/j.foodqual.2021.104193.

Crofton, E. C., Botinestean, C., Fenelon, M., & Gallagher, E. (2019). Potential applications for virtual and augmented reality technologies in sensory science. *Innovative Food Science and Emerging Technologies*, *56*, 102178. Available from https://doi.org/10.1016/j.ifset.2019.102178.

Cummings, J. J., & Bailenson, J. N. (2016). How immersive is enough? A meta-analysis of the effect of immersive technology on user presence. *Media Psychology*, *19*(2), 272−309. Available from http://www.tandf.co.uk/journals/titles/15213269.asp.10.1080/15213269.2015.1015740.

Danner, L., Ristic, R., Johnson, T. E., Meiselman, H. L., Hoek, A. C., Jeffery, D. W., & Bastian, S. E. P. (2016). Context and wine quality effects on consumers' mood, emotions, liking and willingness to pay for Australian Shiraz wines. *Food Research International*, *89*, 254−265. Available from http://www.elsevier.com/inca/publications/store/4/2/2/9/7/0.10.1016/j.foodres.2016.08.006.

Delarue, J., Brasset, A. C., Jarrot, F., & Abiven, F. (2019). Taking control of product testing context thanks to a multi-sensory immersive room. A case study on alcohol-free beer. *Food Quality and Preference, 75*, 78−86. Available from https://doi.org/10.1016/j.foodqual.2019.02.012.

Galiñanes Plaza, A., Delarue, J., & Saulais, L. (2019). The pursuit of ecological validity through contextual methodologies. *Food Quality and Preference, 73*, 226−247. Available from https://doi.org/10.1016/j.foodqual.2018.11.004.

Hartmann, C., & Siegrist, M. (2019). Virtual reality and immersive approaches to contextual food texting. *Context: The effects of environment on product design and evaluation* (pp. 475−500). Woodhead Publishing.

Hathaway, D., & Simons, C. T. (2017). The impact of multiple immersion levels on data quality and panelist engagement for the evaluation of cookies under a preparation-based scenario. *Food Quality and Preference, 57*, 114−125. Available from https://doi.org/10.1016/j.foodqual.2016.12.009.

Hehn, Patrick, Lutsch, Dariah, & Pessel, Frank (2019). *Inducing context with immersive technologies in sensory consumer testing* (pp. 475−500). Elsevier BV. Available from https://doi.org/10.1016/b978-0-12-814495-4.00023-4.

Jennett, C., Cox, A. L., Cairns, P., Dhoparee, S., Epps, A., Tijs, T., & Walton, A. (2008). Measuring and defining the experience of immersion in games. *International Journal of Human Computer Studies, 66*(9), 641−661. Available from https://doi.org/10.1016/j.ijhcs.2008.04.004.

Jeon, M., Fiebrink, R., Edmonds, E. A., & Herath, D. (2019). From rituals to magic: Interactive art and HCI of the past, present, and future. *International Journal of Human Computer Studies, 131*, 108−119. Available from http://www.elsevier.com/inca/publications/store/6/2/2/8/4/6/index.htt.10.1016/j.ijhcs.2019.06.005.

Lichters, M., Möslein, R., Sarstedt, M., & Scharf, A. (2021). Segmenting consumers based on sensory acceptance tests in sensory labs, immersive environments, and natural consumption settings. *Food Quality and Preference, 89*. Available from https://www.journals.elsevier.com/food-quality-and-preference.10.1016/j.foodqual.2020.104138.

Liu, R., Hannum, M., & Simons, C. T. (2019). Using immersive technologies to explore the effects of congruent and incongruent contextual cues on context recall, product evaluation time, and preference and liking during consumer hedonic testing. *Food Research International, 117*, 19−29. Available from http://www.elsevier.com/inca/publications/store/4/2/2/9/7/0.10.1016/j.foodres.2018.04.024.

Sester, C., Deroy, O., Sutan, A., Galia, F., Desmarchelier, J. F., Valentin, D., & Dacremont, C. (2013). Having a drink in a bar: An immersive approach to explore the effects of context on drink choice. *Food Quality and Preference, 28*(1), 23−31. Available from https://doi.org/10.1016/j.foodqual.2012.07.006.

Sinesio, F., Saba, A., Peparaio, M., Saggia Civitelli, E., Paoletti, F., & Moneta, E. (2018). Capturing consumer perception of vegetable freshness in a simulated real-life taste situation. *Food Research International, 105*, 764−771. Available from http://www.elsevier.com/inca/publications/store/4/2/2/9/7/0.10.1016/j.foodres.2017.11.073.

Sinesio, F., Moneta, E., Porcherot, C., Abbà, S., Dreyfuss, L., Guillamet, K., Bruyninckx, S., Laporte, C., Henneberg, S., & McEwan, J. A. (2019). Do immersive techniques help to capture consumer reality? *Food Quality and Preference, 77*, 123−134. Available from https://doi.org/10.1016/j.foodqual.2019.05.004.

Velasco, C., & Obrist, M. (2020). *Multisensory experiences: Where the senses meet technology*, 12932. Available from https://doi.org/10.1093/oso/9780198849629.001.0001.

Wang, Q. J., Escobar, F. B., Mathiesen, S. L., & Mota, P. A. D. (2021). Can eating make us more creative? A multi-sensory perspective. *Foods, 10*(2), 1−17. Available from https://www.mdpi.com/2304-8158/10/2/469/pdf.10.3390/foods10020469.

Witmer, B. G., & Singer, M. J. (1998). Measuring presence in virtual environments: A presence questionnaire. *Presence: Teleoperators and Virtual Environments, 7*(3), 225−240. Available from http://www.mitpressjournals.org/loi/pres.10.1162/105474698565686.

Zandstra, E. H., Kaneko, D., Dijksterhuis, G. B., Vennik, E., & De Wijk, R. A. (2020). Implementing immersive technologies in consumer testing: Liking and Just-About-Right ratings in a laboratory, immersive simulated café and real café. *Food Quality and Preference, 84*, 103934. Available from https://doi.org/10.1016/j.foodqual.2020.103934.

17

Voice-activated technology in sensory and consumer research: a new frontier

Tian Yu[1], Janavi Kumar[2], Natalie Stoer[2], Hamza Diaz[1] and John Ennis[1]

[1]Aigora, Richmond, VA, United States [2]General Mills, Minneapolis, MN, United States

17.1 Advent of voice-activated technology

17.1.1 A brief history

Verbal communication is the primary method of communication among people. However, verbal communication is often accompanied by the need to transform content into written form for record-keeping or circulation among audience. The technology behind this approach is referred to as speech recognition.

Bell and Edison's attempts to record the dictation of notes in the late 19th century, later translated into readable documents by humans, marked the advent of speech recognition. The field has been evolving for over 100 years now. In the mid-20th century, people started to dream about novel interactive methods of using speech recognition technology. In the 1968 movie '2001: A Space Odyssey', the crew talked with an artificial intelligence robot HAL. In 1987 Apple created a concept video about 'Knowledge Navigator' where a Professor used a voice-based assistant combined with a touch screen tablet computer to help him organise his daily tasks and communicate with other colleagues (Apple Knowledge Navigator, 1987).

All the major technology companies have started adopting speech recognition technology in the 21st century. In 2011 Apple introduced *Siri*, a natural language-based voice assistant, in its iPhones. In 2014 Amazon introduced its own voice assistant *Alexa* for its stand-alone smart-speaker device, Echo. In 2016 Google unveiled *Google Assistant*.

The adoption of voice-activated technology by these tech-giants has marked the beginning of a new era of using the technology for day-to-day tasks by individuals and its application for generating novel insights about the improvement of systems and industries. This chapter specifically focuses on the application of the technology to sensory and consumer research.

Digital Sensory Science
DOI: https://doi.org/10.1016/B978-0-323-95225-5.00017-1

17.1.2 Speech recognition basics

As the users of the voice-activated technology, sensory and consumer scientists do not necessarily need to understand the algorithms used in automatic speech recognition, but knowing the elements involved in using this technology would be beneficial for smooth deployment on a larger scale.

Automatic speech recognition has four main components:

1. Signal processing and feature extraction
2. Acoustic model
3. Language model
4. Hypothesis search

The signal processing and feature extraction algorithm takes the human speech as an input and returns the feature vectors as output. The acoustic model takes the feature vectors and predicts which phoneme each vector corresponds to, typically at the subword or character level. The language model guides the acoustic model, providing the context to distinguish between words and phrases that sound phonetically similar. The hypothesis search component combines the acoustic and language models and outputs the word sequence with the highest probability result. Various machine learning models have been extensively used in all of these components (Yu & Deng, 2015).

In the recent decade, as the computational power has increased exponentially and an enormous amount of training data in real-life usage scenarios are more readily available, voice-activated technology has started to penetrate daily lives. A popular example is asking a voice-activated device, such as Amazon's *Alexa*, to play a song, switch off the lights, or share our daily schedules.

17.1.3 Using smart-speakers in sensory consumer science

17.1.3.1 *The smart-speaker market*

According to *IHS Markit*, a market and consumer data agency, as of November 2018, 13.4% of internet households have already equipped themselves with smart gadgets globally. The numbers for selected countries are shown in Fig. 17.1 (Statista, 2018).

In a recent couple of years, the smart-speaker market has been noticeably booming across the globe. Globally, the installed base of smart speakers has reached 320 million units as of 2020, and this number is expected to increase to 640 million by 2024 (Statista, 2021) (Fig. 17.2).

Another interesting point about the smart-speaker market for sensory and consumer scientists is which smart-speaker/voice assistant is the most popular among consumers so that scientists can choose the one that covers the most target population. Also, different platforms/skills are required for implementing customised survey questions (see Section 17.1.3.3) for each device. The answer is country dependent. For example, in a 2021 consumer survey on the United States market (Statista, 2021), 92% of respondents claimed they own a device(s) that has implemented Amazon's *Alexa*, 67% owned device(s) implemented *Google Assistant*, followed by 17% owned Apple HomePod with *Siri*. The detailed survey answers of devices with virtual assistants are shown in Fig. 17.3.

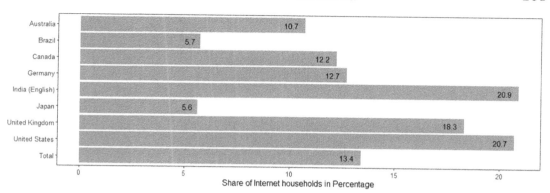

FIGURE 17.1 Worldwide smart-speaker penetration rate among households with internet in 2018. *Source: Modified from Statista (2018). Penetration Rate of Smart Speakers in Internet Households 2018, by Country. https://www. statista.com/statistics/974927/smart-speaker-penetration-rate-by-country/.*

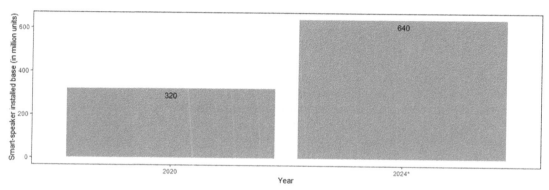

FIGURE 17.2 Worldwide smart-speaker installation in 2020 and 2024 (Statista, 2020). *Source: Modified from Statista (2020). Installed base of smart speakers worldwide in 2020 and 2024 (in million units). https://www.statista.com/ statistics/878650/worldwide-smart-speaker-installed-base-by-country/.*

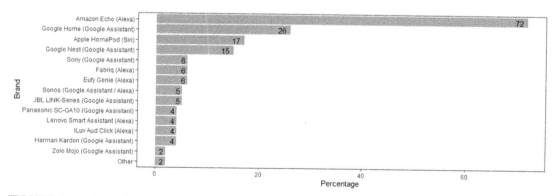

FIGURE 17.3 United States smart speaker ownership by brand in 2021. *Source: From Statista (2021). Statista Dossier about Smart Speakers. https://www.statista.com/study/53329/smart-speakers/.*

The continuing expansion of the smart-speaker market, supported by the constant evolution of the other technologies running on the Internet of Things, will pave the way for sensory and consumer researchers to regularly use smart-speaker devices for market research. This technology has the potential to redefine market research in the future.

17.1.3.2 *The advantages of using smart-speakers in sensory consumer research*

Consumer product goods (CPG) research relies heavily on sensory evaluation. Traditionally, the evaluation was conducted in presetup laboratories with trained respondents where they would evaluate the products and answer the questionnaires through software or on paper. However, some products would require respondents to engage with their hands, such as shampoo and hand soap. Assessing these products is not as easy as evaluating foods and beverages. The panellist has to rely on their memory or the assistance from another person to write down the evaluation details.

Furthermore, there is a growing desire among sensory and consumer scientists to conduct the evaluation and understand the consumer response to the products in natural environments, which has always been a challenge. When consumers use the products and start to answer the online questionnaire, their responses may diverge significantly from their actual experience. Consumer behaviour could also be implicit and challenging to capture in a structured online questionnaire.

Therefore alternative ways for collecting sensory and consumer data would be beneficial, either by simulating the user experience in the laboratory or finding a way to conduct the survey closer to the actual usage scenario, for example, at home.

Voice-activated technology can be a good solution for such use cases. It allows the respondents to free their hands, evaluate products and record the results in real-time and at various locations.

Additionally, from the point of view of neuroscience, product evaluation and answering a visual questionnaire at the same time could result in a high cognitive task load, which could potentially affect the taste perception of the respondents (van der Wal & van Dillen, 2013; Duif et al., 2020). Using smart-speakers for product evaluation would shift some task load from visual to auditory. Although the exact cognitive load of doing visual-based versus auditory-based consumer surveys would need further investigation, it has been reported that visual—auditory dual-task is usually better performing than visual—visual task, when the overall cognitive load is already challenging (Fougnie et al., 2018).

With the birth of the metaverse, virtual experiences will become more commonplace, and future sensory and consumer research must be implemented along with other virtual experiences. Using voice-activated technology in the highly individualised and visual-based metaverse would be a natural solution for both convenience and better use of limited time and attention of the respondents.

17.1.3.3 *The challenges of using smart-speakers in sensory consumer research*

Since voice-activated technology has started gaining traction only recently, sensory and consumer researchers still face several challenges in attempting to use smart speakers for their research projects. The first challenge is from the survey question coding.

Amazon's *Alexa* and Google's *Google Assistant* both have existing platforms for custo-mising or designing conversations in today's market. However, programming skills would be needed to use the platforms.

Alexa allows users to design customised conversations or skills using *Alexa* Software Development Kit (SDK) that requires NodeJS, JSON, or JavaScript programming languages and are deployed as Lambda Functions through Amazon's Alexa Developer Console or Amazon Web Services. In addition, using the Beta testing version, the skills can be limited to a predefined group of users (Skill Beta Testing for Alexa Skills | Alexa Skills Kit, 2022).

The equivalent of AWS Lambda Functions in Google is Google Cloud Functions. Google Assistant also offers a platform for developers to build customised conversations named DialogFlow, which also needs technical skills to make customised surveys for sensory and consumer research use.

Another challenge for sensory and consumer scientists is language limited linguistic diversity provided by these platforms. Currently (2022), Amazon *Alexa* supports nine lan-guages (Arabic, English, French, German, Hindi, Italian, Japanese, Portuguese, and Spanish), including several dialects of English (AU, CA, IN, UK, US), French (CA, FR) and Spanish (ES, MX, US).

On the other hand, *Google Assistant* has different language choices with various devices. For example, Google Nest Mini (2nd gen) offers 15 languages to the users. It provides all the languages *Alexa* offers and includes Danish, Indonesian, Dutch, Norwegian, Korean, Swedish, Mandarin, and Thai.

It should be noted that not all languages have similar performance. As some languages (e.g. English) have more users than others, more data for the AI components of voice-activated technology are available. This has led to more robust voice-activated systems for these languages. Also, from the point of view of the acoustic model, some languages, that is, tonal languages, have extra features. It is expected that speech recognition in these lan-guages is more complex and would require more time to evolve.

Furthermore, an additional challenge in utilizing smart speakers for sensory and con-sumer research is the complexity of survey distribution. When dealing with sensitive survey content that should remain private, researchers might choose to deliver a beta version through users' emails, instructing them to enable the survey themselves. However, given that Amazon accounts are predominantly personal in nature, effectively managing this pro-cess poses a considerable challenge.

While there are some challenges to the application of smart-speaker technology for consumer and sensory research, as it evolves and penetrates deeper into the society, the speech recognition and natural language processing for all languages is poised to improve in the future.

17.2 A case study of smart-speaker in consumer research

17.2.1 Introduction

To fully understand consumer experiences, we used smart-speakers and online surveys to perform a study to compare in-home, real-time, and hands-free consumer testing and consumer testing with online surveys. The smart-speaker survey was adapted to be more friendly to the smart-speaker users. Two types of oats and honey-flavoured snack bars were evaluated in a blinded format. Seventeen respondents were assigned to the smart-speakers and 23 to the traditional surveys.

17.2.2 Experimental procedures/methods

17.2.2.1 Materials

Two types of oats and honey-flavoured snack bars (sample 369 and sample 248) were evaluated in a blinded format. Amazon's Echo (*Alexa*) devices were used as smart-speakers for the hands-free snack bars evaluation.

17.2.2.2 Participants

The respondents were recruited from the *General Mills* employee panel, with a prescreening step to identify which employees already owned *Alexa* devices. Twenty-five respondents who already owned *Alexa* devices were assigned to an *Alexa* Survey, while 25 additional respondents were assigned to a traditional online survey. At the end of the evaluation, 40 respondents completed the survey, with 17 in the *Alexa* survey condition and 23 in the traditional online survey condition. All respondents were followed up with a traditional online survey assessing their experiences after the main survey.

17.2.2.3 Smart-speaker survey development

The traditional online survey questions were adapted to the *Alexa* survey. Briefly, most of the survey questions were rewritten in a close-ended format to align with the normal verbal communication. For rating and just-about-right (JAR) questions, 9- or 5-point scales were used, respectively. Respondents' answers were numbers and recognised by *Alexa*. For Check-All-That-Apply (CATA) questions, a series of 'yes or no' questions were asked. The open-ended questions had the same format as the online survey. The adapted survey was programmed and deployed to respondents through Alexa's 'beta-testing' feature to control who had access to the survey.

17.2.2.4 Data analysis

The transcribed data were captured in real-time and stored in *Google Firebase*. The data were preprocessed for streamlined analysis.

17.2.3 Results

In general, there was an excellent agreement across survey types for preference, while there was a strong differential agreement on overall liking. For CATA responses, both survey types showed strong agreement. However, for open-ended responses, both surveys agreed on the key characteristics, but the smart-speaker survey received fewer responses and word choices than the online survey.

The smart-speaker survey agreed with the online survey on the preference and overall liking of the samples. 82% of the *Alexa* respondents and 83% of the online survey respondents preferred the same product, while 18% of *Alexa* respondents and 13% of online respondents preferred the other product. As the smart-speaker questions were designed for forcing respondents to choose between two samples, there were no 'no preference' respondents. On the other hand, 4% of online respondents answered no preference (see Table 17.1).

For the hedonic ratings, there was strong differential agreement across survey types in our study. As shown in Table 17.2, the differences in the hedonic ratings of the two samples were very consistent, but the responses from the smart-speaker surveys were consistently smaller than those from the online survey.

The CATA responses from smart-speakers were also consistent with the online surveys. For some CATA terms, the frequency of mentions from *Alexa* was slightly higher than from online (Fig. 17.4).

However, the open-ended questions seem to be challenging for the smart-speaker survey. Compared with the online survey, fewer words and less variety of word choices were recorded with *Alexa* in the open-ended question. Current technology and its relatively inflexible format limit the robustness and constrain the responses (Fig. 17.5).

TABLE 17.1 Sample preference comparison with Alexa and online surveys.

Source	248 (%)	No preference (%)	369 (%)
Alexa	18	0	82
Online	13	4	83

TABLE 17.2 Sample hedonic ratings with Alexa and online surveys.

Source	248	369	Difference
Alexa	4.65	6.65	2.00
Online	5.00	7.22	2.22

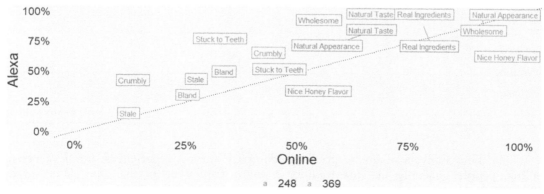

FIGURE 17.4 The Check-All-That-Apply (CATA) responses from Alexa versus online surveys.

4. Immersion technologies, context and sensory perception

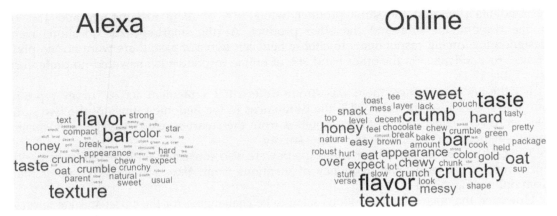

FIGURE 17.5   The word cloud of open-ended questions from Alexa and online surveys.

TABLE 17.3   Alexa versus online surveys liking scores.

Alexa	Online
5.72	7.48

TABLE 17.4   Check-All-That-Apply (CATA) statements percentage comparison between Alexa and online surveys.

Statement	Alexa (%)	Online (%)
Say everything	50	100
Express self	56	100
Recommend	72	100
Easy	78	100
Take again	78	100
Engage	72	87
Fun	67	78
Too short	0	0
Interesting	78	74
Too long	22	0
Hard	28	0

In the meta-analysis where respondents compare survey experiences on *Alexa* versus online, people generally like the traditional online survey better than the *Alexa* survey (Table 17.3). Respondents found the online survey easier, better allowing them to express themselves (Table 17.4).

17.2.4 Learnings and future directions

This study was the first consumer study using voice-activated technology as the data acquisition device. Smart-speakers collected close-ended, quantitative data comparable to the traditional survey. However, the data from open-ended questions collected through smart-speakers are brittle with current technology, where the critical responses are consistent but less numerous than the online survey data.

We have also learned several things from this study to utilise the current technology better. First, to make the survey more conversational and natural, we can design the traditional scaled questions into a series of 'yes or no' questions. For example, 'Do you like it?' followed by 'Do you like/dislike it a lot?' In this way, the liking question is shortened to a 4-point scale, forcing respondents to give an affirmative answer instead of a neutral one.

Second, the length of the smart-speaker survey should be considered carefully. A survey time of around 5—10 min is ideal. A lengthy survey is tedious to respondents and harder to control.

Third, making sure that respondents can access the survey on their own devices presents a practical challenge using Amazon's beta-testing setup. For example, people were not sure which account was associated with the device. The drop-out rate in this study is higher for *Alexa* compared to the traditional one. Collecting and confirming account information before the study would be helpful in reducing the drop-out rates.

As for the open-ended conversational questions, it is challenging for the current technology to understand accents and unusual/long pauses in the sentences. However, with more training data readily available and the advance in speech model and language model algorithms, the application of smart-speaker surveys in sensory and consumer science is bridging the gap between the qualitative and the quantitative tools.

As speech recognition and language translation grow and become sophisticated, the focus group can also become 'smart' in the at-home, in-time scenario.

References

Apple Knowledge Navigator. (1987). Available from https://www.youtube.com/watch?v = umJsITGzXd0&t = 9s.

Duif, I., Wegman, J., Mars, M. M., de Graaf, C., Smeets, P. A. M., & Aarts, E. (2020). Effects of distraction on taste-related neural processing: A cross-sectional fMRI study. *The American Journal of Clinical Nutrition, 111*(5), 950—961. Available from https://doi.org/10.1093/ajcn/nqaa032.

Fougnie, D., Cockhren, J., & Marois, R. (2018). A common source of attention for auditory and visual tracking. *Attention, Perception, & Psychophysics, 80*(6), 1571—1583. Available from https://doi.org/10.3758/s13414-018-1524-9.

Skill Beta Testing for Alexa | Skills Alexa Skills Kit. (2022). Available from https://developer.amazon.com/en-US/docs/alexa/custom-skills/skills-beta-testing-for-alexa-skills.html.

Statista. (2018). Penetration Rate of Smart Speakers in Internet Households 2018, by Country. Available from https://www.statista.com/statistics/974927/smart-speaker-penetration-rate-by-country/.

Statista. (2020). Installed base of smart speakers worldwide in 2020 and 2024 (in million units). Available from https://www.statista.com/statistics/878650/worldwide-smart-speaker-installed-base-by-country/.

Statista. (2021). Smart speaker ownership by brand in the U.S. in 2021. Available from https://www.statista.com/forecasts/997149/smart-speaker-ownership-by-brand-in-the-us.

van der Wal, R. C., & van Dillen, L. F. (2013). Leaving a flat taste in your mouth: Task load reduces taste perception. *Psychological Science, 24*(7), 1277—1284. Available from https://doi.org/10.1177/0956797612471953.

Yu, D., & Deng, L. (2015). *Automatic speech recognition.* Springer. Available from https://doi.org/10.1007/978-1-4471-5779-3.

How to tell powerful digital sensory stories

Shining through the smog: how to tell powerful, purposeful stories about sensory science in a world of digital overload

Sam Knowles

Chief Data Storyteller, The Insight Agents, Lewes, East Sussex, United Kingdom

18.1 Introduction

The digital media landscape promised so much: instant connection between every opinion leader and every data point that matters; a permanent, traceable record of cutting-edge research, capable of standing successive generations of researchers on the shoulders of giants and shortcutting the miss-steps of those who have failed before; and, a permanent, always-evolving repository of what the world knows. If only the reality had lived up to the promise and not a world of information overload, shrouded in an impenetrable smog of data.

To be fair, some of the wide-eyed optimism of Sir Tim Berner's Lee's founding vision of the world wide web has been justified, from PubMed to Google Scholar, from open access journals to open data protocols. Yet there are challenges for digital sensory science just as there are challenges for mass-participation social media. This is less to do with the bad actors and wilful mis- and disinformation that haunt the platforms run by Alphabet, Meta and others, and more to do with a fundamental failure in data storytelling. Data storytelling — bringing together the often fire-and-ice bedfellows of narrative and numbers, stories and statistics — requires special dexterity in the realm of digital and social media platforms. This is because (1) there is so much data available online that serial violators of the principles of better data storytelling will soon be overlooked, and (2) those who respect these principles soon prosper, even if what their research shows ultimately matters less.

18.2 Information overload

Sensory scientists looking to cut through the clutter and make their mark in digital communications — reaching and influencing audiences at scale — should reflect on the sheer volume of noise their data-rich narratives need to cut through if they are to succeed. Viktor Mayer-Schönberger & Cukier — authors of the 2013 book *Big Data: A Revolution That Will Transform How We Live, Work & Think* — have estimated the volume of data created by humanity. From the invention of the printing press in the late 15th century til the end of 2012, they calculated we had generated 1.2 zetabytes — enough data to fill a pile of high-capacity iPads stretching the quarter of a million miles from Earth to the Moon. By the end of 2013? The same again. Recent guesstimates suggest we created 74 zetabytes of data in 2021 alone, over 60 times Mayer-Schönberger and Cukier's figure for total human output less than a decade before. Big Data's capital letters are clearly here to stay.

18.3 Serious data, simply presented, really matters

Because there is so much noise and apparently ever-increasing appetite for more and more data in the world — all connected together via the world wide web — communication via digital media has to work much harder than the relatively closed and self-contained ecosystems of academic journals, conferences and communities. This is why those skilled in data storytelling have the edge over those who look to share all their data with the world — every possible cross-tabulation, every conceivable pivot table, every imaginable analysis of variance.

Covid has underlined the power and impact of great data storytelling — as well as double-underlining the perils of what can happen when data storytelling goes bad. For the first time in more than a century, Governments around the world needed to make evidence-based appeals to shut down liberties citizens had come to take for granted. They did this in part because they were dealing with what former US Defense Secretary, Donald Rumsfeld, famously dubbed 'unknown unknowns', although their motivation was to protect the vulnerable. In free societies not determined to enforce a zero Covid strategy — so we can rule out both China and New Zealand in this analysis — this required an incredibly nuanced balancing act, with Government officials and their scientific advisors walking the tightrope of emotion and logic.

The statistician, Professor Sir David Spiegelhalter, identifies the power and impact of clear, honest and relatively sparse data-driven communication during the pandemic in his 2021 book, *Covid by Numbers*. Contrary to what some politicians on both sides of the Atlantic may have claimed in recent years, we have not 'had enough of experts'. What we have had enough of is experts looking to drown us with data in an attempt to browbeat us into submission. Not to mention self-interested political leaders determined to either overstate or trivialise the severity of particular Covid waves or the efficacy of quack treatments by blinding us with statistics. Again, a classic case of unempathetic information overload.

Equally, there are some occasions or circumstances — circumstances that will be familiar to many working in sensory science — where it is challenging to collect enough data that builds a meaningful narrative, one that is properly evidenced by a sufficiently robust sample to draw meaningful conclusions. Digital technology is proving helpful in these instances, too, enabling sensory scientists to collect validating pools of data via metaanalysis, machine learning, AI simulations and gamification. In these instances digital technology enables us to boost samples and build a more compelling story in a rapid, agile fashion. But whether the challenge sensory scientists face is starting from a position of too much data or too little, the trick in impactful data storytelling is driven by the Goldilocks Principle and the need to get things 'Just Right'. That is partly driven by generations of statistical best practice and partly by humanity and empathy for the audience, and particularly by an appreciation of the likely data tolerance of that audience.

18.4 The psychology of decision-making

What those who misuse data in their storytelling fail to realise − in their unempathetic, inhumane way − is how we make decisions. As the Princeton psychologist, Daniel Kahneman, has demonstrated through his long, illustrious career, we make our decisions emotionally. Decision-making − choosing what to do and what not to do − is mediated using the evolutionarily ancient limbic systems of the reptilian brain, structures that have no access to language, data or facts; quick-and-dirty, energy-lite structures that we share with all other mammals, but also with reptiles and birds. Kahneman calls this System 1 thinking. It is only when we have made our decisions that we go on to justify them rationally, using logic, facts, numbers, statistics. This is known as System 2 thinking in Kahneman's conception, commandeering as it does the uniquely human grey and white matter of the cerebral cortices, evolutionarily recent and not found in other species. System 2 thinking is slow, laborious and energy intensive. Communicators looking to make a persuasive argument with data need to bear this psychological reality of decision-making in mind when building powerful and persuasive stories. It should be a guiding principle for how you construct and tell stories with data.

Telling powerful and purposeful stories with data − in sensory science every bit as much as in public health epidemiology − starts with a fundamental and respectful acknowledgement of the audience you are looking to influence. Data storytelling comes a step before data visualisation, and sensory scientists looking to increase their influence would be well advised to follow the six Golden Rules of better data storytelling. This obtains in formal settings, such as reporting on the real-world impact of their work (consider the United Kingdom's seven-yearly Research Excellence Framework or REF), as well as when looking to engage intelligent lay audiences (in industry, via the popular media, and in how their research is mediated through social and digital platforms). In the rest of this chapter, we will consider the six Golden Rules of better data storytelling.

18.5 The six Golden Rules of data storytelling

18.5.1 Rule 1: know your audience

In the book, film and now the stage play *To Kill A Mockingbird*, the lawyer-hero, Atticus Finch, tells his daughter, Scout: 'You never really understand a person until you consider things from her point of view, until you climb inside of her skin and walk around in it'. Great data storytelling is rooted in the uniquely human ability to feel, understand and express empathy; our capacity to put ourselves in the mind, the mindset, the shoes of those we are looking to influence.

Empathy truly is the superpower that enables scientific and technical experts to read the room — real or virtual; in person, on paper or in pixels — to understand whom they are talking to in advance of opening their mouths or clicking through their first slides. It is what enables them to shape their narratives and infuse them with the right level and complexity of data. Any audience will have its own level of data tolerance and data maturity, but even if you are writing for or speaking to an expert audience of peers with whom you work cheek-by-jowl in the same field, there is no way they can be as familiar with your data as you are. They may know as well as you do the difference between satiety and satiation. Their lab may have explored the dynamics of mouth-feel or the component elements of crunchiness in the same level of granularity as yours. But your data are yours and yours alone, and you should always put yourself in the role of recipient rather than broadcaster when preparing to share your work.

18.5.1.1 *Exercise 1: Pen Portrait*

When I run data storytelling training — for corporate executives as well as for the academic community — one of the first exercises I get them to undertake is called Pen Portrait. I ask them to identify the single most important constituency in an audience they are going to be addressing, in a forthcoming presentation, paper or — increasingly — online forum or platform. Impactful communication in the round is increasingly delivered in the digital ecosystem. Then I get them to spend 15–20 minutes writing a Pen Portrait of the audience — what they are like, what kind of experience with data they have, what their statistical knowledge is likely to be. This will determine how they should structure and then craft their content. What this exercise always does is to underline the importance of tempering numbers in narrative and to encourage statistical moderation. Coming out of the training session, I encourage my workshop participants to do this exercise for every new paper, talk or online forum for which they are preparing. It is best to print out and post Pen Portraits in your line of sight as you create both the form and the detail of your content.

Knowing your audience — through research, experience and crafting Pen Portraits — can also be instructive in helping you to identify which new digital channels, tools and techniques might appeal to them. Assuming that all academics read journals, broadsheet newspapers, and listen to the radio could easily mean — for instance — missing out on the next generation of Early Career Researchers (ECRs). Yet a targeted campaign on TikTok might be exactly the right motivator for inspiring ECRs or an elusive consumer group to engage with your work. And you can only know this for sure if you put the effort in to knowing who they are and what media they consume.

18.5.2 Rule 2: keep it simple — yet smart

The French mathematician Blaise Pascal, the British Prime Minister Winston Churchill, and the Irish poet and playwright Oscar Wilde are all said to have said: 'I would have written you a shorter letter — I just didn't have the time'. In data storytelling — reporting your research results to interest, stimulate and persuade others — the temptation is to share every data set and analysis you have generated in your research. You are passionate about your work and the journey you have taken to reach your conclusions, and you are keen to take others on this journey with you. This is incredibly easy to do in the digital ecosystem as many online forums are effectively unbounded and of infinite capacity. Unlike print publications (which typically set pagination limits) or broadcast (which only have as many hours or minutes as programmes are allocated in a schedule), contributions to the world wide web are often limited only by the size of server farms (i.e. unlimited).

Moving from 'So what?' to 'Now what?'

Yet sharing all your data is not only another failure of empathy; an inability to know your audience and put them first in your quest for influence and impact. It also makes the narrative much more complex than it needs to be. Even our closest friends and colleagues do not want to follow every twist and turn. To make use of your work, they need you to move from 'So what?' — this is what we did, this is what we found about the impact of lavender versus vanilla aroma on mood, for instance, and this is what the data shows — to 'Now what?' — this is what we should do as a result with new product development for bubble bath. The 'So what?... Now what?' mentality keeps things simple — yet smart. And though many researchers feel that simplification makes them simpletons, requiring them to dumb down and trivialise their work, in fact the opposite is the case. Not only does it show care and consideration for those you are looking to influence and persuade. It allows you to pique their interest more quickly and engage in debate more straightforwardly. This is true no matter how junior or senior the storyteller is.

18.5.3 Rule 3: beware the Curse of Knowledge

The Harvard psychologist, Steven Pinker, has broadened his areas of interest over his 25-year career as a writer of popular science best-sellers, from his home turf of linguistics and language acquisition to cognition, nature versus nurture, and knowledge and reasoning. One of his lesser-known works should be required reading for all researchers and scientists, particularly those looking for guidance on how to communicate in the digital media ecosystem. *The Sense of Style: The Thinking Person's Guide to Writing in the 21st Century* from 2014 embodies many of the Golden Rules of data storytelling in how it is written and in the advice it gives. The third Golden Rule is 'Beware the Curse of Knowledge'.

18.5.3.1 Exercise 2: banish the Curse of Knowledge

In *The Sense of Style*, Pinker defines the Curse of Knowledge as: 'The difficulty in imagining what it's like for someone else not to know something that you know'. Even if the phenomenon of the Curse of Knowledge is familiar to you, take 5 minutes now to get a glass of water, make a cup of tea or coffee and reflect on Pinker's definition in the context of your work and how you communicate it. Think particularly of those outside your immediate field of expertise — most people with whom you will be communicating in order to achieve real-world impact — and run through in your mind the extent to which you think you suffer from the Curse.

Pinker picks out five professions he believes are most guilty of the Curse of Knowledge: lawyers, those working in financial services, Government officials, scientists working in business and academics. The Curse of Knowledge is another fundamental failure of empathy, an act of wilful overlooking who makes up your audience, and a refusal to make your arguments — and share your data — in way that invites them in. Because you know your data so much better than those you are sharing it with, you need to give just enough to intrigue them — and no more; just enough to have them sit up, pay attention and ask you follow-up questions. Questions like 'So, did you run this experiment with people who'd never experienced this combination of flavours before?', 'What about participants who actively dislike either flavour in isolation but are unable to detect them in combination?' and 'Did you account for those who are genetically-predisposed to dislike coriander/cilantro?'

It is particularly easy to fall into the trap of the Curse of Knowledge — and it truly is a curse — in online, digital environments. Because of the sheer volume of noise online and the swathes of voices peddling other data sets — both increasing asymptotically, year-on-year — response rates in digital communications can be low. When no response is received, the temptation is to pile on more arguments, more evidence, more data in order to be more convincing. Yet the principles of both rationality and decision-making — particularly, as we saw earlier, Kahneman's paradigm of 'emotional decision-making/rational justification' — suggest that researchers who overcommunicate their data inevitably stray further into the territory of the Curse of Knowledge. Far from shining a light through the smog of digital data overload, this approach is akin to a smoke machine at a glam rock or heavy metal concert. It envelops the audience in confusion and more or less guarantees that they will take refuge elsewhere and listen to someone else who treats them with more empathy and humanity.

18.5.3.2 Case study: a tale of teabag technology

In the mid-1990s, I was part of the team that brought about one of the most important developments in British culture of the late 20th century. I talk, of course, of when teabags moved into the third dimension. When PG Tips stole a march on its competitors by moving from flat and square to three-dimensional and pyramid-shaped, I was working for the PR agency that would be launching the new product innovation. As curious types, we were keen to know what the science was behind the innovation.

We were soon introduced to one of the first — and certainly one of the best — data-driven storytellers I had come across. He showed how fire and ice can work together to produce magic; that telling stories underpinned by complex science — but do not lead with equations or workings-out — really can cut through. Dr. Andrew ('Fred') Marquis was a researcher in thermofluids in the Department of Mechanical Engineering at Imperial College, London. He and his colleagues had been briefed by a PG Tips team, including tea tasters and marketers, supply chain and brand managers, to create something truly revolutionary in teabag technology. Something that would improve the brewing experience given by a teabag and make an even better-tasting cuppa.

When you pour boiling water onto a conventional, square, flat teabag, the pressure of the water on the teabag initially squeezes those delicious tannins and flavonoids from the leaves and into the water. But because the teabag is flat, almost as soon as the flow of water stops when the mug or the teapot is full, teabags tend to clomp up, to use the technical term. The tea leaves stick together and they have no room to move around in the two-dimensional teabag. This reduces the potential for fresh polyphenols to be released from every tea leaf in the teabag, and the brew is suboptimal. To get more taste, people tend to leave the teabag in the cup. But that tends to create a more bitter taste, which some find unpleasant. Or else they grab hold of it and dunk it up and down, and then they usually burn their fingers. The challenge was set.

Rather than dismiss this as marketing fluff or a challenge too trivial, Marquis and his team took up the task with gusto, seeking to build the first teabag that created and sustained a three-dimensional shape when it floated in the mug or teapot. There were some constraints, of course. It had to be of a shape that could be manufactured by Unilever's teabag-making machines with minimum necessary modification. Also, the finished shape needed to be squishable so it could fit into boxes of a similar size to the 40, 80 and 160-teabag boxes PG Tips made at the time. Supermarkets would not suddenly give up twice the space on the shelves just because PG Tips had changed the shape of its teabags.

What is more, having been squashed, the teabags needed to spring back into shape when boiling water was poured over the top. There would be no chance to introduce a new ritual into the tea-making ceremony of − say − shaking out the teabag into its 3D shape before popping it into the cup or mug. It needed to perform as intended and improve brewing performance when anyone was making a cuppa, half-asleep and bleary-eyed.

The thermofluids team got to work. They made teabags of every imaginable three-dimensional shape: cubes, spheres and dodecahedrons. And because of the essential Britishness of tea and teabags − and their waggish sense of humour − they also made top hats and bowler hats. But the shape that won out − the shape that could be made using existing machines, that squashed down into boxes but popped back into shape in the mug, and that gave the best brewing performance − was the tetrahedron or four-faced pyramid.

The way I tell the story is pretty much how Fred told the story. To the PG Tips technical team, to Unilever corporate and to agencies looking for stories. To Radio 4's *Today* Programme and BBC1's *Tomorrow's World*. Here is Fred talking in *The Independent* in 1996: 'The pyramid-shaped teabag tends to naturally float on the surface of the water, allowing the water to flow more freely in and out of the tea bag. It is this extra movement of tea leaves which helps the brewing process. The teabag that works like a teapot with loose leaf tea.'

Of course, there was a wealth of data and statistics underpinning the very complex technology required to produce a tetrahedral teabag. Thermofluids folks also use a lot of equations. But they kept them for their lab books, computer simulations and academic papers published to justify the superior brewing technology the shape provided. But they had the empathy to know who needed to know what. Because of his unusually well-developed empathy he could provide the data-driven ammunition Unilever needed to tell a convincing, evidence-based narrative. He could judge when it was appropriate to be the scientist (with his team and his peers), and when it was appropriate to be the storyteller (with his client and their agencies). He knew his audience, he kept things simple — yet smart — and above all, he avoided the Curse of Knowledge.

18.5.4 Rule 4: find and use relevant data

Great data storytelling demands that the storyteller finds and uses only relevant data. Statisticians — particularly those who work in epidemiology, with their gallows humour — often talk about identifying the 'killer statistics': the sprinkling of data points or handful of numbers that perfectly summarise the story they are looking to tell. Of course it is important not to line up a barrage or battery of numbers and use these to attempt to wear the audience into submission. As we have already seen, this violates the conventions of human and empathetic communication. Good data storytellers should always seek to deliver their narratives with impact.

18.5.4.1 *What type of numbers to use*

More than just finding and using a small amount of the most relevant data, it is also important to choose numbers that have a particular set of qualities that lend themselves to successful narrative. Guidance includes:

- Be accurate but not excessively precise. Data storytelling is not like a mathematics exam; you do not get extra marks for extreme precision or showing your workings-out. Say 'up to', 'about', 'almost' or 'more than', and when your audience asks for clarification and precision, you know they are listening.
- Round numbers up or down, both to avoid unnecessary precision AND to make them more memorable. £350 m is much more memorable and sharable than £346,153,846.
- Avoid really big numbers wherever possible. Many of us find big numbers — billions and especially trillions — really hard to understand. '£350 m a week' is much easier to conceptualise and so commit to memory than the same quantity of money expressed on an annualised basis, '£18bn a year'.
- Use verbs that incorporate numbers — 'halved', 'doubled', 'decimated', 'grew fourfold'.
- Deploy ranking numbers, ordinals (first, second, thirty-fourth) AND quantity numbers, cardinals (1, 2, 34) for variation, as relevant.

18.5.4.2 *How to use the numbers you choose*

As well as choosing the 'what' qualities of their killer statistics with care, data storytellers can also enhance the impact of the numbers they use by paying attention to the 'how'.

- Vary the type of numbers you use — a percentage, a 'one in x', a ratio or a fraction (3/4). Too many of the same sort of numbers jostle and compete with each other, then fall out of working memory.
- Make the first sentence of your first paragraph a number to pique interest. '£350 m a week. That's how much….'
- If you must use big numbers — billions or trillions — set them in context, perhaps with an analogy. '9 million: that's about as many people as live inside the M25 orbital motorway around Greater London' or 'In January 2022, Apple became the first $3tn

company by market capitalisation, making it worth more than all the companies on the London Stock Exchange FTSE 100 Index put together'.

- Repetition aids with memorability. The constant repetition of the (admittedly bogus) statistic '£350m a week to the NHS' by the Vote Leave campaign during the E.U. Referendum in the United Kingdom in 2016 meant that it took on memetic status online and was the defining number of the campaign.

18.5.4.3 Exercise 3: read your work out loud

When you are preparing your data storytelling content – particularly for dissemination online – try reading it aloud to yourself and see how it sounds. Even better, read it aloud to someone else in your company, division or lab. Find someone you know and like and whose judgement you respect but you do not regularly work with. You may be convinced that your latest research has established a critical new distinction between homeostatic and hedonic hunger and you have the data to prove it, but how does it sound to someone else not directly involved in the project at first hearing? Write it as a tweet, a Facebook group post, a short form blog, and road-test the different versions to see which resonates most strongly and why.

Become data storytelling accountability buddies, and read aloud the opening lines, paragraphs or sections of what you are writing. Ask others if they understand what you have said and what the data you have used mean to them. Adjust your content based on their feedback. Then, offer the service back to them. It is amazing how this mutual support service can enhance your data storytelling with very little input from your buddy. Give them a little guidance on the audience who will be reading or listening to the content you share with them, perhaps in the form of a Pen Portrait you have written (see Section 18.5.2).

18.5.5 Rule 5: balance the emotional and the rational

The American business writer, Dan Pink, says in his book *To Sell Is Human*, 'We are all in the moving business' – the business of persuading others to do something different. The purpose of communication is to effect change in attitudes, behaviours, beliefs and ultimately actions. If we are to move others, we need, again, to be cognisant of Kahneman's emotional/rational balance. Usually, those who train storytelling talk about the need to balance the rational and the emotional. Because of the largely unacknowledged primacy of the emotional over the rational, data storytellers are perhaps better advised to balance the emotional with the rational. So, in building data-driven, evidence-based narratives, the smart data storyteller looks to build real, human stories about the impact of their findings on people's lives. Digital communication in particular lends itself to this approach, with talking-heads video the dominant medium for communication on the leading social media platforms run by Alphabet (Google, YouTube) and Meta (Facebook, Instagram, WhatsApp).

1. The three-act story structure

   The fourth century BCE philosopher Aristotle set out two critical frameworks for storytelling that balance the emotional with the rational. In his short, very readable book *The Poetics*, Aristotle established the ground rules for telling stories with impact in the three forms of entertainment on offer to the classical Greeks: comedy, tragedy and epic poetry. In the first framework in the history of literary criticism, he pioneered the three-act story structure, as relevant to Hollywood blockbusters, novels and Netflix boxed sets as it was when the treatise was first published in ~330 BCE. It is a codification of the storytelling structure we humans find endlessly appealing, go looking for to make sense of and navigate the world, and resent if we find is missing.

Aristotle defined the three acts as the thesis, antithesis and synthesis; one perspective, an alternative perspective and the resolution; beginning, middle and end. Between the acts come critical turning points or dramatic junctures. At the culmination of the first act, something big and important happens — a messenger arrives with crucial news, someone dies, a confidence is betrayed. This 'something big and important' is known as the inciting incident in Hollywood, and it propels the protagonist into her journey of discovery in act two. At the end of the second act — which can take up three-quarters of the narrative or more — we reach the second critical turning point, known to dramatists as the climax. To be a satisfying story, this is not the end, and there is often (especially in the case of Marvel films) a lot to get through before the end of the action. And unless you are an absurdist, surrealist or David Lynch, the third act will be used to tidy up loose ends in the story.

2. Pathos, logos, ethos

In a separate work, *The Art of Rhetoric*, Aristotle set out another three-part prescription for better storytelling — in this case for political and forensic speech-making, but as it happens these principles apply well to all varieties of narrative. For a story to move us, it needs to combine pathos (emotion), logos (logic or reason) and ethos (character). We have already addressed the critical importance in narrative construction of empathy, telling real human stories about the impact on real lives of people with whom we can identify — that is the pathos, and it does not necessarily mean what the word translates as literally, which is 'suffering'. To Kahneman's rationale, we start with pathos, to allow the evolutionarily ancient reptilian and limbic brain structures to enable us to make a decision. We then help decisions made to be justified with logic and reason — logos — which also includes facts, statistics and data. But for Aristotle, we need one further component: ethos or 'character', the authority value of the person (or expert) telling the story.

You may be wondering what Greek literature, the first work of literary criticism, Marvel and Salvador Dalí, have to do with presenting the findings of research in sensory science. The answer — as should now be clear — is everything. Reporting on how — say — cues from satiation, the gustatory or olfactory systems affect appetite will be that much more impactful if it follows these principles. This is true of a conference poster, a presentation or keynote, a journal article abstract, a podcast interview or a social media engagement plan.

18.5.6 Rule 6: talk 'Human'

The sixth and final Golden Rule is perhaps both the simplest in essence and the hardest to achieve. 'Human' is a dialect that many in scientific research — academic and industrial — find it very hard indeed to speak. Many researchers use jargon and technical language, acronyms and initialisms — as well as armfuls of comparatively raw, unfiltered statistics — as if this makes them sound somehow more authoritative and intellectual. In fact, apart from a relatively narrow clique of colleagues and other specialists working in exactly the same field, what this approach achieves is confusion, misunderstanding and alienation, even among an intelligent lay audience.

In recent years, there have been two principal drivers of researchers starting to move away from the unhuman, inhumane dialect often favoured by research scientists — and those in the humanities too. The first is the growing requirement for academics to report on the real-world impact of their work, when applying for funding, when reporting to grant-awarding bodies, and especially in exercises like UK Research & Innovation's REF. REF impact case studies are reviewed by intelligent lay readers, and research funding is allocated on the basis of research that has demonstrable impact that is well and clearly reported. Most major universities now have impact officers and teams who work cross-departmentally to up the clarity of their submissions, a critical element of which is data storytelling. After three rounds of the seven-yearly REF, this rigour is starting to filter its way through to academic writing more generally.

The second reason for the pivot to 'Human' is the increasing role of the digital media ecosystem — including web repositories and social media platforms — as a primary means of dissemination of research findings. It is too easy to ignore content that is not written in a clear, succinct fashion, with just the right level of data as part of the narrative. If Institution or Company X cannot tell you clearly what their work has found — if its reporting generates stifled yawns, distraction and a feeling of TLDR — then Institution or Company Y is just a couple of clicks away. X will be ignored, not approached for collaboration, and have fewer links made to it online, making it fall down the natural search algorithm; the opposite will happen to Y, and in part because those who work at Y are better data storytellers.

18.5.6.1 *Five top tips for talking 'Human'*

The dialect of 'Inhuman' not only features jargon and technical language, abbreviations, acronyms and initialisms but also includes many, many verbs in the passive voice. It takes us much longer to process 'the mouse was chased by the cat' than 'the cat chased the mouse', and not just because the latter has fewer words. 'Inhuman' also favours high levels of abstract nouns (concept, framework, matrix) over concrete (actual, physical things). Zombie nouns, made from a verb stem and the ubiquitous suffix '-isation', are especially time- and energy-inefficient in terms of processing. You should use all of these less in your data storytelling — and, indeed, do precisely the opposite:

1. Rule out abbreviations.
2. Use the active voice.
3. Upweight your use of concrete nouns.
4. Tell stories about real people and the emotional consequences.
5. Use your narrative to share the implications of your research on the potential of people's real lives.

18.6 Conclusions and future perspectives

The brothers, Chip and Dan Heath, said memorably in their book *Made to Stick: Why Some Ideas Survive & Others Die*: 'After a presentation, 63% of attendees remember stories. Only 5% remember statistics'. I love the suspicious accuracy of 63%, which may well be a self-consciously spurious statistic. What is undoubtedly true is that many more recall stories than statistics. This does not mean as a data storyteller — particularly a researcher in the field of sensory science, looking to communicate with impact in the increasingly digital media ecosystem — you need to avoid numbers altogether. Far from it.

If you want to be a better data storyteller and shine through the smog, follow these six Golden Rules.

1. Know your audience and write a Pen Portrait of who it is you are trying to influence. Assess their likely data tolerance and maturity and be sure you always create content with them in mind.
2. Keep it simple — yet smart. Take the time to use only the data you really need to include, pruning the excess. Harness your inner Pascal, Churchill or Wilde and 'write a shorter letter'. It is well worth the time.
3. Beware the Curse of Knowledge. The Curse of Knowledge is real and it is a Curse. All of the Golden Rules are related to empathy, but none more so than Rule 3.
4. Find and use only truly relevant data. Make the numbers you choose accurate but not excessively precise, round numbers up and down, and remember that your working life is not a mathematics exam.
5. Balance the emotional and the rational, both to mimic human decision-making and to observe Aristotle's two rules of three: thesis-antithesis-synthesis and pathos-logos-ethos.
6. Talk 'Human' — that rarest of dialects for research scientists writing online.

None of these Golden Rules is complex, difficult to achieve or revolutionary, in and of itself. But take them together and rigorously apply them to your data storytelling, and you will notice deeper and richer interactions with those you are looking to influence. The defining equation of communications impact in the always-on, digitally mediated knowledge economy is simply this: Analytics + Storytelling = Influence. Your role is to give your data meaning. As Nate Silver said in the introduction to his 2011 book, *The Signal and the Noise*: 'The numbers have no way of speaking for themselves. We speak for them. We imbue them with meaning'.

Capturing and holding an audience's attention is one of the principal attributes investors look for in entrepreneurs and start-ups looking for funding. Every start-up accelerator and incubator has its own template for a 60-second or elevator pitch, and coaches put in hours with entrepreneurs behind the scenes to ensure that they become pitch perfect. This entrepreneurial spirit is also of very great advantage to sensory scientists seeking to communicate what they have found, what it means and what should change as a result, particularly when their messages are mediated via the digital media ecosystem. So when you are preparing your next data read-out on the impact of aroma on satiety or the multisensory implications of 'crunchiness' versus 'crispiness', put yourself in the mindset of an entrepreneur pitching on *Dragon's Den* or *Shark Tank*.

References

Aristotle (~330 BCE). (1991). *The art of rhetoric*. Penguin Classics.

Aristotle (~330 BCE). (1996). *The poetics*. Penguin Classics.

Mayer-Schönberger, V., & Cukier, K. (2013). *Big data: A revolution that will transform how we live, work, and think*. John Murray.

Further reading

Berners-Lee, T. (1989). Information management: A proposal. http://www.w3.org/History/1989/proposal.html.

Heath, C., & Heath, D. (2008). *Made to stick: Why some ideas take hold and others come unstuck*. Arrow.

Kahneman, D. (2011). *Thinking, fast & slow*. Penguin.

Knowles, S. K. Z. (2018). *Narrative by numbers: How to tell powerful & purposeful stories with data*. Routledge.

Knowles, S. K. Z. (2022). *Asking smarter questions: How to be an agent of insight*. Routledge.

Lee, H. (1960). *To kill a mockingbird*. J.B. Lippincott & Co.

McKee, R. (1999). *Story: Substance, structure, style, and the principles of screenwriting*. Methuen.

Pinker, S. (2014). *A sense of style: The thinking person's guide to writing in the 21st century*. Allen Lane.

Silver, N. (2013). *The signal & the noise*. Penguin.

Spiegelhalter, D. (2019). *The art of statistics: Learning from data*. Pelican.

Spiegelhalter, D., & Masters, A. (2021). *Covid by numbers: Making sense of the pandemic with data*. Pelican.

Index

Note: Page number followed by "*f*" and "*t*" refer to figures and tables, respectively.

CPI Antony Rowe
Eastbourne, UK
August 31, 2023